D1735698

*Edited by*
*Chang Lu*

**Chemical Cytometry**

## *Further Reading*

D. Anselmetti (Ed.)

**Single Cell Analysis**

**Technologies and Applications**

2009

ISBN: 978-3-527-31864-3

T. Yanagida, Y. Ishii (Eds.)

**Single Molecule Dynamics in Life Science**

2009

ISBN: 978-3-527-31288-7

X.-H. N. Xu (Ed.)

**New Frontiers in Ultrasensitive Bioanalysis**

**Advanced Analytical Chemistry Applications in Nanobiotechnology, Single Molecule Detection, and Single Cell Analysis**

2007

ISBN: 978-0-471-74660-7

C.A. Mirkin, C.M. Niemeyer (Eds.)

**Nanobiotechnology II**

**More Concepts and Applications**

2007

ISBN: 978-3-527-31673-1

R.S. Marks, C.R. Lowe, D.C. Cullen, H.H. Weetall, I. Karube (Eds.)

**Handbook of Biosensors and Biochips**

**2 Volume Set**

2007

ISBN: 978-0-470-01905-4

*Edited by Chang Lu*

# Chemical Cytometry

## Ultrasensitive Analysis of Single Cells

WILEY-VCH Verlag GmbH & Co. KGaA

**The Editor**

***Prof. Chang Lu***
Virginia Tech
Department of Chemical Engineering
Blacksburg
Virginia 24061
USA

**Cover**
The cover image is based on the work by Ana Valero and Albert van den Berg.



**Library of Congress Card No.:** applied for

**British Library Cataloguing-in-Publication Data**
A catalogue record for this book is available from the British Library.

**Bibliographic information published by the Deutsche Nationalbibliothek**
The Deutsche Nationalbibliothek lists this publication in the Deutsche Nationalbibliografie; detailed bibliographic data are available on the Internet at http://dnb.d-nb.de.



Printed in the Federal Republic of Germany
Printed on acid-free paper

**Cover Design** Formgeber, Eppelheim
**Typesetting** Laserwords Private Limited, Chennai, India
**Printing and Binding** betz-druck GmbH, Darmstadt

**ISBN:** 978-3-527-32495-8

# Contents



*Chemical Cytometry*. Edited by Chang Lu

ISBN: 978-3-527-32495-8

# Preface

Chemical cytometry has grown into a mature field after decades of development. Although the exact definition of chemical cytometry varies among different researchers, most chemical cytometry studies involve breaching the cell membrane for probing intracellular molecules. As one will find from this book, chemical cytometry and related methods have provided a powerful set of tools that can be used to tackle diverse problems over a wide range of topics. Chemical analysis of cells at the single-cell level provides several distinct advantages: first, it allows probing the heterogeneity of a cell population; second, it is possible to look into the functions/behaviors of a cell or a subcellular component as a biological unit; third, the sensitivity permits investigation of scarce biological samples such as those derived from primary materials. The state-of-the-art sensors provide the capability for examining single cells by optical, electronic, and other spectroscopic means. However, when a single-cell study is conceived, both the feasibility and the necessity need to be intensively debated: Is the molecule of interest abundant enough for the detection method of choice? Will the single-cell data yield unique information that is not available from bulk studies? Is heterogeneity in the cell population expected? How many cells need to be examined in order to generate conclusions with statistical significance for the population?

The goal of this book is to provide a survey of the field of chemical cytometry and single-cell analysis. It covers important topics such as analysis of proteins and nucleic acids from single cells, common approaches such as capillary electrophoresis and microfluidics, and widely-used detection methods such as optical, electrochemical, and mass spectrometric tools. The collection does not mean to be exhaustive. The hope is that, by knowing the works by some of the best researchers in the field and understanding the rationale behind them, scientists can assess the potential utility of chemical cytometry approach in their own work.

I would like to express my gratitude to all the authors who contributed to the book, and to Frank Weinreich and Heike Noethe at Wiley-VCH for their assistance.

West Lafayette, Indiana, November 2009 *Chang Lu*

*Chemical Cytometry*. Edited by Chang Lu

ISBN: 978-3-527-32495-8

## List of Contributors

*Nancy L. Allbritton*
University of North Carolina at
Chapel Hill
Department of Chemistry
Chapman Hall
Chapel Hill
NC 27599-3290
USA

*Ning Bao*
Purdue University
Department of Agricultural and
Biological Engineering
225 South University Street
West Lafayette
IN 47907
USA

*Michael Brenner*
University of North Carolina at
Chapel Hill
Department of Chemistry
Chapman Hall
Chapel Hill
NC 27599-3290
USA

*Yan Chen*
California Institute of Technology
Department of Bioengineering &
Electrical Engineering
1200 E California Boulevard
Pasadena
CA 91125
USA

*Daniel T. Chiu*
University of Washington
Department of Chemistry
Bagley Hall
Seattle
WA 98195-1700
USA

*Yan Dong*
The Pennsylvania State University
Department of Chemistry
104 Chemistry Building
University Park
PA 16802
USA

*Norman J. Dovichi*
University of Washington
Department of Chemistry
Bagley Hall Room 36
Seattle
WA 98195-1700
USA

*Chemical Cytometry*. Edited by Chang Lu

ISBN: 978-3-527-32495-8

***David Essaka***
University of Washington
Department of Chemistry
Bagley Hall Room 36
Seattle
WA 98195
USA

***Andrew G. Ewing***
Göteborg University
Department of Chemistry
Kemivägen 10
SE-41296 Göteborg
Sweden

*and*

The Pennsylvania State University
Department of Chemistry
104 Chemistry Building
University Park
PA 16802
USA

***Michael L. Heien***
The Pennsylvania State University
Department of Chemistry
104 Chemistry Building
University Park
PA 16802
USA

***Ole Hindgaul***
Carlsberg Laboratory
Gamle Carlsberg Vej 10
DK-2500
Valby Copenhagen
Denmark

***Wenrui Jin***
Shandong University
School of Chemistry and
Chemical Engineering
27 Shanda Nanlu
Jinan 250100
China

***Ann M. Knolhoff***
University of Illinois at
Urbana-Champaign
2317 Beckman Institute
405 N Mathews Avenue
Urbana
IL 61801
USA

***Sumith Kottegoda***
University of North Carolina at
Chapel Hill
Department of Chemistry
Chapman Hall
Chapel Hill
NC 27599-3290
USA

***Michael E. Kurczy***
The Pennsylvania State University
Department of Chemistry
104 Chemistry Building
University Park
PA 16802
USA

***Ni Li***
University of California
Department of Chemistry
501 Big Springs Road
Riverside
CA 92521
USA

***Robert M. Lorenz***
University of Washington
Department of Chemistry
Bagley Hall
Seattle
WA 98195-1700
USA

***Chang Lu***
Virginia Tech
Department of Chemical
Engineering
Blacksburg
Virginia 24061
USA

***Monica P. Palcic***
Carlsberg Laboratory
Gamle Carlsberg Vej 10
DK-2500
Valby
Copenhagen
Denmark

***Ryan M. Phillips***
University of North Carolina at
Chapel Hill
Department of Pharmacology
120 Mason Farm Road
Chapel Hill
NC 27599-7365
USA

***Jillian Prendergast***
The Johns Hopkins School of
Medicine
Department of Pharmacology
725-N. Wolfe Street
Baltimore
MD 21205
USA

***Angela Proctor***
University of North Carolina at
Chapel Hill
Department of Chemistry
Chapman Hall
Chapel Hill
NC 27599-3290
USA

***Kangning Ren***
Hong Kong University of Science
and Technology
Department of Chemistry
Clear Water Bay
Kowloon
Hong Kong
China

***Stanislaw S. Rubakhin***
University of Illinois at
Urbana-Champaign
2317 Beckman Institute
405 N Mathews Avenue
Urbana
IL 61801
USA

***Ronald L. Schnaar***
The Johns Hopkins School of
Medicine
Department of Pharmacology
725-N. Wolfe Street
Baltimore
MD 21205
USA

*and*

The Johns Hopkins School of
Medicine
Department of Neuroscience
725-N. Wolfe Street
Baltimore
MD 21205
USA

***Christopher E. Sims***
University of North Carolina at
Chapel Hill
Department of Chemistry
Chapman Hall
Chapel Hill
NC 27599-3290
USA

***Jonathan V. Sweedler***
University of Illinois at
Urbana-Champaign
2317 Beckman Institute
405 N Mathews Avenue
Urbana
IL 61801
USA

***Ana Valero***
Ecole Polytechnique Federale de
Lausanne
LMIS4
Microsystem Laboratory 4
BM 3124
Station 17
CH-1015 Lausanne
Switzerland

***Albert van den Berg***
University of Twente
Mathematics and Computer
Science
Faculty of Electrical Engineering
BIOS Lab-on-a-chip group
Veldmaat 10
7500 AE Enschede
The Netherlands

***David R. Walt***
Tufts University
Department of Chemistry
62 Talbot Avenue
Medford
MA 02155
USA

***Hsiang-Yu Wang***
National Cheng Kung University
Department of Chemical
Engineering
No. 1 University Road
Tainan City, 701
Taiwan

***Jun Wang***
Purdue University
Department of Agricultural and
Biological Engineering
225 South University Street
West Lafayette
IN 47907
USA

***Ragnhild D. Whitaker***
Tufts University
Department of Chemistry
62 Talbot Avenue
Medford
MA 02155
USA

***Colin D. Whitmore***
University of Washington
Department of Chemistry
Bagley Hall Room 36
Seattle
WA 98195
USA

***Hongkai Wu***
Hong Kong University of Science
and Technology
Department of Chemistry
Clear Water Bay
Kowloon
Hong Kong
China

***Shan Yang***
University of North Carolina at
Chapel Hill
Department of Chemistry
Chapman Hall
Chapel Hill
NC 27599-3290
USA

***Jiang F. Zhong***
University of Southern California
Department of Pathology
Keck School of Medicine
2025 Zonal Avenue
Los Angeles
CA 90033
USA

***Wenwan Zhong***
University of California
Department of Chemistry
501 Big Springs Road
Riverside
CA 92521
USA

# 1
# Origin, Current Status, and Future Perspectives of Chemical Cytometry

*Norman J. Dovichi*

## 1.1
## The Cell and Cytometry

Cells are the organizing units of life. However, not all cells are alike; individual cells can differ significantly from their neighbors. These differences can be due to genetic, environmental, cell cycle, or stochastic effects, and these differences can have profound effects upon the behavior of a cellular population. A knowledge of the function and composition of individual cells will lead to deeper understanding of the response to disease, development, and environmental insult.

Most biological techniques are incapable of determining cellular heterogeneity. Instead, most biological techniques begin with the homogenization of many millions of cells, and those techniques simply report the average composition and response of the population. These methods are inherently blind to the distribution of the cellular population.

There is a simple reason why most biological measurements are performed on millions of cells. Cells are small, and very few biological techniques are capable of analyzing the minute amount of material contained within a cell. Consider a typical cell from a higher eukaryote. That cell might be 15 μm in diameter, corresponding to a volume of 1 pl and a mass of 1 ng. The cell is roughly 15% protein, corresponding to 150 pg; roughly 7% nucleic acids, corresponding to 70 pg; and roughly 2.5% carbohydrates and 2.5% lipids, corresponding to 25 pg. Analysis of this minute amount of material is a formidable challenge.

## 1.2
## Flow and Image Cytometry

Cytometry refers to the instrumental characterization of individual cells. Classical cytometry methods were developed in collaboration with physicists, and they employed sophisticated optical and electronic means to determine the minute amount of material within a cell. Howard Shapiro's classic text *Practical Flow Cytometry*, which is now in its fourth edition, provides a delightful view both of the

*Chemical Cytometry*. Edited by Chang Lu

ISBN: 978-3-527-32495-8

distant history of cytometric methods and of the current status of flow cytometry, which is arguably the most powerful and popular classic method of cytometry [1].

In flow cytometry, a dilute suspension of cells is passed through a tightly focused laser beam [2]. Each cell generates optical signals that provide information on the cellular composition, organization, and size. For example, fluorescently labeled antibodies provide information on cell-surface markers, intercalating dyes provide information on DNA content, and fluorogenic reagents provide information on enzymatic activity within the cell. Near-angle light scattering provides information on cell size, while right-angle scattering provides information on subcellular components. The impedance change that occurs as a cell occludes a narrow aperture in the flowing stream can be used to estimate the cell size.

Flow cytometry has a number of powerful attributes. First, it can process tens of thousands of cells per second, providing outstanding statistics on the distribution of cellular properties. Second, it can operate in multiparameter mode, simultaneously monitoring a dozen cellular properties. Third, it is a nondestructive technology. Fourth, cells can be automatically selected in flow-sorting instruments. Sorters essentially incorporate ink-jet printer heads to deflect drops containing the desired cells into receiving vessels. Perhaps the most powerful commercial flow sorter is manufactured by Cytopeia, a company founded by Ger van den Engh in Seattle and now part of Beckton–Dickinson. Their Influx instrument is specified to sort 25 000 cells per second, which is near the fundamental limit set by hydrodynamics [3].

Image cytometry is the second form of classic cytometry. In this case, a fluorescent microscope is equipped with a camera, which records the image of the field of view. Image processing is then used to characterize the cellular population. Like flow cytometry, image cytometry can process very large numbers of cells and is a nondestructive technology. Also like flow cytometry, image cytometry can only monitor a handful of parameters from each cell.

Both flow and image cytometry have important limitations. First, only a dozen components can be analyzed from a single cell, limited by the spectral resolution of the instrument and the overlap of fluorescent reagents. Second, conventional cytometry employs affinity probes against known targets; the unexpected is invisible.

## 1.3 Chemical Cytometry

Conventional cytometry methods rely on physical methods to characterize cells. Chemical cytometry, by contrast, employs powerful methods of analytical chemistry to characterize cells. These methods, in general, are destructive – the cell must be lysed before analysis. These methods inevitably are much slower than classic cytometry; obtaining large amounts of information from a cell inevitably takes time, and analyzing large numbers of cells tends to be a heroic effort. However, these methods can provide extraordinarily rich information on the composition of a cell; this rich information ultimately motivates development of the technology.

This book provides a snapshot of the technology in the first decade of the twenty-first century. This time is exciting. Chemical cytometry is escape from its infancy, and the field is ripe for rapid advancement.

This chapter focuses on some of the history of chemical cytometry. The remainder of the book provides a glimpse into its future.

### 1.3.1 Prehistory of Chemical Cytometry – Microchemistry, Microspectroscopy, and Microseparations

The field of microchemistry developed roughly 100 years ago. At that time, microanalysis referred to analysis of microgram or smaller amounts of material. Most of the methods developed during this time were simply miniaturized versions of classic wet-chemical analysis, employing clever designs for burettes, pipettes, and sensitive balances. However, these measurements predated electronic means of measurement and required great manual dexterity to obtain quantitative results.

Perhaps reflecting the influence of physicists who moved into the field after World War II, or perhaps reflecting familiarity with microscopy, biologists appear to have been much more receptive to the use of spectroscopic methods for microanalysis than analytical chemists in the first half of the twentieth century. The collaboration between biologists and physicists resulted in early versions of microspectrometry and cytometry that employed remarkably sophisticated fluidics, optics, and electronics, culminating in the earliest flow and image cytometry methods. As one example, Gastone Matioli performed pioneering spectroscopic measurements of hemoglobin from individual erythrocytes as an early example of image cytometry in 1962 [4]. The authors were able to resolve the absorbance spectra of hemoglobin A and F. However, other forms of hemoglobin were not resolved by spectral difference, which suggests that a separation method is required to characterize these components.

Electrophoresis and chromatography are the preeminent methods for separation of biological molecules. Although electrophoresis was known since the 1800s, it was not widely used until Arne Tiselius developed moving boundary electrophoresis in 1937 [5]; a decade later, Pauling and Itano employed the technique to measure the pH dependence of electrophoretic mobility of normal and sickle-cell anemia variants of hemoglobin [6]. They demonstrated that the disease was associated with a significant difference in the isoelectric point of the two forms of the protein, reflecting differences in the structure of the two variants.

Moving boundary electrophoresis is a cumbersome technology, which severely limited its application. A set of papers appeared in the late 1940s that described the use of silica gel and agarose as supporting media for electrophoretic separations [7, 8]. Another set of papers appeared in 1950 describing paper as a supporting medium for electrophoresis, the most highly cited of which was authored by Tiselius [9]. Finally, in 1959, Raymond published a description of the use of polyacrylamide as a separation matrix [10].

### 1.3.2
### Early History of Chemical Cytometry – Before Ultrasensitive Detection

The inauguration of chemical cytometry can be traced to three papers by J. Edström. The first, published in 1953, demonstrated the electrophoretic separation of picograms of nucleotides. In this method, a 1-cm-long single silk fiber was used for the separation. A mixture of nucleotides was pipetted onto the fiber, and electrophoresis was performed at an electric field of 120 V $cm^{-1}$. After 2 h, the nucleotides had traveled 500 μm and were resolved into two bands. Detection was through ultraviolet (UV) microscopy at 275 nm. This paper was quickly followed by two others that described the application of the technique to the analysis of the nucleotides from single neurons [11–13]. Interestingly, the latter separation was performed at a field strength of 5000 V $cm^{-1}$ but nevertheless required half an hour to achieve a 0.5 mm separation.

As noted before, Matioli recognized that spectroscopy was not sufficient to resolve hemoglobins from single cells. Inspired by Edström's work, he began to use electrophoresis for the separation of hemoglobin at the single-cell level. Initial attempts employed a cellulose fiber for separation of hemoglobin variants from single erythrocytes. However, those efforts failed, and Matiolli instead used a polyacrylamide fiber to support the separation [14]. Erythrocytes were dispersed in the polymerizing acrylamide solution, which was pulled into a 10-μm-thick, 100-μm-wide ribbon. Erythrocytes were randomly distributed within the ribbon. Electrophoresis was performed at 1000 V $cm^{-1}$, and resolution of hemoglobin variants was achieved after 5 min. Detection was through bright-field microscopy at 420 nm.

Fiber-based separation methods are reminiscent of the early microchemical methods developed by analytical chemists, in that both required great dexterity and patience, and were only practiced in a few labs.

As a much more practical format, several groups began to perform polyacrylamide gel electrophoresis within glass capillaries; Neuhoff has written a valuable review of the field [15]. These capillary-based systems, the predecessors of modern capillary electrophoresis, employed precision-bore capillaries for the separation, often using tubes manufactured by the Drummond Corporation as disposable microliter-scale pipettes. These capillaries are still manufactured, and my research group employs them as precision spacers in the construction of some of our instrumentation.

The smallest of these capillaries were 100 μm i.d. and 30 mm long, although 500-μm-diameter tubes were much more commonly used. Because of the relatively large inner diameter, these early capillary electrophoresis systems were operated at modest electric fields (20–50 V $cm^{-1}$), which produced slow separations that required half an hour or more for completion. These early capillary electrophoresis systems suffered from one serious limitation; they did not employ on-column detectors. Instead, the gel was forced from the capillary and then stained, for example, by Coomassie blue to visualize proteins.

Despite limitations associated with detection and separation efficiency, these systems found use for the characterization of single cells. In one notable example,

Marchalonis and Nossal isolated the antibodies secreted by a single immunologically active cell. They labeled the antibodies with $^{125}$I, separated proteins using cellulose acetate electrophoresis at 35 V cm$^{-1}$, and detected the labeled proteins using autoradiography. Over 125 cells were successfully analyzed, and a fairly large heterogeneity was observed in the electrophoretic mobility of the labeled components. This data was used to support the hypothesis that a single cell produces a homogeneous antibody [16].

Both Wilson and Ruchel employed electrophoresis for the separation of proteins from single giant neurons of *Aplysia* [17, 18]. These neurons are ~100 μm in diameter and contain perhaps 100 ng of proteins. These quite large cells provided enough material to generate 2–20 separations. In an impressive experiment, cells were passed through sequential extractions to isolate different fractions. For example, cytoplasmic components could be isolated by osmotic shock, followed by treatment with ethylene glycol and a second osmotic shock, followed by treatment with 2 N NaCl to isolate a third fraction, followed by extraction with 8 M urea, and a final solubilization with SDS. In addition to proteins, low-molecular weight peptides were observed from the neurons. These peptides underwent changes with time, and there were hints of secretary granules observed in some electropherograms.

In 1975, O'Farrell introduced two-dimensional gel electrophoresis [19]. Shortly thereafter, miniaturized versions were reported [20]. In the miniaturized methods, isoelectric focusing (IEF) was performed in a Drummond capillary. After focusing, the capillary's contents were carefully extruded onto a miniature SDS-PAGE slab gel, and separated at right angles to the IEF gel. Neukirchen and colleagues employed a miniaturized two-dimensional gel electrophoresis separation method and silver staining to characterize proteins separated from a single *Drosophila* egg [21]. Several hundred proteins were resolved from the micrograms of protein contained in the egg. However, like most two-dimensional electrophoresis methods, reproducibility and throughput were very limited, and the technology required careful manual manipulations.

### 1.3.3
### Prehistory of Chemical Cytometry – The Origins of Capillary-based Separations

The second half of the 1970s saw developments that were of fundamental importance in the development of modern chemical cytometry. Capillary-based separations, including capillary liquid chromatography and capillary electrophoresis, were developed. Ishii, Tsuada, and Novotny developed technology for performing capillary liquid chromatography [22, 23], and Jorgenson published several highly important papers on capillary electrophoresis [24, 25].

Both capillary liquid chromatography and capillary electrophoresis are instrumental methods of analysis, where a high-performance separation column is reused for dozens or hundreds of analyses. This ability to reuse the separation column allows for automation where samples can be loaded from an autosampler for high-throughput and reproducible analysis. These capillary separations have typically employed fused silica capillaries with 10–50 μm i.d. and can easily handle nanoliter or smaller sample volumes, which is well matched to the volume of a cell.

### 1.3.4
### Prehistory of Chemical Cytometry – The Origins of Ultrasensitive Detection

Although lasers became commercially available in the later part of the 1960s, it took another decade before they became common excitation sources for fluorescence detection. In their pioneering work, Zare and coworkers employed lasers as detectors for the chromatographic separation of aflatoxins in the mid-1970s [26, 27]. These papers reported subpicogram detection based on the native fluorescence of these carcinogenic natural products. They were shortly followed by two other examples in the early 1980s [28, 29], which again employed native fluorescence for femtogram detection. The high spatial coherence of the laser allowed the beam to be focused to a high-intensity spot in the fluorescence detector, resulting in much improved detection limits compared with conventional lamp-based excitation.

I had the good pleasure of doing a postdoctoral fellowship with Dick Keller at Los Alamos Scientific Laboratory in 1981–1982. Through a series of happy accidents, I was able to spend half of my time working in the flow cytometry laboratory with Jim Jett and John Martin. The Los Alamos flow cytometry laboratory had made a number of very important contributions to the field, perhaps the most important being the development of the first flow sorter by Mack Fulwyler [30]. I worked with Jett and Martin to investigate the performance of the sheath-flow cuvette, a key component of a flow cytometer, as a fluorescence detector for neat solutions. We obtained attogram detection limits for a fluorescent dye, which represented a two order of magnitude improvement in detection limit over earlier work. We also mapped an approach to use laser-induced fluorescence for single-molecule detection, which represents the ultimate detection limit [31, 32].

In 1988, Yung-Fong Cheng, a graduate student in my laboratory, coupled capillary electrophoresis with a fluorescence detector based on the sheath-flow cuvette [33]. The combination produced both outstanding electrophoretic separation and ultrasensitive fluorescence detection. He achieved low zeptomole detection limits for a number of fluorescein thiohydantoin derivatives of amino acids, which represented a 10 000-fold improvement in detection limit for these biological compounds. Optimization of the laser power resulted in an order of magnitude improvement in detection limit [34]. Zare and Jorgenson also published papers in 1988 that coupled capillary electrophoresis with laser-induced fluorescence, in both cases achieving attomole detection limits [35, 36].

### 1.3.5
### Prehistory of Chemical Cytometry – The Polymerase Chain Reaction and Fluorescence-based DNA Sequencing

The development of the polymerase chain reaction in 1985 provided an extraordinarily powerful tool for nucleic acid analysis from minute samples [37].

Similarly, the development of fluorescent-based DNA sequencing based on Sanger's chain-terminating reaction provided a powerful means of identifying nucleic acids [38, 39]. When coupled with capillary electrophoresis, the genomics community had two powerful tools for the rapid characterization of the genetic content of single cells [40–43].

### 1.3.6
### Chemical Cytometry: Protein Analysis – The First Experiments

Modern chemical cytometry had its genesis in a series of papers published by Jim Jorgenson, Andy Ewing, and Ed Yeung toward the end of the 1980s. In each case, capillary separations were employed as powerful tools to resolve components from single cells. These papers addressed the important issues in chemical cytometry: cell lysis, labeling, separation, and detection.

Jorgenson published the first modern chemical cytometry paper in 1987 in a somewhat obscure journal [44]. That paper was followed by two higher profile papers [45, 46]. In all cases, giant neurons were used for analysis. Cell lysis was primitive: the cell was placed within a microcentrifuge tube and manually homogenized. The homogenate was labeled with naphthalene-2,3-dicarboxaldehyde (NDA) for fluorescence detection. An aliquot was injected into a capillary liquid chromatographic column or a capillary electrophoresis column for separation. Components were detected by voltammetry or, in the *Science* paper, by lamp-based fluorescence. Jorgenson later collaborated with Mark Wightman to employ electrochemistry to detect catecholamines either within or secreted from individual cells [47, 48].

Capillary liquid chromatography is not ideal for single-cell analysis. It requires transfer of the lysed cellular components to an injection loop for introduction to the separation column. Although it is possible to perform the transfer with the relatively large cells employed in the earliest papers, such manipulations with a typical eukaryote cell would result in unacceptable levels of dilution and poor signal-to-noise. Little work has been done since then using liquid chromatography for chemical cytometry.

Capillary electrophoresis is a much more convenient tool for single-cell analysis because injection is much simpler and the capillary's inner lumen can act as a convenient microscale reactor for component labeling. Both Jorgenson and Ewing used capillary electrophoresis for single-cell analysis in the 1980s [49]. Like Jorgenson, Ewing also worked with single giant neurons. However, Ewing used an etched capillary tip to puncture the neuron's membrane and to sample a portion of the cell's cytoplasm. Capillary electrophoresis was used to separate the sampled components, and amperometry was used to detect easily oxidizable components.

### 1.3.7
### Chemical Cytometry: Protein Analysis – Cell Lysis

These papers laid the foundation for the use of modern separation tools to characterize cellular components. However, they required manual maceration of the cell (Jorgenson) or puncture of the cell membrane (Ewing) to obtain material for analysis.

Yeung published the first description of cell lysis within an electrophoresis column before analysis [50]. Erythrocytes were treated in suspension with monobromobimane to fluorescently label glutathione. After labeling, the intact, labeled cell was aspirated within a capillary. The labeled erythrocytes were lysed by passage of a 1% SDS solution, a digitonin solution, or the separation buffer. After lysis, electrophoresis was used to resolve the components. By lysing the cell within the capillary, the labeled contents were contained within the capillary, which eliminated sample loss during handling.

In some cases, cells can undergo metabolism during the period between lysis and analysis. To minimize such changes, Allbritton developed several tools for rapid cell lysis. In her earliest experiments, a suspension of cells was dropped onto a microscope slide. A pulsed Nd-YAG laser was focused near a cell on the slide. Absorbance of the beam by the slide produced an acoustic wave that lysed the nearby cell. Simultaneously, an aspiration pulse was applied to the distal end of the capillary, drawing the cellular contents within the capillary for subsequent separation [51]. This laser-pipette allowed for the very rapid sampling of cellular contents, which minimized perturbations to the cellular contents associated with the handling of the cells. Allbritton and Ramsey collaborated to employ a localized high voltage to induce rapid cell lysis of cells flowing through a microfabricated device [52].

### 1.3.8
### Chemical Cytometry: Protein Analysis – Native Fluorescence

Yeung employed a UV laser to excite native fluorescence from proteins and other components within a single cell. The first example revisited Matioli's characterization of hemoglobin in single erythrocytes, this time using a UV laser to excite native fluorescence from the high-abundance protein [53]. Similar technology was used for detection of catecholamines from single adrenal chromaffin cells, serotonin in neurons, and proteins from neurons [54–58]. Sweedler also employed capillary electrophoresis to study neurotransmitters present in single neurons [59, 60].

Unfortunately, native fluorescence usually requires lasers that operate in the UV portion of the spectrum; these lasers tend to be expensive and have limited life span, which has discouraged their application for detection of native fluorescence in chemical cytometry.

In addition, there has been interest in using much less expensive and much more reliable lasers operating in the visible portion of the spectrum to monitor expression of green fluorescent protein (GFP) from single eukaryote [61] and single prokaryote cells [62]. GFP is excited in the blue portion of the spectrum and can be detected with very high sensitivity. GFP-fusion proteins can be prepared by genetic engineering, which represents a highly precise means of fluorescently labeling a specific protein. However, since it requires genetic engineering, a significant effort is required to perform the labeling.

### 1.3.9
### Chemical Cytometry: Protein Analysis – On-column Labeling

Laser-induced fluorescence of labeled components is likely the most widely used detection method for chemical cytometry. Low cost and reliable lasers that operate in the visible portion of the spectrum are well matched to a wide range of derivatization chemistries and can easily produce detection limits in the zeptomole range. However, labeling adds an extra step to the analysis and places limits on the separation buffer, which must also be compatible with the labeling chemistry.

Ewing reported an on-column lysis and labeling scheme for chemical cytometry. In this experiment, the cell was injected into the capillary, lysed, and its contents labeled with a fluorogenic reagent. By performing the labeling chemistry within the capillary, manipulations were dramatically simplified, excessive dilution was avoided, and the entire reaction product was available for subsequent electrophoretic separation [63].

On-column labeling usually requires a high concentration of derivatization reagent. The excess reagent, and fluorescent impurities present within the reagent, can create a sea of impurity peaks that can swamp the signal from the analyte of interest. Several fluorogenic reagents have been commercialized. These reagents undergo a dramatic bathochromic shift upon reaction, which usually shifts the absorbance spectrum from the UV for the reagent to be visible for the product. The product is excited without interference from the unreacted reagent. For many years, we employed the fluorogenic reagent 3-(2-furoyl) quinoline-2-carboxaldehyde (FQ) as a fluorogenic reagent. The reagent reacts with primary amines, such as the $\varepsilon$-amine of lysine residues, to produce a product that is excited in the blue portion of the spectrum. However, FQ converts the cationic group into a neutral product. Unfortunately, it is very difficult to convert all lysine residues into labeled products for most proteins. The reaction instead converts a single protein into a set of labeled products; there are $2^N - 1$ possible fluorescent products where $N$ is the number of primary amines on the target [64]. These products produce a charge ladder, where the charge of the product is related to the number of labels that were incorporated [65]. These products have different electrophoretic mobility and generate very complex electropherograms. We discovered that incorporation of an anionic surfactant, such as SDS, collapses the electropherogram into a sharp electrophoretic peak with very high separation efficiency, which makes FQ useful for capillary sieving electrophoresis and micellar electrokinetic capillary chromatography, both of which employ charged surfactants in the separation [66].

Unfortunately, IEF is not compatible with charged surfactants, and FQ produces very complex electropherograms from standard proteins [67]. Fortunately, Wolfbeiss has recently developed a fluorogenic reagent that produces a cationic product upon reaction with an amine [68]. We have employed the reagent to label proteins for capillary IEF [69]. Our initial attempts were plagued by a very high background signal from fluorescent impurities present within commercial ampholytes. We then employed a bank of high-intensity photodiodes to photobleach those impurities, which dramatically decreased the background signal and allowed us to obtain zeptomole mass detection limits [70]. We are in the process of employing this reagent to applications of IEF in chemical cytometry.

### 1.3.10
### Chemical Cytometry: Protein Analysis – Two-dimensional Capillary Electrophoresis

A one-dimensional separation method is able to resolve perhaps 100 components from a complex protein mixture obtained from a single cell. The protein content of single cells can be simplified by differential extraction [18]. However, such manipulations require large cells such as giant neurons, and require cumbersome manual manipulations. Instead, it is valuable to perform two-dimensional separations on the proteins from single cells. In two-dimensional separations, the contents of a cell undergo separation in a first dimension capillary. Once the fastest moving components reach the distal end of that capillary, fractions are repetitively transferred to a second capillary for subsequent separation. Our initial work employed the FQ reagent to label proteins and various forms of capillary electrophoresis for separation [71–73]. While these forms of electrophoresis provided outstanding separation of biogenic amines, the resolution of proteins was disappointing [74]. Fortunately, Wolfbeiss's fluorogenic dyes allow IEF as a separation mode. We are developing a two-dimensional capillary electrophoresis separation based on IEF in the first dimension and the capillary equivalent to SDS-PAGE in the second; this combination should provide outstanding resolution of proteins from single cells.

## 1.4
## Chemical Cytometry of DNA and mRNA

There has been quite a bit of activity in the analysis of genomic and mRNA expression in single cells. The availability of commercial primer sets for whole-genome PCR amplification has proven to be particularly powerful for such studies. In the experiments, a single cell is lysed in a nanoliter container. After whole-genome amplification, any sequencing method can be employed to characterize nucleic acids. In the original approach, pyrosequencing was employed [75]. For mRNA studies, it is necessary to incorporate a reverse transcriptase step to produce cDNA, which is then amplified and analyzed by western blotting, hybridization to an oligonucleotide array, or direct sequencing [76–79].

## 1.5
## Metabolic Cytometry

We coined the term *metabolic cytometry* to refer to the analysis of metabolites in single cells [80]. Metabolites consist of peptides, lipids, carbohydrates, biogenic amines, and other low-molecular weight components. A number of fluorogenic reagents have been developed to monitor the activity of specific hydrolytic enzymes. These reagents are taken up by cells and converted into membrane impermeable and highly fluorescent products; reagents are available to assay for esterase, phosphatase, glucosidase, and other enzymes. These reagents are commonly used in flow cytometry. However, they can only provide information on the activity of a single enzyme; more general metabolic activity requires more powerful chemical cytometry methods.

### 1.5.1 Directed Metabolic Cytometry

Metabolic studies tend to fall into two classes. In directed metabolic cytometry, cells are treated with a fluorescent substrate. Conversion of the substrate to product results in a change in electrophoretic behavior that is detected as a mobility shift. Allbritton's work on kinase activity is an example of directed metabolic cytometry [81–83]. In these studies, cells are treated with fluorescently labeled peptides that are substrates to different kinases. After incubation, cells are lysed and the products analyzed by capillary electrophoresis with laser-induced fluorescence detection. Incorporation of a phosphate group results in a shift in the peptide's mobility. By careful adjustment of the peptide structure and electrophoretic conditions, it is possible to simultaneously monitor the activity of a number of kinases.

We have developed another form of directed metabolic cytometry. In our case, we incubate cells with a fluorescent oligosaccharide or glycolipid [80, 84–86]. Cells take up the substrate and a cascade of enzymes convert it to a series of catabolic and anabolic products. However, as long as the fluorescent tag remains, any metabolic product can be detected with yoctomole detection limits [87]. These metabolic cascades can become extremely complex, particularly for carbohydrate metabolism, and the resolution of very complex mixtures of metabolic products can be a challenge. Chapter 2 of this book presents some recent results on the metabolism of sphingolipids in single neurons.

As a final example of metabolic cytometry, Kennedy has developed a powerful microfluidic system for monitoring insulin production from single islets of Langerhans [88]. This technology employs repeated sampling from the islet followed by competitive immunoassay at rates of 5400 assays per hour, which provides extraordinary temporal resolution of insulin production.

### 1.5.2 Shotgun Metabolic Cytometry

The directed metabolic cytometry methods are designed to monitor activity of specific enzymes. This situation is not useful when surveying the metabolism of single cells. In this case, it is useful to employ shotgun metabolic cytometry that provides a broad survey of metabolites. Sweedler has worked extensively in this area, developing powerful mass spectrometric methods to characterize peptides and small-molecule neurotransmitters in single neurons and organelles [89–93].

## 1.6 Future Perspectives on Instrumentation for Chemical Cytometry

The technology employed in chemical cytometry is maturing rapidly. The procedures for cell culture, isolation, lysis, and labeling are mostly developed. Separation of very complex samples, such as single-cell protein lysates, remains challenging, but recent advances in ultrasensitive capillary IEF provides a powerful new tool.

Microfluidic devices play an important role in manipulating cells; when coupled with capillary array electrophoresis, they provide an opportunity for high-throughput and high-resolution analysis [94].

### 1.6.1 Chemical Cytometry of Primary Cells Is the Wave of the Future

In our experience, analysis of cultured cells can produce highly reproducible results for chemical cytometry of proteins, carbohydrates, and glycolipids. As a result, cultured cells are wonderful reagents to investigate the inherent precision of chemical cytometry. However, such analyses ultimately are boring. If cells generate reproducible chemical cytometry patterns, then there is little need to employ chemical cytometry for their analysis. It is much simpler to analyze a cellular homogenate using conventional analytical methods.

There are some examples when analysis of cell cultures may be of interest. For example, cells in culture pass through the cell cycle as they grow and divide, and chemical cytometry can be used to characterize the evolution in cellular composition during the cell cycle [80, 95–97]. We have used image cytometry on cells treated with an intercalating stain to segregate cells into the G1 and G2/M phases of the cell cycle before analysis [80, 95], and we have sorted ethanol fixed cells to isolate cells at a specific phase of the cell cycle before chemical cytometry [96, 97].

However, more interesting results can be obtained from analysis of primary cells. We have investigated a number of examples of chemical cytometry of primary cells. In each case, we observe a much more heterogeneous behavior than the corresponding cultured cells. Chapter 2 of this book presents some recent results on sphingolipid metabolism in single neurons, which indeed is much more complex than similar experiments performed on cultured cells. We envision that chemical cytometry will be a particularly powerful tool for characterizing the changes that accompany the development of an embryo [98, 99] and differentiation of a stem cell.

### 1.6.2 Challenges – Tissue Dissociation Can Introduce Changes in a Cell's Composition

Often it is the details that bedevil the applications for chemical cytometry. For example, dissociation of primary cells from tissues can be remarkably challenging. While trypsin and related proteases can be used to dissociate cells, the activity of those enzymes undoubtedly induces chemical and biochemical responses. For this reason, hematopoietic cells are tempting targets for chemical cytometry because they do not require dissociation from a tissue.

### 1.6.3 Challenges – Cells Need to Be Fixed to Prevent Changes Associated with Sample Handling

Similarly, it is necessary to consider the changes induced while handling cells. Viable cells that sit on ice for several hours before analysis will likely undergo stress response, changing the characteristics of the cell. We have decided to fix cells as soon as possible to minimize changes during subsequent handling. Formalin is valuable for fixing cells before metabolic cytometry since it does not react with the target compounds. This fixing is of great convenience. The data provided in Chapter 2 were generated from cells that were dissociated and incubated at Johns Hopkins University in Baltimore and then shipped to the University of Washington for analysis, which is at a distance of 4000 km. Such an arrangement would not be possible with unfixed cells, in which case the chemical cytometry instrument would need to be in very close proximity to the animal facility.

Unfortunately, formalin cross-links proteins and nucleic acids, and that reagent cannot be used to fix cells before chemical cytometry of those components. Hence, we have been investigating the use of ethanol as a fixative [74]. Ethanol appears to efficiently denature enzymes, which should preserve the cellular composition, without introduction of cross-links. However, ethanol has several problems. It can make membranes permeable, which can result in the loss of low-molecular weight components. Treated cells tend to be much more sticky than their untreated counterparts, which can introduce challenges in the manipulation of the fixed cells. Ethanol causes proteins to precipitate; their subsequent resolubilization can require the use of surfactants, which can interfere in some analyses.

### 1.6.4 Challenges – The Connection Between the Real World and the World of Chemical Cytometry

Finally, there needs to be a connection between the world of chemical cytometry and the analysis of large-scale lysates. The former provides wonderful detail on cellular composition and response to environmental changes while the latter provides ample material for conventional analytical measurements.

Identification of components observed in single-cell analysis is fraught with challenges, particularly for fluorescence-based assays. The presence of a peak in an electropherogram simply means that some component was present. As noted nearly 50 years ago by Matioli *et al.* [4], spectral information provides very limited information for the identification of components.

Ideally, mass spectrometry would be used for identification of components in chemical cytometry. As demonstrated by Sweedler, the current generation of mass spectrometers has sufficient sensitivity to monitor peptide and metabolite expression in single neurons [89–93, 100]. Unfortunately, the sensitivity of those instruments is insufficient to detect any but the most abundant proteins in a single cell.

We have employed purified proteins as standards, which are used to spike electropherograms; those proteins are isolated from gel electrophoresis of large-scale homogenates [101]. The isolation of proteins is very tedious, and alternative methods are required for component identification. Fortunately, capillary electrophoresis can be very reproducible, so that analysis of large homogenates by capillary electrophoresis with mass spectrometry detection may provide tentative identification of the components in single cells.

## Acknowledgments

I acknowledge support from the National Institutes of Health (P50HG002360, R01NS061767, and R33CA122900).

## References

1 Shapiro, H. (2003) *Practical Flow Cytometry*, Wiley-Liss, Hoboken.
2 Steinkamp, J.A. (1984) Flow cytometry. *Rev. Sci. Instrum.*, **55**, 1375–1400.
3 Ibrahim, S.F. and van den Engh, G. (2003) High-speed cell sorting: fundamentals and recent advances. *Curr. Opin. Biotechnol.*, **14**, 5–12.
4 Matioli, G., Thorell, B., and Brody, S. (1962) Microspectrophotometric determination of differentially extracted haemoglobin in single erythrocytes. *Acta Haematol.*, **28**, 73–85.
5 Tiselius, A. (1937) A new apparatus for electrophoretic analysis of colloidal mixtures. *Trans. Faraday Soc.*, **33**, 524–530.
6 Pauling, L., Itano, H.A., Wells, I.C., Schroeder, W.A., Kay, L.M., Singer, S.J., and Corey, R.B. (1950) Sickle cell anemia hemoglobin. *Science*, **111**, 459–459.
7 Gordon, A.H., Keil, B., and Sebesta, K. (1949) Electrophoresis of proteins in agar jelly. *Nature*, **164**, 498–499.
8 Consden, R., Gordon, A.H., and Martin, A.J.P. (1946) Ionophoresis in silica jelly – a method for the separation of amino-acids and peptides. *Biochem. J.*, **40**, 33–41.
9 Cremer, H.D. and Tiselius, A. (1950) Elektrophorese von eiweiss in filtrierpapier. *Biochem. Z.*, **320**, 273–283.
10 Raymond, S. and Weintraub, L. (1959) Acrylamide gel as a supporting medium for zone electrophoresis. *Science*, **130**, 711.
11 Edstrom, J.E. (1953) Ribonucleic acid mass and concentration in individual nerve cells – a new method for quantitative determinations. *Biochim. Biophys. Acta*, **12**, 361–386.
12 Edstrom, J.E. (1953) Nucleotide analysis on the cyto-scale. *Nature*, **172**, 809.
13 Edstrom, J.E. and Hyden, H. (1954) Ribonucleotide analysis of individual nerve cells. *Nature*, **174**, 128–129.
14 Matioli, G.T. and Niewisch, H.B. (1965) Electrophoresis of hemoglobin in single erythrocytes. *Science*, **150**, 1824–1826.
15 Neuhoff, V. (2000) Microelectrophoresis and auxiliary micromethods. *Electrophoresis*, **21**, 3–11.
16 Marchalonis, J.J. and Nossal, G.J. (1968) Electrophoretic analysis of antibody produced by single cells. *Proc. Natl. Acad. Sci. U.S.A.*, **61**, 860–867.
17 Wilson, D.L. (1971) Molecular weight distribution of proteins synthesized in

single, identified neurons of Aplysia. *J. Gen. Physiol.*, **57**, 26–40.

18 Ruchel, R. (1976) Sequential protein analysis from single identified neurons of Aplysia californica. A microelectrophoretic technique involving polyacrylamide gradient gels and isoelectric focusing. *J. Histochem. Cytochem.*, **24**, 773–791.

19 OFarrell, P.H. (1975) High-resolution 2-dimensional electrophoresis of proteins. *J. Biol. Chem.*, **250**, 4007–4021.

20 Ruchel, R. (1977) 2-dimensional micro-separation technique for proteins and peptides, combining isoelectric-focusing and gel gradient electrophoresis. *J. Chromatogr.*, **132**, 451–468.

21 Neukirchen, R.O., Schlosshauer, B., Baars, S., Jackle, H., and Schwarz, U. (1982) Two-dimensional protein analysis at high resolution on a microscale. *J. Biol. Chem.*, **257**, 15229–15234.

22 Ishii, D., Asai, K., Hibi, K., Jonokuchi, T., and Nagaya, M. (1977) Study of micro-high-performance liquid-chromatography. 1. Development of technique for miniaturization of high-performance liquid-chromatography. *J. Chromatogr.*, **144**, 157–168.

23 Tsuda, T. and Novotny, M. (1978) Packed Microcapillary columns in high-performance liquid-chromatography. *Anal. Chem.*, **50**, 271–275.

24 Jorgenson, J.W. and Lukacs, K.D. (1981) Zone electrophoresis in open-tubular glass-capillaries. *Anal. Chem.*, **53**, 1298–1302.

25 Jorgenson, J.W., and Lukacs, K.D. (1983) Capillary zone electrophoresis. *Science*, **222**, 266–272.

26 Diebold, G.J. and Zare, R.N. (1977) Laser fluorimetry – subpicogram detection of aflatoxins using high-pressure liquid-chromatography. *Science*, **196**, 1439–1441.

27 Berman, M.R. and Zare, R.N. (1975) Laser fluorescence analysis of chromatograms – sub-nanogram detection of aflatoxins. *Anal. Chem.*, **47**, 1200–1201.

28 Sepaniak, M.J. and Yeung, E.S. (1980) Determination of adriamycin and daunorubicin in urine by high-performance liquid-chromatography with laser fluorometric detection. *J. Chromatogr.*, **190**, 377–383.

29 Folestad, S., Johnson, L., Josefsson, B., and Galle, B. (1982) Laser-induced fluorescence detection for conventional and microcolumn liquid-chromatography. *Anal. Chem.*, **54**, 925–929.

30 Fulwyler, M.J. (1965) Electronic separation of biological cells by volume. *Science*, **150**, 910–911.

31 Dovichi, N.J., Martin, J.C., Jett, J.H., and Keller, R.A. (1983) Attogram detection limit for aqueous dye samples by laser-induced fluorescence. *Science*, **219**, 845–847.

32 Dovichi, N.J., Martin, J.C., Jett, J.H., Trkula, M., and Keller, R.A. (1984) Laser-induced fluorescence of flowing samples as an approach to single-molecule detection in liquids. *Anal. Chem.*, **56**, 348–354.

33 Cheng, Y.F. and Dovichi, N.J. (1988) Subattomole amino-acid analysis by capillary zone electrophoresis and laser-induced fluorescence. *Science*, **242**, 562–564.

34 Wu, S.L. and Dovichi, N.J. (1989) High-sensitivity fluorescence detector for fluorescein isothiocyanate derivatives of amino-acids separated by capillary zone electrophoresis. *J. Chromatogr.*, **480**, 141–155.

35 Roach, M.C., Gozel, P., and Zare, R.N. (1988) Determination of methotrexate and its major metabolite, 7-hydroxymethotrexate, using capillary zone electrophoresis and laser-induced fluorescence detection. *J. Chromatogr.*, **426**, 129–140.

36 Nickerson, B. and Jorgenson, J.W. (1988) High-speed capillary zone electrophoresis with laser-induced fluorescence detection. *J. High Resolut. Chromatogr. Chromatogr. Commun.*, **11**, 533–534.

37 Saiki, R.K., Scharf, S., Faloona, F., Mullis, K.B., Horn, G.T., Erlich, H.A., and Arnheim, N. (1985) Enzymatic

amplification of beta-globin genomic sequences and restriction site analysis for diagnosis of sickle cell anemia. *Science*, **230**, 1350–1354.

38 Smith, L.M., Sanders, J.Z., Kaiser, R.J., Hughes, P., Dodd, C., Connell, C.R., Heiner, C., Kent, S.B.H., and Hood, L.E. (1986) Fluorescence detection in automated DNA-sequence analysis. *Nature*, **321**, 674–679.

39 Prober, J.M., Trainor, G.L., Dam, R.J., Hobbs, F.W., Robertson, C.W., Zagursky, R.J., Cocuzza, A.J., Jensen, M.A., and Baumeister, K. (1987) A system for rapid DNA sequencing with fluorescent chain-terminating dideoxynucleotides. *Science*, **238**, 336–341.

40 Swerdlow, H. and Gesteland, R. (1990) Capillary gel-electrophoresis for rapid, high-resolution DNA sequencing. *Nucleic Acids Res.*, **18**, 1415–1419.

41 Drossman, H., Luckey, J.A., Kostichka, A.J., Dcunha, J., and Smith, L.M. (1990) High-speed separations of DNA sequencing reactions by capillary electrophoresis. *Anal. Chem.*, **62**, 900–903.

42 Cohen, A.S., Najarian, D.R., and Karger, B.L. (1990) Separation and analysis of DNA-sequence reaction-products by capillary gel-electrophoresis. *J. Chromatogr.*, **516**, 49–60.

43 Swerdlow, H., Wu, S., Harke, H., and Dovichi, N.J. (1990) Capillary gel-electrophoresis for DNA sequencing - laser-induced fluorescence detection with the sheath flow cuvette. *J. Chromatogr.*, **516**, 61–67.

44 Kennedy, R.T., St Claire, R.L., White, J.G., and Jorgenson, J.W. (1987) Chemical-analysis of single neurons by open tubular liquid-chromatography. *Mikrochim. Acta*, **2**, 37–45.

45 Kennedy, R.T., Oates, M.D., Cooper, B.R., Nickerson, B., and Jorgenson, J.W. (1989) Microcolumn separations and the analysis of single cells. *Science*, **246**, 57–63.

46 Kennedy, R.T. and Jorgenson, J.W. (1989) Quantitative-analysis of individual neurons by open tubular liquid-chromatography with voltammetric detection. *Anal. Chem.*, **61**, 436–441.

47 Cooper, B.R., Jankowski, J.A., Leszczyszyn, D.J., Wightman, R.M., and Jorgenson, J.W. (1992) Quantitative determination of catecholamines in individual bovine adrenomedullary cells by reversed-phase microcolumn liquid chromatography with electrochemical detection. *Anal. Chem.*, **64**, 691–694.

48 Cooper, B.R., Wightman, R.M., and Jorgenson, J.W. (1994) Quantitation of epinephrine and norepinephrine secretion from individual adrenal medullary cells by microcolumn high-performance liquid chromatography. *J. Chromatogr. B*, **653**, 25–34.

49 Olefirowicz, T.M. and Ewing, A.G. (1990) Capillary electrophoresis in 2 and 5 microns diameter capillaries: application to cytoplasmic analysis. *Anal. Chem.*, **62**, 1872–1876.

50 Hogan, B.L. and Yeung, E.S. (1992) Determination of intracellular species at the level of a single erythrocyte via capillary electrophoresis with direct and indirect fluorescence detection. *Anal. Chem.*, **64**, 2841–2845.

51 Sims, C.E., Meredith, G.D., Krasieva, T.B., Berns, M.W., Tromberg, B.J., and Allbritton, N.L. (1998) Laser-micropipet combination for single-cell analysis. *Anal. Chem.*, **70**, 4570–4577.

52 McClain, M.A., Culbertson, C.T., Jacobson, S.C., Allbritton, N.L., Sims, C.E., and Ramsey, J.M. (2003) Microfluidic devices for the high-throughput chemical analysis of cells. *Anal. Chem.*, **75**, 5646–5655.

53 Lee, T.T. and Yeung, E.S. (1992) Quantitative determination of native proteins in individual human erythrocytes by capillary zone electrophoresis with laser-induced fluorescence detection. *Anal. Chem.*, **64**, 3045–3051.

54 Lillard, S.J. and Yeung, E.S. (1996) Analysis of single erythrocytes by injection-based capillary isoelectric focusing with laser-induced native fluorescence detection. *J. Chromatogr. B*, **687**, 363–369.

55 Tong, W. and Yeung, E.S. (1997) On-column monitoring of secretion of catecholamines from single bovine adrenal chromaffin cells by capillary

electrophoresis. *J. Neurosci. Methods*, **76**, 193–201.

56 Ho, A.M. and Yeung, E.S. (1998) Capillary electrophoretic study of individual exocytotic events in single mast cells. *J. Chromatogr. A*, **817**, 377–382.

57 Parpura, V., Tong, W., Yeung, E.S., and Haydon, P.G. (1998) Laser-induced native fluorescence (LINF) imaging of serotonin depletion in depolarized neurons. *J. Neurosci. Methods*, **82**, 151–158.

58 Yeung, E.S. (1999) Study of single cells by using capillary electrophoresis and native fluorescence detection. *J. Chromatogr. A*, **830**, 243–262.

59 Lapainis, T., Scanlan, C., Rubakhin, S.S., and Sweedler, J.V. (2007) A multichannel native fluorescence detection system for capillary electrophoretic analysis of neurotransmitters in single neurons. *Anal. Bioanal. Chem.*, **387**, 97–105.

60 Zhang, X. and Sweedler, J.V. (2001) Ultraviolet native fluorescence detection in capillary electrophoresis using a metal vapor NeCu laser. *Anal. Chem.*, **73**, 5620–5624.

61 Hu, K., Ahmadzadeh, H., and Krylov, S.N. (2004) Asymmetry between sister cells in a cancer cell line revealed by chemical cytometry. *Anal. Chem.*, **76**, 3864–3866.

62 Turner, E.H., Lauterbach, K., Pugsley, H.R., Palmer, V.R., and Dovichi, N.J. (2007) Detection of green fluorescent protein in a single bacterium by capillary electrophoresis with laser-induced fluorescence. *Anal. Chem.*, **79**, 778–781.

63 Gilman, S.D. and Ewing, A.G. (1995) Analysis of single cells by capillary electrophoresis with on-column derivatization and laser-induced fluorescence detection. *Anal. Chem.*, **67**, 58–64.

64 Zhao, J.Y., Waldron, K.C., Miller, J., Zhang, J.Z., Harke, H., and Dovichi, N.J. (1992) Attachment of a single fluorescent label to peptides for determination by capillary zone electrophoresis. *J. Chromatogr.*, **608**, 239–242.

65 Gao, J., Gomez, F.A., Härter, R., and Whitesides, G.M. (1994) Determination of the effective charge of a protein in solution by capillary electrophoresis. *Proc. Natl. Acad. Sci. U.S.A.*, **91**, 12027–12030.

66 Lee, I.H., Pinto, D., Arriaga, E.A., Zhang, Z., and Dovichi, N.J. (1998) Picomolar analysis of proteins using electrophoretically mediated microanalysis and capillary electrophoresis with laser-induced fluorescence detection. *Anal. Chem.*, **70**, 4546–4548.

67 Richards, D.P., Stathakis, C., Polakowski, R., Ahmadzadeh, H., and Dovichi, N.J. (1999) Labeling effects on the isoelectric point of green fluorescent protein. *J. Chromatogr. A*, **853**, 21–25.

68 Wetzl, B.K., Yarmoluk, S.M., Craig, D.B., and Wolfbeis, O.S. (2004) Chameleon labels for staining and quantifying proteins. *Angew. Chem. Int. Ed. Engl.*, **43**, 5400–5402.

69 Ramsay, L.M., Dickerson, J.A., and Dovichi, N.J. (2009) Attomole protein analysis by CIEF with LIF detection. *Electrophoresis*, **30**, 297–302.

70 Ramsay, L.M., Dickerson, J.A., Dada, O., and Dovichi, N.J. (2009) *Anal. Chem.*, **81**, 1741–1746.

71 Michels, D.A., Hu, S., Schoenherr, R.M., Eggertson, M.J., and Dovichi, N.J. (2002) Fully automated two-dimensional capillary electrophoresis for high sensitivity protein analysis. *Mol. Cell Proteomics*, **1**, 69–74.

72 Michels, D.A., Hu, S., Dambrowitz, K.A., Eggertson, M.J., Lauterbach, K., and Dovichi, N.J. (2004) Capillary sieving electrophoresis-micellar electrokinetic chromatography fully automated two-dimensional capillary electrophoresis analysis of Deinococcus radiodurans protein homogenate. *Electrophoresis*, **25**, 3098–3105.

73 Hu, S., Michels, D.A., Fazal, M.A., Ratisoontorn, C., Cunningham, M.L., and Dovichi, N.J. (2004) Capillary sieving electrophoresis/micellar electrokinetic capillary chromatography for two-dimensional protein fingerprinting of single mammalian cells. *Anal. Chem.*, **76**, 4044–4049.

74 Kraly, J.R., Jones, M.R., Gomez, D.G., Dickerson, J.A., Harwood, M.M., Eggertson, M., Paulson, T.G., Sanchez, C.A., Odze, R., Feng, Z., Reid, B.J., and Dovichi, N.J. (2006) Reproducible two-dimensional capillary electrophoresis analysis of Barrett's esophagus tissues. *Anal. Chem.*, **78**, 5977–5986.

75 Marcy, Y., Ouverney, C., Bik, E.M., Losekann, T., Ivanova, N., Martin, H.G., Szeto, E., Platt, D., Hugenholtz, P., Relman, D.A., and Quake, S.R. (2007) Dissecting biological "dark matter" with single-cell genetic analysis of rare and uncultivated TM7 microbes from the human mouth. *Proc. Natl. Acad. Sci. U.S.A.*, **104**, 11889–11894.

76 Brady, G., Barbara, M., and Iscove, N.N. (1990) Representative in vitro cDNA amplification from individual hemopoietic cells and colonies. *Methods Mol. Cell. Biol.*, **2**, 17–25.

77 Dulac, C. and Axel, R. (1995) A novel family of genes encoding putative pheromone receptors in mammals. *Cell*, **83**, 195–206.

78 Tietjen, I., Rihel, J.M., Cao, Y.X., Koentges, G., Zakhary, L., and Dulac, C. (2003) Single-cell transcriptional analysis of neuronal progenitors. *Neuron*, **38**, 161–175.

79 Klein, C.A., Seidl, S., Petat-Dutter, K., Offner, S., Geigl, J.B., Schmidt-Kittler, O., Wendler, N., Passlick, B., Huber, R.M., Schlimok, G., Baeuerle, P.A., and Riethmuller, G. (2002) Combined transcriptome and genome analysis of single micrometastatic cells. *Nat. Biotechnol.*, **20**, 387–392.

80 Krylov, S.N., Zhang, Z., Chan, N.W.C., Arriaga, E., Palcic, M.M., and Dovichi, N.J. (1999) Correlating cell cycle with metabolism in single cells: the combination of image and metabolic cytometry. *Cytometry*, **37**, 15–20.

81 Lee, C.L., Linton, J., Soughayer, J.S., Sims, C.E., and Allbritton, N.L. (1999) Localized measurement of kinase activation in oocytes of Xenopus laevis. *Nat. Biotechnol.*, **17**, 759–762.

82 Meredith, G.D., Sims, C.E., Soughayer, J.S., and Allbritton, N.L. (2000) Measurement of kinase activation in single mammalian cells. *Nat. Biotechnol.*, **18**, 309–312.

83 Sims, C.E. and Allbritton, N.L. (2003) Single-cell kinase assays: opening a window onto cell behavior. *Curr. Opin. Biotechnol.*, **14**, 23–28.

84 Krylov, S.N., Starke, D.A., Arriaga, E.A., Zhang, Z., Chan, N.W., Palcic, M.M., and Dovichi, N.J. (2000) Instrumentation for chemical cytometry. *Anal. Chem.*, **72**, 872–877.

85 Le, X.C., Tan, W., Scaman, C.H., Szpacenko, A., Arriaga, E., Zhang, Y., Dovichi, N.J., Hindsgaul, O., and Palcic, M.M. (1999) Single cell studies of enzymatic hydrolysis of a tetramethylrhodamine labeled trisaccharide in yeast. *Glycobiology*, **9**, 219–225.

86 Whitmore, C.D., Hindsgaul, O., Palcic, M.M., Schnaar, R.L., and Dovichi, N.J. (2007) Metabolic cytometry. Glycosphingolipid metabolism in single cells. *Anal. Chem.*, **79**, 5139–5142.

87 Zhao, J.Y., Dovichi, N.J., Hindsgaul, O., Gosselin, S., and Palcic, M.M. (1994) Detection of 100 molecules of product formed in a fucosyltransferase reaction. *Glycobiology*, **4**, 239–242.

88 Dishinger, J.F., Reid, K.R., and Kennedy, R.T. (2009) Quantitative monitoring of insulin secretion from single islets of Langerhans in parallel on a microfluidic chip. *Anal. Chem.*, **81**, 3119–3127.

89 Li, L., Romanova, E.V., Rubakhin, S.S., Alexeeva, V., Weiss, K.R., Vilim, F.S., and Sweedler, J.V. (2000) Peptide profiling of cells with multiple gene products: combining immunochemistry and MALDI mass spectrometry with on-plate microextraction. *Anal. Chem.*, **72**, 3867–3874.

90 Rubakhin, S.S., Garden, R.W., Fuller, R.R., and Sweedler, J.V. (2000) Measuring the peptides in individual organelles with mass spectrometry. *Nat. Biotechnol.*, **18**, 172–175.

91 Rubakhin, S.S., Churchill, J.D., Greenough, W.T., and Sweedler, J.V. (2006) Profiling signaling peptides in single mammalian cells using mass spectrometry. *Anal. Chem.*, **78**, 7267–7272.

92 Rubakhin, S.S. and Sweedler, J.V. (2007) Characterizing peptides in individual mammalian cells using mass spectrometry. *Nat. Protoc.*, **2**, 1987–1989.

93 Lapainis, T., Rubakhin, S.S., and Sweedler, J.V. (2009) Capillary electrophoresis with electrospray ionization mass spectrometric detection for single-cell metabolomics. *Anal. Chem.*, **81**, 5858–5864.

94 Boardman, A.K., McQuaide, S.C., Zhu, C., Whitmore, C.D., Lidstrom, M.E., and Dovichi, N.J. (2008) Interface of an array of five capillaries with an array of one nanoliter wells for high-resolution electrophoretic analysis as an approach to high-throughput chemical cytometry. *Anal. Chem.*, **80**, 7631–7634.

95 Hu, S., Le, Z., Krylov, S., and Dovichi, N.J. (2003) Cell cycle-dependent protein fingerprint from a single cancer cell: image cytometry coupled with single-cell capillary sieving electrophoresis. *Anal. Chem.*, **75**, 3495–3501.

96 Sobhani, K., Fink, S.L., Cookson, B.T., and Dovichi, N.J. (2007) Repeatability of chemical cytometry: 2-DE analysis of single RAW 264.7 macrophage cells. *Electrophoresis*, **28**, 2308–2313.

97 Harwood, M.M., Bleecker, J.V., Rabinovitch, P.S., and Dovichi, N.J. (2007) Cell cycle-dependent characterization of single MCF-7 breast cancer cells by 2-D CE. *Electrophoresis*, **28**, 932–937.

98 Harwood, M.M., Christians, E.S., Fazal, M.A., and Dovichi, N.J. (2006) Single-cell protein analysis of a single mouse embryo by two-dimensional capillary electrophoresis. *J. Chromatogr. A*, **1130**, 190–194.

99 Hu, S., Lee, R., Zhang, Z., Krylov, S.N., and Dovichi, N.J. (2001) Protein analysis of an individual Caenorhabditis elegans single-cell embryo by capillary electrophoresis. *J. Chromatogr. B*, **752**, 307–310.

100 Li, L., Garden, R.W., Romanova, E.V., and Sweedler, J.V. (1999) In situ sequencing of peptides from biological tissues and single cells using MALDI-PSD/CID analysis. *Anal. Chem.*, **71**, 5451–5458.

101 Hu, S., Le, Z., Newitt, R., Aebersold, R., Kraly, J.R., Jones, M., and Dovichi, N.J. (2003) Identification of proteins in single-cell capillary electrophoresis fingerprints based on comigration with standard proteins. *Anal. Chem.*, **75**, 3502–3505.

# 2
# Metabolic Cytometry – The Study of Glycosphingolipid Metabolism in Single Primary Cells of the Dorsal Root Ganglia

*Colin D. Whitmore, Jillian Prendergast, David Essaka, Ole Hindgaul, Monica P. Palcic, Ronald L. Schnaar, and Norman J. Dovichi*

## 2.1
## Introduction

### 2.1.1
### Glycosphingolipids Play an Important Role in Neuronal Membranes

The surfaces of animal cells are covered with glycoproteins and glycolipids. The sugar chains, termed *oligosaccharides*, face the outside of the cells where they can act as recognition molecules in a very wide array of biological events [1–3]. Both glycoproteins and glycolipids are present in cell membranes. The brain is unique among vertebrate tissues because >80% of the conjugated saccharides are glycolipids rather than glycoproteins. Glycosphingolipids are the most common glycolipids. In these compounds, the lipid portion consists of ceramide, an *N*-acyl sphingosine. The sugar chains of 1 to over 20 residues can take on very complex structures. The compounds are not static within the cell but undergo continual internalization and recycling [2]. It appears that the expression of these compounds is tightly regulated with biosynthesis and biodegradation in careful balance.

These compounds play important roles in transmembrane signal transduction, apoptosis, Alzheimer's, diabetes, and neuronal functions. There is a rich literature describing the changes in structures of glycolipids, which accompany embryonic development and tumor progression. A set of devastating diseases arises from inborn errors of metabolism involving enzymes in the biosynthetic and biodegradation pathways for glycolipids; the best known of these are Tay-Sachs and Gaucher diseases.

### 2.1.2
### The Metabolism of Glycosphingolipids Is Quite Complex

Glycosphingolipid metabolism has been worked out in detail (*http://www.genome.jp/kegg/pathway.html*). Figure 2.1 presents a portion of that metabolism, beginning with ceramide and adding sugars to create larger and more complex carbohydrate

*Chemical Cytometry*. Edited by Chang Lu

ISBN: 978-3-527-32495-8

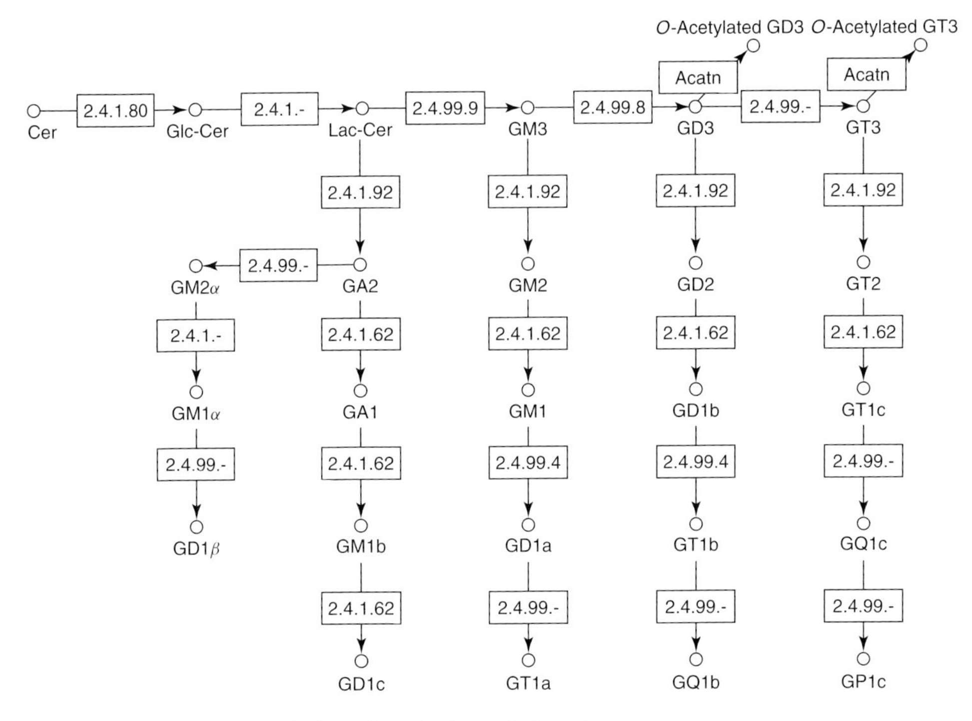


**Figure 2.1** Partial glycosphingolipid metabolic pathway. The circles are different glycolipids and the squares contain the EC number for the anabolic enzyme responsible for the production of that compound.

head groups. The distribution of glycosphingolipids is the result of anabolic enzymes that add successive sugars and catabolic enzymes that remove those sugars. The numbers in each box is the EC number for the anabolic enzymes. The corresponding catabolic enzyme is not shown.

### 2.1.3
### Expression of Glycosphingolipids Is Quite Heterogeneous in Primary Tissues

Labeled antibodies and glycolipid-binding toxins have been used to monitor expression of specific glycosphingolipids in various primary tissues. These studies demonstrate extraordinary heterogeneity in the distribution of these compounds. That distribution presumably reflects differences in the function of individual neurons. The hypothalami of mice with Tay-Sachs disease, in which the ganglioside $G_{M2}$ accumulates excessively, were stained to examine $G_{M2}$ storage. The staining revealed that accumulation occurred unevenly among the neurons [4]. The effects of gangliosides are also heterogeneous. Mice that could not synthesize lipids larger than $G_{M3}$ and $G_{D3}$ were engineered. They had axon degeneration at 43 times the rate of control animals, but this degeneration took place only in a minority of cells [5].

The dorsal root ganglion (DRG) is one of the several groups of nerve cells clustered together near the spinal column, comprising part of the peripheral nervous system. DRGs contain the cell bodies of neurons that carry sensory impulses along a single long axon extending from their distal termini (e.g., in the skin) to the spinal cord. Antibody stains were used to characterize the expression of different glycosphingolipids in DRGs, and heterogeneity was clearly demonstrated. Gong *et al.* showed significant variation in the types of gangliosides present in the various neurons of DRGs [6].

### 2.1.4 Metabolic Cytometry Has Been Used to Characterize Glycosphingolipid Metabolism in Model Cell Systems

The use of affinity reagents (antibodies, toxins) to monitor glycosphingolipids has several limitations. First, reagents are not available for all glycosphingolipids. Second, the affinity reagents can exhibit cross-reactivity. Third, while different fluorescent stains can be used to visualize several targets simultaneously, correcting for spectral cross talk is difficult, which limits the dynamic range of the measurement. Finally, it is difficult to simultaneously monitor more than three components due to spectral overlap.

Compounds labeled with highly fluorescent dyes can be separated with high resolution by capillary electrophoresis (CE) and detected with great sensitivity by laser-induced fluorescence [7–11]. If a fluorescent label is attached to a substrate along a metabolic pathway, metabolites that retain their label can be detected at very low levels.

We employed chemical cytometry to monitor glycosphingolipid metabolism in cultured AtT20 cells; these pituitary tumor cells have some resemblance to neurons and provide a convenient model system. In those studies, $G_{M1}$ was tagged with tetramethylrhodamine, a highly fluorescent dye that does not prevent uptake or metabolism of the substrate [11]. Pagano and coworkers have used similar fluorescent compounds to examine glycosphingolipid localization and transport [12, 13]. By complexing these compounds with defatted bovine serum albumin, the authors were able to prompt cells to uptake their probes.

Figure 2.2 presents the structure and metabolic transformations expected for tetramethyl rhodamine (TMR)-$G_{M1}$. The substrate can undergo catabolic transformation, losing terminal sugar groups to successively produce the simpler $G_{M2}$, $G_{M3}$, and LacCer structures. LacCer can undergo loss of the terminal galactose and glucose, resulting in the fluorescently labeled ceramide. The LacCer can also undergo anabolic transformation with the addition of sugars to create the asialo $G_{A2}$ and $G_{A1}$, where asialo refers to the absence of a sialic acid. $G_{M3}$, a branch point in the biosynthesis of the more abundant gangliosides can undergo addition of sugars to create larger structures such as $G_{D3}$, $G_{D1b}$, and $G_{T1b}$. Finally, $G_{M1}$ can itself accept additional sugars to become $G_{D1a}$ or $G_{T1a}$.

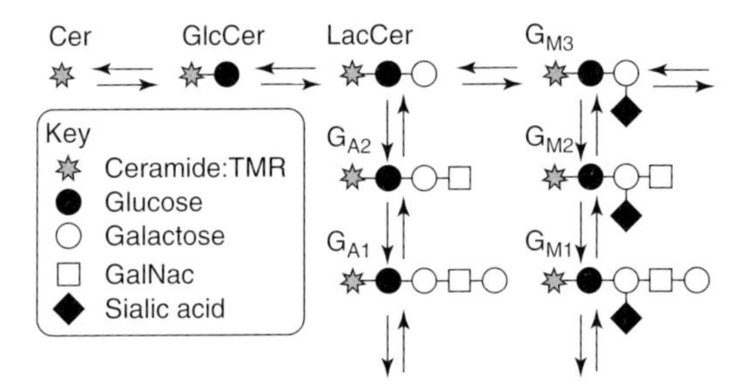


**Figure 2.2** Glycosphingolipid metabolic pathway. Cer is ceramide with a covalently attached tetramethyl rhodamine.

We adapted Pagano's method to induce an immortalized cell line to take up our fluorescent $G_{M1}$ analog. The cells metabolized the lipid into products that retained the tetramethylrhodamine tag. After incubation, cells were selected, injected onto a capillary, and lysed by contact with the separation buffer. The cell's contents were then separated by CE and detected by laser-induced fluorescence [14]. Relatively little variation existed among AtT20 cells. There is a limited amount of relevant biological data that can be gleaned from studying a model cell line. Therefore, it is crucial to adapt our technique to allow for the interrogation of primary nerve (and other) cells freshly isolated from animals.

In this chapter, we present a method for the analysis of glycosphingolipid metabolism in individual, primary cells. We apply this method to the conversion of the ganglioside $G_{M1}$ by single neurons isolated from rat DRG.

## 2.2 Material and Methods

### 2.2.1 Reagents

The synthesis of tetramethylrhodamine-labeled $G_{M1}$, $G_{A1}$, $G_{M2}$, $G_{A2}$, $G_{M3}$, LacCer, GlcCer, and Cer has been previously described [15]. The preparation involved acylation of the homogeneous C18 lyso-forms of $G_{M1}$, LacCer, GlcCer, and Cer with the *N*-hydroxysuccinimide ester of a $\beta$-alanine tethered 6-tetramethylrhodamine derivative. The resulting glycosphingolipids were treated with galactosidase, sialidases, and sialytranserase enzymes to create the other standards.

### 2.2.2 Cells and Cell Culture

DRG neurons were prepared from four- to five-day-old rat pups using the protocol described [16], with the following changes. Collagenase for DRG dissociation was increased to 10 mg ml$^{-1}$. An additional purification step was used after dissociation and cell filtering. A suspension of cells in medium (~1.5 ml) was pipetted over a two-step gradient of 25 and 50% Percoll in Hepes-buffered saline (3 ml each; Sigma–Aldrich). After centrifugation (800g, 25 min) cells at the 25%/50% interface were collected, washed by centrifugation in saline, resuspended in growth medium, and plated on a 35-mm cell culture dish coated with 10 μg $ml^{-1}$ laminin (Cultrex/Trevigen, Gaithersburg, MD) for 1 h at 37 °C. Nonadherent cells were collected and plated at 200 000 cells/well on 24-well plates (Costar, Corning, NY) precoated with 125 μg $ml^{-1}$ poly-D-lysine (Sigma–Aldrich) and 10 μg $ml^{-1}$ laminin. Cells were cultured in complete (serum containing) medium for 24 h, washed, cultured in serum-free growth medium for 24 h, then TMR-$G_{M1}$ added.

### 2.2.3 Uptake of Fluorescent $G_{M1}$

Fluorescent $G_{M1}$ was complexed to defatted bovine serum albumin in 1 : 1 ratio using a procedure adapted from Pagano [12]. The plated neurons were grown in this media for 24 h.

### 2.2.4 Homogenate and Single Cell Preparation

To prepare cellular homogenates, cultured DRG neurons were washed with phosphate-buffered saline (Dulbecco's formulation, PBS) then were dislodged using trypsin–EDTA solution (2.5 mg $ml^{-1}$ trypsin, 1 mM EDTA; Invitrogen). After 5 min at 37 °C, cells were gently triturated from the culture surface and transferred to a glass tube. Soybean trypsin inhibitor (2.5 mg $ml^{-1}$; Sigma–Aldrich) was added, the cells collected by centrifugation, and washed repeatedly with PBS. Finally, the cell pellet was lysed by addition of 1% SDS in PBS and the lysate was kept frozen until analysis.

For single cell analyses, cultured DRG neurons were washed with PBS and collected as described above. The washed cell pellet was resuspended in a solution of 4% paraformaldehyde in PBS. After 12 min at ambient temperature the fixed cells were collected by centrifugation (200g, 3 min) and the cells incubated in 10 mM glycine in PBS to quench the fixative, then washed repeatedly with the same solution. Finally, cells were resuspended in the same solution and shipped to University of Washington on ice prior to analyses.

### 2.2.5 Capillary Electrophoresis

The CE system is similar to others used in this group [7–11, 17, 18]. An uncoated, fused-silica capillary (35 cm long, 20 μm i.d., 150 μm o.d.) was used for the separation. The distal tip of the capillary was placed in a sheath-flow cuvette for postcolumn fluorescence detection. A 10 mW frequency-doubled Nd-YAG laser operating at 532 nm was focused in the cuvette to excite fluorescence. Emission was imaged by a 0.45 NA microscope objective through a 580 DF40 band-pass filter onto a gradient index-lens coupled fiber optic. The optic transferred the light to a single-photon counting avalanche photodiode that was interfaced to a data acquisition card. The signal was recorded on a PC using Labview and data were processed using Matlab running on a Macintosh computer.

The cells were lysed in the capillary. Injection was accomplished by exposing the sheath flow at the terminus of the capillary to a 1 m drop for 1 s, which created a siphon to inject a small volume. In the procedure, a plug of 2% Triton X-100 in water was injected. An inverted microscope was then used to locate a cell on an uncoated, Teflon-printed 21-well slide. The capillary was centered over the cell with a set of micromanipulators, and the cell was aspirated into the capillary. A second plug of Triton X-100 was added, and the cell was allowed to lyse for 5 s. The capillary was then placed in a running buffer composed of 10 mM sodium tetraborate, 35 mM sodium deoxycholate, and 5 mM methyl-$\beta$-cyclodextrin [19]. Separation was performed at 25 kV.

Data were treated with a 7-point median filter to remove spikes due to bubbles and cellular debris and then convoluted with a 44-point Gaussian function with 5 point standard deviation. Peak alignment was performed using a two-point procedure based on the $G_{M1}$ and Cer peak positions.

## 2.3 Results and Discussion

### 2.3.1 Dorsal Root Ganglia Homogenate

We first analyzed a homogenate prepared from primary DRG neurons (Figure 2.3). The largest peak is that of the substrate, $G_{M1}$. This data demonstrates that the dissociated, plated DRG neurons are capable of taking up $G_{M1}$ from media, and of catabolizing it into the anticipated products. The metabolic profile produced by these cells is very similar to that produced by the AtT20 cells [14]. There is significant accumulation of $G_{M2}$ and ceramide, smaller amounts of $G_{M3}$, lactosylceramide and glucosylceramide, and several peaks that correspond to metabolites for which we do not have standards.

### 2.3.2 DRG Single Cells

The homogenate of DRG neurons had a glycosphingolipid profile very similar to that from an immortalized cell line, while the single cells were more diverse (Figure 2.4).

First, there was great variation in total labeled lipid content from cell to cell. This variation can be seen in confocal microscopy (Figure 2.5). The variation is more striking in our CE data. There was a three order of magnitude range in summed lipid peaks (Figure 2.6). The uptake reached very high levels for some cells. In

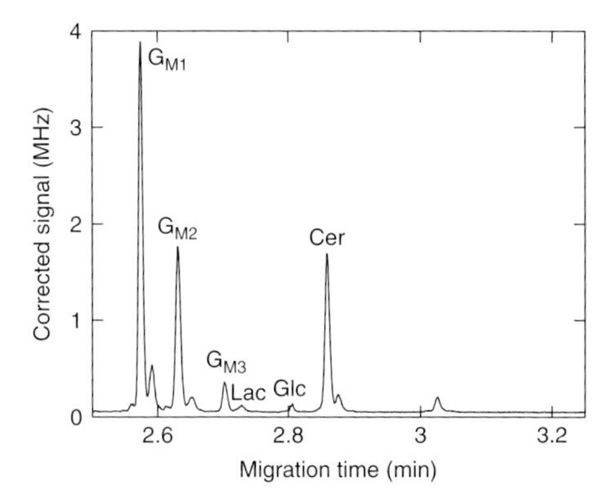


**Figure 2.3** Electropherogram of a homogenate of primary DRG neurons. Cells were harvested, plated, and then incubated with a fluorescent analog of $G_{M1}$.

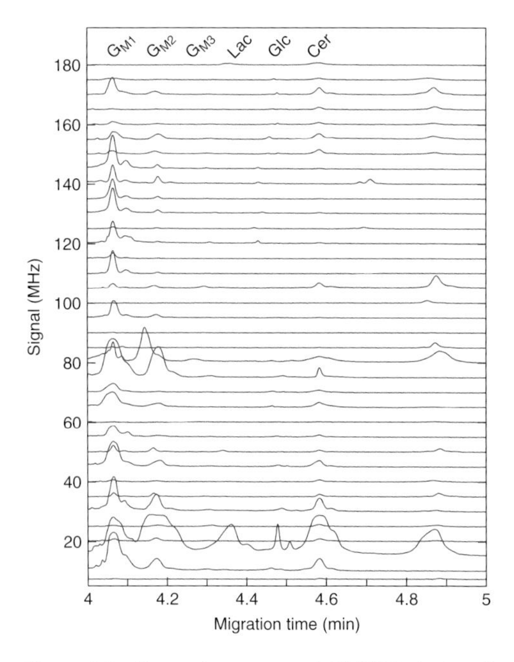


**Figure 2.4** Electropherograms of 36 DRG single cells. Note the variation in peak intensity; some traces are nearly featureless at this scale, while others are above the detector's linear range.

those extreme cases, the signal exceeded the dynamic range of our detector, which resulted in severely distorted electrophoresis peaks; the estimate of uptake for those cells is certainly underestimated in Figure 2.6.

Second, primary neurons also showed much more variation in their metabolic profiles than did immortalized cells. We can measure this heterogeneity as deviation

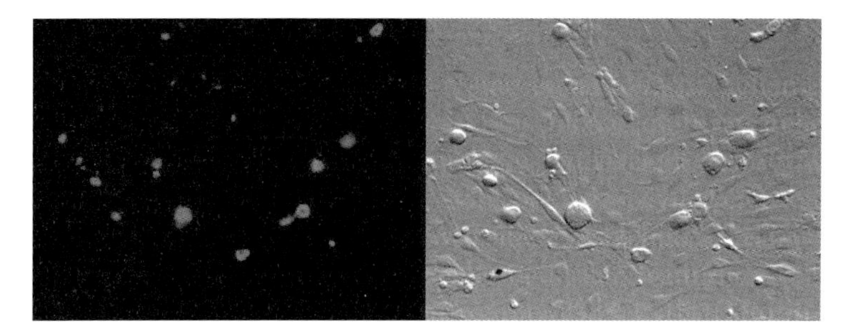

**Figure 2.5** Fluorescence and bright-field images of plated DRG neurons grown in the presence of $G_{M1}$ (TMR). Note that cells that appear similar in bright field vary considerably in fluorescence intensity.

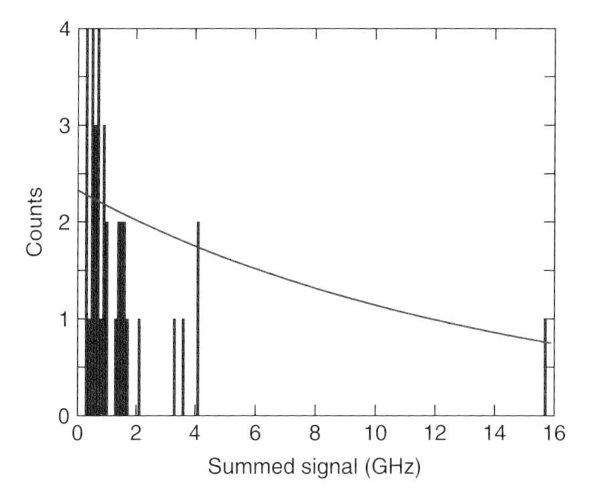


**Figure 2.6** Histogram of substrate uptake. Substrate uptake was estimated by summing the signal from lipid peaks present in a single cell electropherogram. The smooth curve is the least squares fit of an exponential decay to the histogram.

of normalized peak areas of the three major peaks, $G_{M1}$, $G_{M2}$, and ceramide, which produce signals that are large enough to be quantified in every cell, and account for approximately 80% of total lipid signal. Normalization is accomplished by dividing each peak's height by the summed intensity of all three peaks. In the AtT20 cell line, the peak heights of $G_{M1}$, $G_{M2}$, and ceramide had relative standard deviations of 9, 16, and 40%, respectively. In the primary DRG neurons, the relative deviations are a good deal larger: 38, 45, and 81%. The data provided hints of population differences. Plotting the level of ceramide versus $G_{M1}$, (Figure 2.7), suggests the presence of two populations; one in which ceramide is linearly dependent on $G_{M1}$, and the other in which ceramide is at a consistent, lower level which is independent of the substrate's quantity.

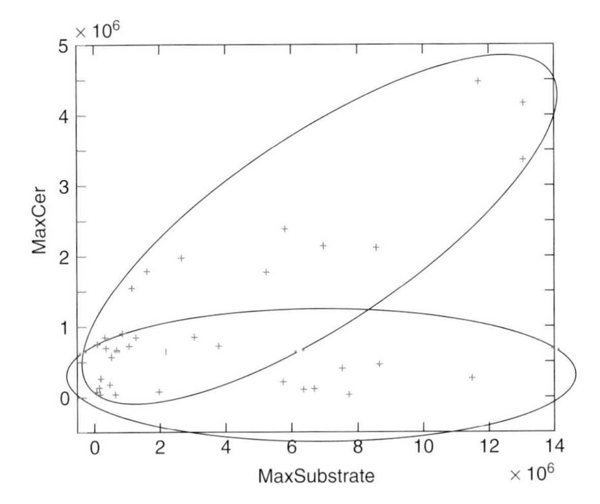


**Figure 2.7** Scatter plot of $G_{M1}$ versus ceramide. The distribution suggests the presence of two populations, one in which ceramide is dependent on substrate concentration and the other in which it is independent.

## 2.4 Conclusions

We have demonstrated a method of harvesting primary DRG neurons and inducing them to take up a fluorescent glycosphingolipid analog. We have shown that these neurons metabolize that analog, and that we can fix them, ship them 4000 km between laboratories, and analyze the fluorescent lipid content of single cells. We have shown that, as might be expected for a diverse neuronal cell population, there is a great cell-to-cell heterogeneity in the uptake and metabolism of this lipid.

## Acknowledgments

We gratefully acknowledge support from the National Institutes of Health (5R01NS61767). CW also acknowledges a fellowship from the Analytical Division of the American Chemical Society. JP was supported by NIH training grant P32GM07445.

## References

1 Tettamanti, G. (2004) Ganglioside/glycosphingolipid turnover: new concepts. *Glycoconj. J.*, **20**, 301–317.

2 Colombaioni, L. and Garcia-Gil, M. (2004) Sphingolipid metabolites in neural signalling and function. *Brain Res. Brain Res. Rev.*, **46**, 328–325.

3 Ohtsubo, K. and Marth, J.D. (2006) Glycosylation in cellular mechanisms of health and disease. *Cell*, **126**, 855–867.

4 Platt, F.M., Neises, G.R., Reinkensmeier, G., Townsend, M.J., Perry, V.H., Proia, R.L., Winchester, B., Dwek, R.A., and Butters, T.D. (1997) Prevention of lysosomal storage in Tay-Sachs mice treated with N-butyldeoxynojirimycin. *Science*, **276**, 428–431.
5 Sheikh, K.A., Sun, J., Liu, Y., Kawai, H., Crawford, T.O., Proia, R.L., Griffin, J.W., and Schnaar, R.L. (1999) Mice lacking complex gangliosides develop Wallerian degeneration and myelination defects. *Proc. Natl. Acad. Sci. U.S.A.*, **96**, 7532–7537.
6 Gong, Y., Tagawa, Y., Lunn, M.P., Laroy, W., Heffer-Lauc, M., Li, C.Y., Griffin, J.W., Schnaar, R.L., and Sheikh, K.A. (2002) Localization of major gangliosides in the PNS: implications for immune neuropathies. *Brain*, **125**, 2491–2450.
7 Krylov, S.N., Zhang, Z., Chan, N.W.C., Arriaga, E., Palcic, M.M., and Dovichi, N.J. (1999) Correlating cell cycle with metabolism in single cells: the combination of image and metabolic cytometry. *Cytometry*, **37**, 15–20.
8 Cheng, Y.F. and Dovichi, N.J. (1988) Subattomole amino-acid analysis by capillary zone electrophoresis and laser-induced fluorescence. *Science*, **242**, 562–564.
9 Wu, S.L. and Dovichi, N.J. (1989) High-sensitivity fluorescence detector for fluorescein isothiocyanate derivatives of amino-acids separated by capillary zone electrophoresis. *J. Chromatogr.*, **480**, 141–155.
10 Zhao, J.Y., Dovichi, N.J., Hindsgaul, O., Gosselin, S., and Palcic, M.M. (1994) Detection of 100 molecules of product formed in a fucosyltransferase reaction. *Glycobiology*, **4**, 239–242.
11 Whitmore, C.D., Olsson, U., Larsson, E.A., Hindsgaul, O., Palcic, M.M., and Dovichi, N.J. (2007) Yoctomole analysis of ganglioside metabolism in PC12 cellular homogenates. *Electrophoresis*, **28**, 3100–3104.
12 Pagano, R.E. and Martin, O.C. (1988) A series of fluorescent N-acylsphingosines: synthesis, physical properties, and studies in cultured cells. *Biochemistry*, **27**, 4439–4445.
13 Marks, D.L., Singh, R.D., Choudhury, A., Wheatley, C.L., and Pagano, R.E. (2005) Use of fluorescent sphingolipid analogs to study lipid transport along the endocytic pathway. *Methods*, **36**, 186–195.
14 Whitmore, C.D., Hindsgaul, O., Palcic, M.M., Schnaar, R.L., and Dovichi, N.J. (2007) Metabolic cytometry. Glycosphingolipid metabolism in single cells. *Anal. Chem.*, **79**, 5139–5142.
15 Larsson, E.A., Olsson, U., Whitmore, C.D., Martins, R., Tettamanti, G., Schnaar, R.L., Dovichi, N.J., Palcic, M.M., and Hindsgaul, O. (2007) Synthesis of reference standards to enable single cell metabolomic studies of tetramethylrhodamine-labeled ganglioside GM1. *Carbohydr. Res.*, **342**, 482–489.
16 Mehta, N.R., Lopez, P.H., Vyas, A.A., and Schnaar, R.L. (2007) Gangliosides and Nogo receptors independently mediate myelin-associated glycoprotein inhibition of neurite outgrowth in different nerve cells. *J. Biol. Chem.*, **282**, 27875–27886.
17 Dovichi, N.J. and Hu, S. (2003) Chemical cytometry. *Curr. Opin. Chem. Biol.*, **7**, 603–608.
18 Kraly, J.R., Jones, M.R., Gomez, D.G., Dickerson, J.A., Harwood, M.M., Eggertson, M., Paulson, T.G., Sanchez, C.A., Odze, R., Feng, Z., Reid, B.J., and Dovichi, N.J. (2006) Reproducible two-dimensional capillary electrophoresis analysis of Barrett's esophagus tissues. *Anal. Chem.*, **78**, 5977–5986.
19 Zhang, L., Hu, S., Cook, L., and Dovichi, N.J. (2002) Analysis of aminophospholipid molecular species by methyl-beta-cyclodextrin modified micellar electrokinetic capillary chromatography with laser-induced fluorescence detection. *Electrophoresis*, **23**, 3071–3077.

# 3
# Cell Signaling Studied at the Single-cell Level

*Angela Proctor, Shan Yang, Sumith Kottegoda, Michael Brenner, Ryan M. Phillips, Christopher E. Sims, and Nancy L. Allbritton*

## 3.1
## Introduction

Prior to having the ability to analyze signal transduction events in individual cells, signaling was thought to occur on a graduated scale. However, as technological improvements have enabled cells to be individually probed, it has become increasingly apparent that this is not always the case. It is now known that in many instances a cell may react to an incoming stimulus in an all-or-none fashion [1]. This important information is lost in a homogenous assay format by virtue of signal averaging across a bulk cell population. For example, in a cell population where only a select few cells show abnormal signaling, an average may indicate "normal" behavior, whereas a single-cell interrogation process would reveal abnormal cells. Tumors are an example of this situation, where the vast majority of cells within the tumor may be normal, and only a minority of cells demonstrates abnormal signaling behavior. It is in such instances where the ability to study individual cells gives valuable insight into the true behavior of signal transduction events.

Protein kinases are central to many signal transduction pathways. These enzymes add the terminal phosphate group from adenosine triphosphate (ATP) to a serine, threonine, or tyrosine residue on a target protein, which results in a modification of the target's bioactivity [2]. Protein phosphatases remove this phosphate group from the protein, returning it to the prephosphorylated form. This phosphorylation/dephosphorylation of proteins acts as a switch in many signal transduction pathways, with the phosphorylated form generally signaling the "on" state, and the dephosphorylated form signaling the "off" state [1]. The ability to monitor the amount of phosphorylation by a kinase in a single cell can give an indication of how the signaling pathway is functioning.

This chapter describes our group's efforts in the development and application of analytical technologies for the study of normal and aberrant signaling pathways through monitoring kinase activity within single cells. The techniques rely on microelectrophoresis, a technique that has been in use for many years for high-sensitivity chemical separations to characterize the contents of single or small

*Chemical Cytometry*. Edited by Chang Lu

ISBN: 978-3-527-32495-8

numbers of cells, a field known as *chemical cytometry* [3]. Our method utilizes an integrated platform comprising an inverted microscope, pulsed laser, and micro-electrophoresis system, either in a capillary or microfluidic format. To perform a single-cell biochemical assay, cells are loaded with fluorescently labeled substrates that function as reporters of the activity of the target enzyme. For kinases, the substrate is typically a fluorescently labeled peptide with a serine, threonine or tyrosine residue or a small lipid molecule. At the desired time after reporter loading, a cell is lysed by a focused laser pulse. Dilution of cytoplasmic molecules by diffusion and turbulent mixing terminates cellular reactions with sufficient rapidity to prevent measurement artifacts from cell lysis [4]. Following lysis, the cellular contents are immediately loaded into a separation channel and separated by electrophoresis. Owing to their different chemical properties, the substrate and product forms of the reporter are readily separated, detected using laser-induced fluorescence (LIF), and identified by their characteristic migration times. The ratio of the peak areas of the substrate and product are then used as a measure of the targeted enzyme's activation. Both fluorescent peptides and lipids have been used successfully as substrates for kinases [5–9]. The phosphorylated products can also be dephosphorylated by phosphatases; therefore, the measurement reflects the dynamic equilibrium of the kinase and phosphatase activity in the single cell [5–7]. An additional advantage of this approach is the ability to measure the activation of multiple enzymes simultaneously in a single cell by loading the cell with reporters for a variety of enzymes. At present, three kinases have been assayed at the same time in single cells, although the very high separation efficiency of microelectrophoretic methods makes it likely that a much larger number is possible [7]. A second advantage of this capillary electrophoresis (CE)-based method is that it is fully compatible with light microscopy so that the cells can be monitored using a variety of optical tools prior to analysis by microcolumn electrophoresis. Included in this chapter is a discussion of which signaling pathways have been interrogated; the reporters utilized to study these pathways; how the reporters are loaded into cells; the techniques for cell lysis; the electrophoretic and detection modes used in these determinations; and the high-throughput screening technology developed thus far for single-cell analyses.

## 3.2 Analytes Examined and Reporters Used

### 3.2.1 Reporter Properties

In the context of this chapter, a reporter can be any molecule, such as a fluorescently tagged peptide, lipid, or carbohydrate, which is detectable by CE [10]. When choosing a reporter, there are several parameters that must be examined. First of all, the reporter must be able to be loaded into the cell. The reporter must be specific for the process of interest and must undergo a change detectable by CE, achieved by altering the electrophoretic mobility of the reporter. The chosen reporter must be distinguishable from cellular components and must be compatible with the chosen detection method. Fluorescently labeled substrates are often used as reporters because of the very low detection limits afforded by LIF detection. Finally, the reporter availability and storage options must be considered. Some reporters are commercially available (derivatization may be necessary), while others may be synthesized in the lab. Long-term storage of the reporter may necessitate special treatment, such as storing peptide reporters at −70 °C to prevent degradation or storing lipid reporters under nitrogen or argon to prevent oxidation.

### 3.2.2 Probing System Function

When first setting up a system for single-cell component analysis, the abilities of the system must be explored under standardized conditions. CE and microfluidic systems can be tested by loading cells with two fluorescent dyes, usually fluorescein and Oregon green [11–15]. There are several reasons why these two dyes are utilized. First, both are small molecules that can be easily loaded into cells. An argon ion laser, commonly used in LIF, provides excitation at 488 nm, close to the maximum excitation wavelength for both compounds. These compounds have different electrophoretic mobilities and can be distinguished from one another as well as from other nonfluorescent cellular components by CE. Finally, reporter molecules are frequently labeled with fluorescein or fluorescein derivatives, and hence the use of these dyes provides some data on separation and detection limits of labeled reporters under similar conditions. Once the new system's functionality has been verified with these fluorescent molecules, more complicated analyses can be undertaken.

### 3.2.3 Peptides as Reporters

Kinases are enzymes that play an integral role in nearly all signal transduction pathways by phosphorylation of target substrates [1]. To better understand the roles that specific kinases play in signaling pathways, analysis of kinase activity within a single cell is critical. Single-cell assays have been used to measure activity of several kinases implicated in cancer, including protein kinase A (PKA) [7, 16], protein kinase B (PKB or Akt) [6, 9], protein kinase C (PKC) [7, 16], calcium-calmodulin activated kinase II (CamKII) [6, 7, 9, 16], and cdc2 protein kinase (cdc2K) [7, 16]. Fluorescently tagged peptide substrates have been used in numerous studies of kinase activity in single cells [6, 7, 9, 16]. The peptides specific to the desired kinase are purchased commercially or synthesized in the lab, loaded into the cell and monitored by CE to determine the activity of the kinase in single cells.

There are many benefits in using peptide substrates as reporters for kinase activity, but also many drawbacks. Peptide substrates are easily synthesized by utilizing solid phase-peptide synthesis with Fmoc-protected amino acids [10]. They are readily labeled with fluorescein or other organic dyes by reacting with the free amine of the resin-bound peptide. The first obstacle arises when the reporter must be loaded into the cell. However, several methods exist to accomplish this task, either by physical means or by making the reporters membrane permeable [1]. Initially, the reporter is a nonphosphorylated substrate of the desired kinase, which is then phosphorylated by the kinase on a serine, threonine, or tyrosine residue. Addition of the phosphate group from ATP changes the net charge on the peptide, thus altering the electrophoretic mobility of the phosphorylated substrate and allowing for separation from the starting material. In addition, due to the varying electrophoretic mobilities of the substrate reporters and their phosphorylated counterparts, it is feasible to simultaneously monitor the activity of multiple kinases in a single cell. CE has excellent resolving power, allowing for the separation of multiple components in a single sample. An additional benefit of using sensitive LIF detection with CE is that the reporter concentration loaded into cells can be much lower than native substrate concentrations, which eliminates competitive inhibition [1].

A drawback to using peptide reporters is the possibility that the substrate will be phosphorylated by a nontarget kinase, usually one closely related to the target kinase or one recognizing a similar consensus sequence. Care must be taken in substrate design to avoid this problem and a screen against potentially interfering kinases may be necessary to more accurately understand what is being reported. In addition, the substrate reporters have been shown to have lower $K_M$ and $V_{max}$ values than protein substrates, something that may also affect the assay and the interpretation of the results [1]. A final drawback when using peptide reporters is their susceptibility to degradation by proteases that cleave and digest peptides. There is no guarantee that the reporter will behave as expected once it is cleaved and the results cannot be trusted once this occurs. Attempts to prevent degradation of the reporters to increase stability and lifetime in the cell are as important as ensuring kinase specificity.

Reporter loading difficulties can limit the utility of fluorescein-labeled peptide reporters [17]. Already in use in microscopic cell biology studies, biarsenical dyes have recently been used to fluorescently label recombinant proteins and synthesized peptides, which can then be used as reporters. This method requires a genetically encoded motif of —C—C—X—X—C—C— (where X is any noncysteine amino acid) that binds two arsenic atoms with high affinity [18]. FlAsH [4′,5′-bis(1,3,2-dithioarsolan-2-yl)fluorescein] has two As(III) subunits that pair to the cysteine thiol groups located in the motif. FlAsH is a fluorescent dye that permeates the cell membrane and increases in fluorescence once it binds to the tetracysteine motif. FlAsH has an excitation maximum at 508 nm, making it compatible with CE–LIF. One of the main benefits of this approach is that the reporter is synthesized by the cell and does not need to be loaded into the cell. The FlAsH dye then binds to the reporter, providing a fluorescently labeled reporter inside the cell. Owing to the differences in electrophoretic mobility of the free and bound dye, these analytes can be easily separated by CE, and hence removal of the unbound dye prior to analysis is unnecessary. Also of note is that labeling with FlAsH has been shown to have minimal influence on the substrate properties of the reporter [17].

### 3.2.4
### Lipids as Reporters

In addition to peptide reporters, lipid-based substrates have been used as reporters for the lipid kinases. These include fluorescently labeled sphingosine and phosphatidyl inositol phosphates [5, 8]. Sphingosine is phosphorylated by sphingosine kinase 1 and 2 (SK1 and SK2) and dephosphorylated by the phosphatases SPP1 and SPP2 [5]. Although the balance between these pathways is yet to be determined, SK1 is thought to be oncogenic and SK2 is involved in immune response. Monitoring of the amount of sphingosine and sphingosine-1-phosphate in the cell can give an idea of the activity of these kinases and phosphatases. Sphingosine labeled with fluorescein remains cell permeant, so that it may be passively loaded into the cells, where CE–LIF has been used to observe the relative amounts of the phosphorylated and nonphosphorylated sphingosine reporter in cells. Other lipid reporters include phosphatidyl inositol 4,5-bisphosphate ($PIP_2$) and phosphatidyl inositol 3,4,5-trisphosphate ($PIP_3$) [8]. Mwongela *et al.* used Bodipy-Fl tagged $PIP_2$ and $PIP_3$ to monitor the *in vitro* activity of the enzyme phosphatidylinositide 3-kinase (PI 3-K), which is a known tumor promoter and converts $PIP_2$ to $PIP_3$. These reporters were also used to monitor the reverse conversion, catalyzed by the lipid phosphatase PTEN (phosphatase and tensin homolog detected on chromosome 10), which is a known tumor suppressor. The electrophoretic mobilities of Bodipy-Fl labeled $PIP_2$ and $PIP_3$ are different enough to show separation by CE with good resolution, heralding a promising method for monitoring the activities of PI 3-K and PTEN in cells.

### 3.2.5
### Secondary Reporters

Use of CE in series with a biological detector cell has allowed for quantitative determination of inositol 1,4,5-trisphosphate ($IP_3$) in *Xenopus laevis* oocytes [19–22]. In these works, CE was used to isolate and deliver the analyte $IP_3$ to a detector cell loaded with the $Ca^{2+}$ sensitive indicator mag-fura-2. Presence of $IP_3$ triggered $IP_3$-induced $Ca^{2+}$ release, in turn causing a fluorescence increase in mag-fura-2. By using this reporter system, physiologic $IP_3$ concentrations [19], spatial concentrations [21], and fertilization-dependent concentrations of $IP_3$ in the *X. laevis* oocytes were determined [22].

## 3.3
## Cell Preparation and Reporter Loading

Effective loading of fluorescent reporters into cells is critical for their use in single-cell analysis. The success of a given loading method depends on the properties of the reporter itself and the cell being studied. A variety of such techniques are described here, each with unique advantages as well as potential pitfalls, Figure 3.1.

### 3.3.1
### Disruptive Loading Methods

Microinjection is a method by which reporters can be directly injected into the cells of interest using a fine glass capillary. It is a relatively labor-intensive technique which, while allowing for the introduction of nearly any reporter, also subjects the cells to high, often lethal, levels of stress. Because the cell membrane must be punctured by the capillary to introduce the reporter, cells that must be grown in suspension, as well as cells with low cytoplasmic or nuclear volume, are not generally compatible with this technique [23]. Furthermore, this technique is not well suited to experiments in which stress-related pathways are to be interrogated, as the shock of microinjection itself can activate signaling mechanisms related to stress response. Nevertheless, microinjection has been used for introduction of fluorescent reporters into a variety of cells [4, 6, 7, 15, 16].

Another disruptive loading technique is pinocytic loading. Here, cells are first subjected to a hypertonic solution containing the reporter molecule, inducing the formation of fluid-filled plasma membrane invaginations known as *pinosomes*. Next, the cells are treated with a hypotonic solution, resulting in rupture of the pinosomes and release of the reporter into the cytoplasm. Finally, cells are washed with extracellular buffer (ECB) prior to CE [6]. Again, this technique involves a relatively high degree of shock to the cells of interest and thus is not ideally suited for interrogation of pathways that would be obscured or confounded by stress response or membrane repair mechanisms.

### 3.3.2
### Nondisruptive Loading Methods

While the aforementioned techniques can be highly effective, certain applications of fluorescent reporters require methods of loading that are less stressful to the cells being analyzed. There are a number of options available for reporter loading based on diffusion or cellular transport mechanisms that require no physical or osmotic manipulation of the cells.

Some commonly utilized fluorescent dyes, including fluorescein and Oregon green, are commercially available in an acetyl ester form. The addition of acetyl groups to the carboxylic acid moieties of these dyes increases their hydrophobicity substantially and permits diffusion across the plasma membrane. Once in the cytoplasm, the ester linkages are cleaved by ubiquitous esterases, rendering the dye charged and thereby trapping it inside the cell [24]. Fluorescein diacetate (mixed isomers) and Oregon green 488 carboxylic acid diacetate 6-isomer have been successfully loaded into rat basophilic leukemia (RBL) cells [12, 24, 25], Ba/F3 (murine leukemia) [11], and Jurkat (nonadherent human leukemia) cells [14]. In each case, simply incubating the cells in a solution of the diacetate dye provided effective loading.

Peptide-based reporters are frequently charged, hydrophilic molecules that will not pass through a plasma membrane unassisted. However, the conjugation of a peptide reporter to a TAT sequence, taken from amino acid residues 48–57 of the TAT protein of HIV, has been demonstrated to allow entry of the reporter into cells through a mechanism not yet fully elucidated. The TAT peptide (RKKRRQRRR) has been used to load substrates for PKB and CamKII into HT1080, RBL, and NIH 3T3 cells with efficacy similar to microinjection, but without the need for physical manipulation of the cells [9]. The TAT sequence has been shown to effectively aid in reporter loading through attachment via either a disulfide bond or photolabile linker. In the latter example, cells are exposed to UVA (ultraviolet A) light after reporter loading and the TAT tag is spontaneously removed by linkage degradation. It should be noted that certain cells, including B lymphocytes, do not show effective uptake of TAT-conjugated compounds [26].

Another valuable tool for peptide reporter loading, particularly in cells that are incompatible with endocytosis-based strategies such as TAT conjugation, is the coupling of a hydrophobic lipid tag to the peptide reporter. The conjugation of a C14 myristate group to otherwise cell-impermeant peptides has been shown to promote efficient loading into cells [26], though the exact mechanism for membrane transport of these tagged peptides has not been fully established. Furthermore, conjugation of myristate to the peptide reporter via a disulfide bond allows for spontaneous release of the reporter once in the cytoplasm. In the above work, a number of peptides were synthesized using standard solid-phase technique, with the addition of a myristate-conjugated lysine and an N-terminal fluorescein to each. Ba/F3 cells were prepared for loading by centrifuging the cells at 800 g for 2 min, discarding the supernatant, and resuspending in an ECB. The cells were then incubated in a 20 μM solution of reporter in ECB for 30 min at 37 °C.

Lipid molecules are naturally membrane-permeant, a quality that simplifies the loading of lipid-based reporters. This principle was illustrated in the measurement of SK activity using sphingosine fluorescein (SphFl) as a reporter [5]. In this work, Ba/F3 cells and sheep erythrocytes were loaded with SphFl by incubation with a solution of the reporter in ECB at 37 °C for 15–20 min.

## 3.4 Cell Lysis and Sampling Techniques

Cell lysis is an integral step in single-cell analysis because it releases the compounds for CE analysis. The cell is a complex mixture of cell contents and lysis techniques must be compatible with downstream analysis; thus, the method of cell lysis must be chosen very carefully [27]. The foremost methods to date for single-cell lysis are discussed below.

### 3.4.1 Chemical Lysis

Chemical lysis is typically performed with a detergent. The detergent solubilizes the lipids and proteins of the cell membrane to create pores and eventually lyse the cells.

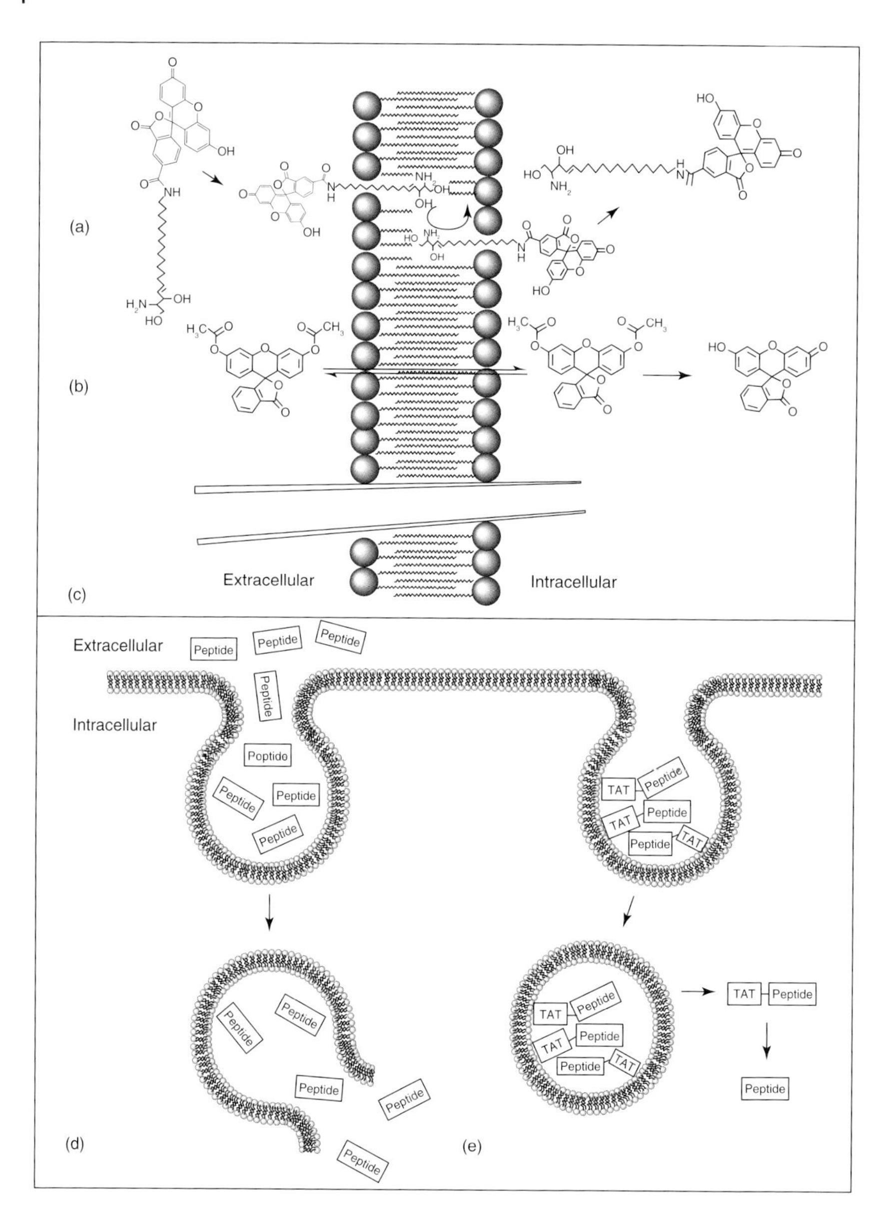


**Figure 3.1** Reporter loading methods. (a) A proposed mechanism for the transport of myristoylated peptides into cells involves insertion of the lipid tail into the bilayer, followed by flipping across the membrane, and ultimately cleavage of the linker releasing the reporter. (b) Passive diffusion of diacetylated fluorescein across the membrane. (c) Depiction of micropipette insertion into the membrane (not to scale), through which reporters can be directly injected. (d) Through pinocytic loading, pinosomes are induced and allowed to lyse intracellularly. (e) TAT conjugation is thought to cause adsorption onto the membrane surface followed by entry into the cell through endocytosis.

The cell lysis buffer contains either strong ionic detergents such as sodium dodecyl sulfate (SDS), milder nonionic detergents such as Triton X-100, or zwitterionic detergents such as 3-[(3-cholamidopropyl)dimethylammonio]-1-propanesulfonate. Lysis by detergent has been well developed for bulk cell lysis and has also been translated to the single-cell level [28]. This method is easy to use with low cost. However, it takes seconds to permeabilize the cell membrane and terminate chemical reactions within the cell, which is too long a time frame for observations of many signaling pathways. This delay allows the cell time to initiate signal transduction pathways involved in membrane repair and stress response, rendering it difficult to accurately assess the cell's true biochemical activity. Also of concern is the remaining detergent in the cell lysate, which can influence the downstream analysis [10].

### 3.4.2 Hypotonic Lysis

The cell membrane is completely permeable to water. When the cell is exposed to a hypotonic solution lower in salt concentration than the cytoplasm, water from the hypotonic solution moves into the cell via osmosis, causing it to swell and rupture. Li *et al.* used distilled water as a hypotonic solution to lyse cells. A cell was immersed in the hypotonic solution for 10 s and was observed to swell, destroying the cell membrane. The hypotonic solution was then immediately exchanged for ECB and sampling and electrophoresis of the cell contents was performed [4]. This method does not contaminate the lysate because only distilled water is introduced; however, the results of these studies showed that cellular membrane repair mechanisms may activate a variety of enzyme systems in response to the cellular damage during this slow lysis process.

### 3.4.3 Laser Lysis

For decades, it has been known that pulsed laser radiation focused at high numerical apertures (pulsed laser microbeams) can create damage on the cellular or subcellular level [29]. More recently, highly focused laser microbeams with pulses on femtosecond to nanosecond timescales are increasingly used for lysis of single cells. Rau *et al.* did critical work in describing the process of lysing cells with a pulsed laser. As shown in Figure 3.2, a directed nanosecond pulse from a laser through a high numerical aperture objective lens was focused onto a small spot near the cell to be lysed. After the pulse, localized plasma occurred at the focus point, generating a cavitation bubble. The cavitation bubble expanded and contracted within microseconds, resulting in the formation of a shock wave, which ruptured the cell membrane. Thus, a cell or cells located near the focusing point have been observed to be lysed either during the expansion of the cavitation bubble ($<1\,\mu s$ after the pulse) or collapse of the bubble (approximately $30\,\mu s$ after the pulse). The zone of lysis could be controlled within a small area by altering the laser energy [30].

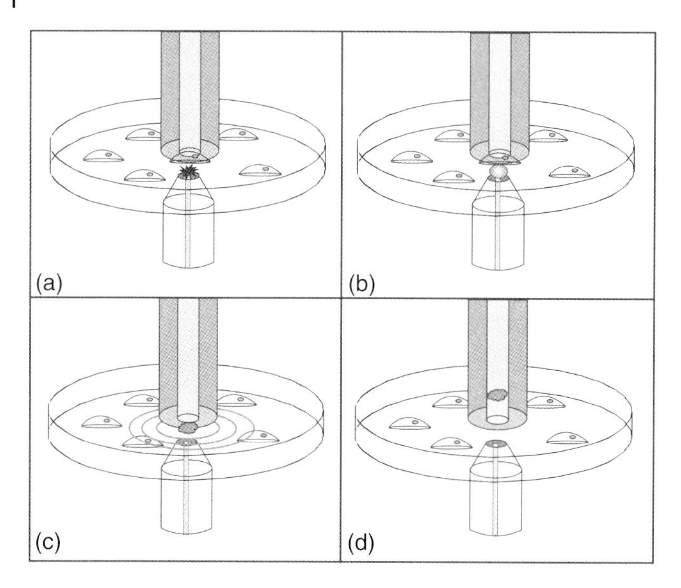


**Figure 3.2** Schematic of laser-induced cell lysis. (a) A pulsed laser microbeam is focused near a targeted cell, generating the localized plasma. (b) A cavitation bubble is formed in the center around the focused laser pulse beam. (c) The expansion of the cavitation bubble generates the shock waves (gray ellipses) and ruptures the cell. (d) The cells are selectively lysed and aspirated into the capillary.

Laser lysis is a rapid way to lyse the cell. Many biochemical reactions happen within seconds or less; thus, the time to terminate cellular reactions (the temporal resolution) becomes critical when cellular properties such as enzymatic activity are being analyzed [25]. For such biochemical analyses, accurate measurement requires complete cell lysis with termination of biochemical reactions in subsecond time periods. Thus, fast laser lysis makes it possible to study highly dynamic processes in cells. For example, Li *et al.* utilized a pulsed laser microbeam for cell lysis and was able to investigate the activity of important cell signaling kinases PKC and CamKII [31].

Laser lysis is also very suitable for and compatible with a microfluidic measurement platform because of the optical permeability of the microfluidic system. The fast lysis time and the very limited zone of lysis area (less than 5 μM) enable coupling of the laser lysis method to a microfluidic chip. Lai *et al.* have demonstrated the use of a pulsed laser microbeam for single-cell lysis followed by electrophoretic separation of cellular analytes in a poly(dimethylsiloxane) (PDMS) microfluidic device [11].

### 3.4.4
### Electrical Lysis

Although laser lysis has many advantages, it also has some drawbacks. Not only are pulsed lasers expensive, but also manipulation of the beam is required, bringing

**Table 3.1** Properties of commonly used methods for single-cell lysis.

| Lysis mode | Temporal resolution | Protein denaturation | Cost | Compatibility with microchips |
|---|---|---|---|---|
| Chemical | Seconds | High | Low | Good |
| Hypotonic | Seconds | Low | Low | Good |
| Laser | Microseconds to nano-seconds | None | High | Excellent |
| Electrical | Milliseconds | None | Moderate | Very good |

up safety concerns and a need for training prior to use. Thus, finding an alternative method for rapid cell lysis that is inexpensive and easy to use is very attractive.

When the cell is exposed to an electric field, transmembrane potentials are generated, causing a rupture of the lipid bilayer. Pores are formed that will lead to cell lysis if there is a high enough electric field strength and enough time. Han *et al.* have also developed and demonstrated rapid lysis of single, adherent cells with a voltage pulse of microsecond to millisecond duration followed by efficient collection and separation of the cellular contents in a capillary [25]. A comparison of electrical lysis to laser-based lysis was also made, which showed that the lysis time and sampling efficiencies of the two techniques were comparable.

The cell for single-cell analysis may be lysed either before or after entry into the capillary or microchannel by one of the several methods mentioned above. Depending on different applications, certain methods are preferable to others [27]. A comparison of commonly used methods for single-cell lysis is listed in Table 3.1.

### 3.4.5 Sampling Techniques

Efficient sampling techniques are highly challenging to the application of CE and microchip analysis for the study of single cells not only because of the extremely small amounts and volumes of cell contents, but also because of the required detachment of adherent cells from their growth surface. Sampling techniques to date have been classified into two categories: whole cell sampling and subcellular sampling. Generally, either part of a cell (subcellular sampling) or the entire cell's contents (whole cell sampling) can be loaded by hydrodynamic or electroosmotic fluid flow into a capillary or microchannel for electrophoretic analysis.

### 3.4.6 Whole Cell Sampling

Whole cell sampling for CE or microchannel electrophoresis typically includes the loading of a single cell suspended in solution or a cell attached to a substrate surface into a capillary or microchannel. Sims *et al.* developed a laser-based cell lysis technique to sample and analyze a cell using CE [24]. An individual cell attached to the coverslip of a cell chamber was put 10–20 μm beneath one end of a capillary mounted on the stage of a microscope. After the cell was lysed by a focused laser pulse, a potential difference was applied across the capillary to electrokinetically load the cell's contents into the capillary in 33 ms, possibly much faster. Dilution of cytoplasmic molecules by diffusion and turbulent mixing terminates the cellular reactions on a similar time scale. The cellular contents were then separated and analyzed by CE–LIF.

Marc *et al.* has also developed a fast electrical lysis procedure for single-cell analysis as a lower cost alternative, with fewer safety concerns than the laser-based system [12]. In this setup, cells are grown on a gold-coated coverslip in a trap. A gold-coated capillary with a tapered tip was positioned above the cell chosen for analysis and a voltage pulse was applied between the capillary and the coverslip. The applied voltage pulse passed through the cell membrane, lysed the cell and loaded the cellular contents into the capillary for separation.

### 3.4.7
### Subcellular Sampling

Subcellular sampling refers to the analysis of only the cellular region of interest instead of whole cell contents. Many biochemical processes occur in localized spatially discrete regions of cells. Subcellular sampling could minimize interference from extracellular components, and offers the possibility to sample specific subcellular compartments. By this sampling method, the function of different compartments can be studied and thus simplify the information obtained, giving a more basic understanding of the mechanisms occurring within cellular processes. Various methods for subcellular sampling have been discussed in the literature. These methods include direct insertion of a capillary tip into a cell followed by hydrodynamic or electrokinetic loading of cytoplasm into a capillary for electrophoretic analysis or the use of a laser to detach a cellular process for further analysis [27].

Lee *et al.* have demonstrated spatially localized measurement of kinase activation in the oocyte of the frog, *X. laevis* [16]. The large size of the *X. laevis* oocytes and eggs make them extremely useful as model cells for biomedical science investigations. One end of a capillary was directly inserted into the targeted region of the cell and a vacuum force was utilized to aspirate the subcellular contents into the capillary. Next, a voltage was applied across the capillary for separation analysis. With this sampling technique, it is possible to define a spatial resolution depending on the volume sampled, the capillary's internal diameter and the insertion depth of the capillary into the oocyte. Typical achievable spatial resolutions are from 50 to 250 μm. Such micron-sized spatial resolution enables the identification of gradients of metabolites across the 3 mm circumference of the oocyte. However, only giant cells (>1 nl) that are much greater in size than the typical mammalian cell (1 pl) are suitable for this method. Significant perturbations to the cell at the time of sampling are also limitations to this method.

Li *et al.* were able to show a successful method for lysis of portions of a single neuronal process pheochromocytoma (PC 12 cells) without damaging the neuron body by using the laser-based cell lysis technique [4]. The laser energy was carefully adjusted from 1 to 3 μJ to find an appropriate laser energy level to generate an attenuated shock wave, which could selectively lyse the process while leaving the cell body and intervening neuronal process intact. Lysis of the cellular tip occurred in less than 33 ms. In addition, by modulating the laser pulse energy, the size of the region of cellular disruption could also be controlled. Simultaneously, with the laser lysis process, subcellular contents were loaded into a capillary for electrophoretic separation. Targeting molecules in subcellular compartments was performed to demonstrate the feasibility of performing measurements by the subcellular sampling technique described above. A reporter localized to the nucleus was detected on the electropherogram following laser-mediated disruption of the cell and the nucleus.

## 3.5 Electrophoresis Separation Conditions

Electrophoresis-based separations are a viable method for chemical analysis of single cells [8, 11, 17, 21] and this separation method has been extensively characterized and successfully used to separate a wide variety of molecules [5, 8, 16, 32]. Separations are based on differences in electrophoretic mobilities as a result of mass, size, shape, and charge differences in the analytes [33]. Migration times can be optimized by altering the buffer and changing the conditions used for the electrophoretic separation [8, 31]. Both capillary and microfluidic electrophoresis have proven useful in the interrogation of biological pathways in mammalian cells and *X. laevis* oocytes [7, 24].

### 3.5.1 Buffer Exchange

The use of electrophoresis for *in vivo* studies requires that the separation buffer be biologically compatible with the cells [31]. Generally, electrophoretic buffers are not biologically compatible because their composition is detrimental to the living cell [34]. Unless the separation is to be conducted in a physiologic buffer such as ECB, the cell buffer must be exchanged for electrophoretic buffer prior to separation. This requirement of having to use two different buffers is commonly referred to as the *buffer exchange problem*.

Lee *et al.* demonstrated a successful buffer exchange when monitoring the activity of SK in intact sheep erythrocytes [5]. The cells were kept in a cell-compatible ECB during handling and were chemically lysed when exposed to the electrophoretic separation buffer. Bodipy-labeled sphingosine and spingosine-1-phosphate from the cells were then separated and detected with CE–LIF. Separations were performed in a fused-silica capillary (50 μm i.d., 360 μm o.d.) with an effective length of 18 cm and a total length of 41 cm. An applied field strength of $+487\ \mathrm{V\,cm^{-1}}$ was used for the separation. The electrophoretic buffer yielding the best resolution and efficiency was 100 mM Tris, 10 mM sodium deoxycholate (SDC), 20% 1-propanol and 5% of a dynamic coating, EOTrol LR polymer solution, from Target Discovery (Palo Alto, CA). The addition of the surfactant SDC improved the efficiency of the separation and reduced capillary wall interactions, shortening elution time to less than 10 min. 1-Propanol was used as an organic modifier to improve separation efficiency and the dynamic coating was used to reduce interactions with the capillary wall.

Lee *et al.* investigated the activity of cdc2K, PKA, and PKC in *X. laevis* oocytes utilizing a slightly different technique [16]. A positively charged coating of poly-ethyleneimine was utilized in the fused-silica capillary to reduce adsorption of analytes to the capillary surface. Triton X-100, a surfactant used in the separation buffer (45 mM NaCl, 1.6 mM KCl, 0.6 mM $MgCl_2$, 1% Triton X-100, 10 mM HEPES, pH 7.4), is toxic to living cells, so a buffer exchange was required prior to the commencement of the electrophoretic separation. The oocytes were kept in an ECB and a capillary was utilized to puncture the membrane for the loading of the cytoplasm. The capillary inlet was then translated to the electrophoretic buffer reservoir and CE immediately initiated.

Another demonstration of a successful buffer exchange was shown by Marc *et al.* in 2007 [13]. This work utilized a coaxial sheath around the capillary to provide electrophoretic buffer for the separation while keeping this toxic buffer away from viable cells. The adherent cells were located in a PDMS channel and surrounded in an ECB appropriate for viable cells until the moment of lysis. The separation capillary with the surrounding sheath was positioned 10–40 μm over the cell to be sampled. Sheath flow and voltage were applied simultaneously, serving to lyse the cell and commence electrophoretic separation of the cell contents. The capillary was then translated upstream to the next viable cell, where the process was repeated. This sheath adaptor on the capillary and constant fluid flow across the cells allowed for cells upstream of the capillary to remain viable in the ECB buffer while the separation occurred in an appropriate electrophoretic buffer. Figure 3.3a–c depicts this instrument design.

### 3.5.2
### No Buffer Exchange

Another way to avoid the problem of cytotoxicity from the separation buffer is to use an electrophoretic buffer that is compatible with living cells [31]. This technique eliminates the buffer exchange problem and reduces the number of steps and time required for analysis. Meredith *et al.* utilized CE without a buffer exchange to monitor both pharmacologically and physiologically stimulated PKA, PKC, CamKII, and cdc2 kinase activity in RBL and mouse Swiss 3T3 cells [7]. The physiologic buffer of 135 mM NaCl, 5 mM KCl, 10 mM HEPES, 1 mM $MgCl_2$, 1 mM $CaCl_2$, pH 7.4 was also used as the electrophoretic buffer. The use of this buffer for CE was possible because the separation did not require a dynamic coating, surfactant or other organic modifier – all compounds that are cytotoxic. It is also possible to add low concentrations of biologically compatible buffer additives such betaine to the ECB. Betaine-containing buffers have been shown to separate lysozyme and alpha-chymotrypsinogen utilizing CE [35].

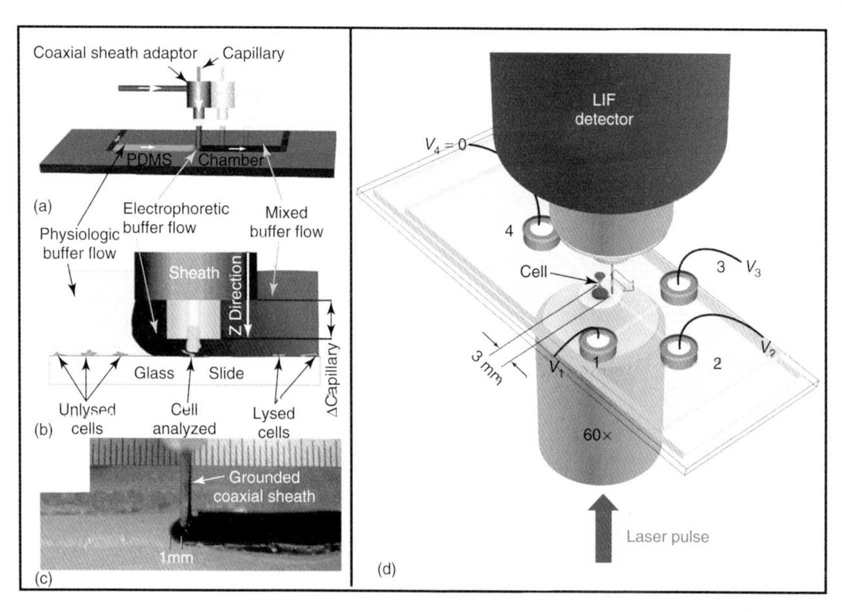


**Figure 3.3** Schema for capillary- and microfluidic-based serial analysis of cell analytes. (a) A capillary encased in a sheath adaptor remains stationary while the microchamber is translated to position the cell under the capillary. (b) A close-up, side view of the capillary and coaxial sheath. (c) Photograph of the coaxial system. The coaxial sheath fluid flow containing a dye is dark blue. Note that this fluid extends only 1 mM upstream of the capillary position. (d) Schematic of a microfluidic chip for separation of cellular contents. A laser pulse lyses the cell and the contents are separated in the microchannel and detected with LIF. (Figure (a–c) were reprinted with permission from [13], Copyright 2007 by The American Chemical Society; (d) was reprinted with permission from [11].)

### 3.5.3 Microchannel Electrophoretic Separations

In addition to the capillary electrophoretic techniques, electrophoresis in microchannels has also been used to probe single cells. In a feasibility study to look at separations in coated microchannels, Phillips *et al.* demonstrated baseline resolution of fluorescein and Oregon green separated electrophoretically on a PDMS microchip with 30 μm × 30 μm phosphatidylcholine-coated channels with a separation distance of 2 mm and an applied field strength of $+133\,\text{V}\,\text{cm}^{-1}$ [36]. In this same work, a high concentration ECB buffer (135 mM NaCl, 5 mM KCl, 10 mM HEPES, 1 mM $MgCl_2$, 1 mM $CaCl_2$, pH 7.4) was utilized to separate the phosphorylated and nonphosphorylated forms of a peptide substrate reporter for Abelson tyrosine kinase (Abl). The two peaks representing the phosphorylated and nonphosphorylated forms of the substrate were nearly baseline separated in 10 s with good efficiency and migration time reproducibility. This same buffer system was also utilized previously for peptide separations utilizing a capillary; however, the efficiency and reproducibility were poor, with unstable migration times and peak tailing [31].

In another study, Lai and coworkers separated the contents of whole Ba/F3 cells in a microfluidic channel using a separation buffer isotonic to the ECB [11]. The isotonic buffer consisted of 1 mM $KH_2PO_4$, 4 mM $Na_2HPO_4$, 245 mM D-mannitol and 40 mM poly(ethylene glycol), pH 7.4, and was used only for electrophoresis. A 3-mm-long, 50 μm × 50 μm polymer-coated PDMS channel was utilized for the electrophoretic separation of fluorescein and Oregon green in the cells. The cells were lysed with a 4.2 μJ laser pulse and the cell contents were injected into the separation channel by altering the voltage applied to the channels on the microchip. A field strength of $+175\ V\,cm^{-1}$ was utilized for the separation and detection occurred via LIF near the end of the separation channel. See Figure 3.3d for a schematic of the setup. While the isotonic buffer utilized in this study gave an adequate separation, it was not as good as a previous separation performed with an alternative electrophoretic buffer (25 mM Tris and 192 mM glycine, pH 8.4) [37, 38], indicating that the isotonic buffer was not an optimal electrophoretic buffer.

## 3.6 Detection

Detection is a very important step in any chemical analysis. Advancements in detection techniques have had a major impact on chemical cytometry. The cellular environment is chemically complex and may contain only a few hundred copies of target analyte distributed in the picoliter-to-nanoliter volume of the cell [3]. Therefore, an ultrasensitive detection method is necessary for single-cell chemical analyses. LIF is the most sensitive of the techniques for detection in CE and microchip devices [39]. The high photon flux and spatial coherence of laser light provide excellent properties for a fluorescence excitation source. This section will discuss CE–LIF detection on custom-built instruments for single-cell analysis.

### 3.6.1 Instrumentation

Fluorescence detection is extremely dependent on the instrumental design. Because of the complexity and state of the art of chemical cytometry, CE–LIF systems are built in-house. Sims and coworkers first introduced a CE–LIF system coupled with a laser-based cell lysis method (Figure 3.4) [24]. Over time, this setup was continuously upgraded for enhanced sensitivity as well as to meet experimental demands [17, 25].

#### 3.6.1.1 Excitation

Selection of fluorescently labeled analytes and fluorescent dyes used in most investigations to date (refer to Table 3.2) were mainly based on availability of an

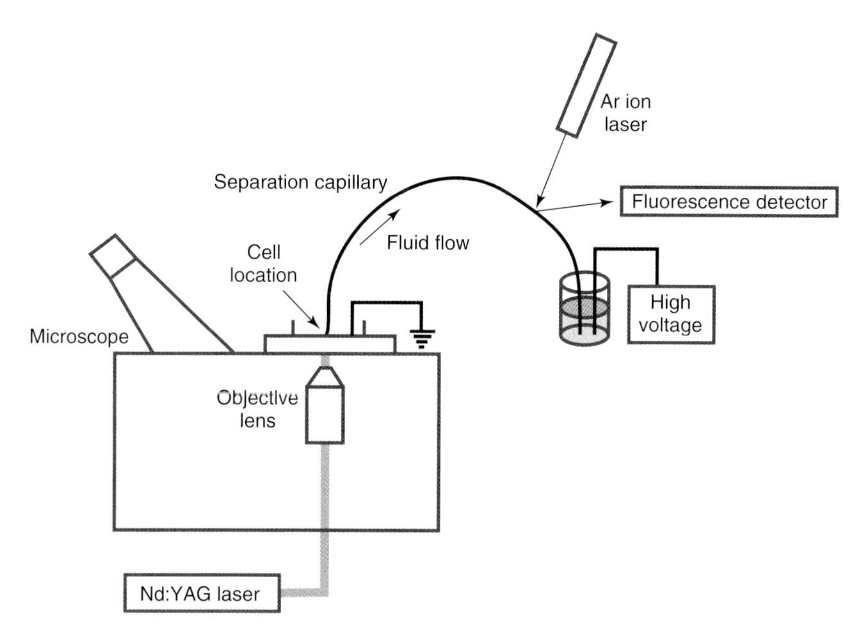


**Figure 3.4** Schematic of the CE–LIF system coupled to a laser-based cell lysis technique.

**Table 3.2** A summary of the reporters used, processes that are interrogated with these reporters, and the various methods for loading the reporters into cells.

| Reporter | Process interrogated | Loading methods | References |
|---|---|---|---|
| Fluorescein and Oregon green | Electrophoretic system function | Microinjection, pinocytic loading, diacetate incubation | [11–15] |
| Fluorescein-labeled peptides | Protein kinases | Microinjection, pinocytic loading, TAT conjugation, C18 conjugation | [6, 7, 9, 16] |
| FlAsH-labeled peptides | Protein kinases | Genetically engineered tetracysteine motif; incubated with FlAsH dye | [17] |
| Fluorescein-labeled sphingosine | SK1, SK2, SPP1, and SPP2 | Incubation | [5, 8] |
| Bodipyl-Fl phosphatidyl inositol phosphates | PI3-K and PTEN | Incubation | [8] |

excitation light source. In most cases, green fluorescence (535 ± 20 nm) is detected; therefore, either an argon ion laser (488 nm; Uniphase, San Jose, CA) [24] or a solid state laser (473 nm; Lasermate Group Inc., Pomona, CA) [5] was used as an excitation source. A laser line filter was placed in the excitation pathway when multiline argon ion lasers (488, 514 nm) were used as the light source. To allow the capillary lumen to be interrogated by the focused laser beam, an optical window in the capillary was created by burning off the polyimide coating and the laser beam was focused down to a size less than the inner diameter of the capillary by a focusing lens.

#### 3.6.1.2 Fiber-optic Cable-based Excitation Pathway

Continual improvements to the custom-built CE–LIF system have been made to increase the robustness and improve the detection limits. Kottegoda and coworkers replaced the open space excitation pathway with a single-mode fiber-optic cable that coupled a focusing lens of 3.5 cm focal length to the end of the fiber (Oz Optics, Ottawa, ON, Canada) [17]. This new design provides a sharper spot size on the capillary than was previously obtained. For a 50 μm i.d. capillary, a 40 μm spot size was maintained, minimizing the laser scattering from the glass–liquid interface parallel to the optical pathway. Owing to the fiber-optic cable setup, the laser was positioned outside the main body of the CE–LIF system, improving the stability of the excitation pathway.

#### 3.6.1.3 Emission

The measured fluorescence intensity is linearly dependent on the excitation power of the light source. For excitation, 2 mW of laser power was maintained at the capillary detection window and fluorescence was collected with a microscope objective (40 × 0.75 N.A., Plan Fluor, Nikon, Melville, NY) at a right angle to the capillary and the laser beam. The emission light was collected and measured with a photomultiplier tube (PMT; R928, Hammamatsu, Bridgewater, NJ) or an intensified charge couple device (CCD) camera (PhotometricsCoolSNAP*fx;* Roper Scientific Inc., Tucson, AZ) [11] after filtering it with a 488 nm holographic notch filter and band pass filter, 535 ± 50 nm. The broad spectral response (400–900 nm) and high sensitivity of the PMT give an amplification of a low light signal with little noise and fast response times. However, optical filtering is an essential part of wavelength selection and assists in the reduction of the background noise. The holographic notch filter and the spatial filter are used to eliminate most of the laser scatter that contributes to the background noise.

#### 3.6.1.4 Data Collection and Analysis

A custom program was written in TestPoint (Keithley Instruments, Inc., Cleveland, OH) for data collection and hardware control. This program also controls the Nd:YAG laser firing for cell lysis and the CE by controlling the high-voltage power supply. The PMT current was amplified and converted to a voltage with a preamplifier (model 1212, DL Instrument, Dryden, NY). The fluorescence signal and current reading from the CE were constantly monitored and digitized by a data acquisition board (KPCI 3100, Keithley Instrument). Upon run completion, the data was stored in ASCII format for later manipulation in a graphic software program. Many graphic software programs are available for use and allow for peak identification and peak area calculations.

### 3.6.2
### Limit of Detection (LOD)

After the initial setup of the detection and electrophoresis system, it must be tested and adjusted to achieve the detection limits and sensitivity necessary for single-cell experiments. To detect single cells loaded with nanomolar concentrations of the fluorescently labeled reporter, the detection limit of the system must be on the order of $10^{-20}$ mol [1, 24]. To determine the limit of detection (LOD), fluorescent standards were injected at successively lower concentrations until the electrophoretic peak was three times larger than the baseline. To accurately calculate the mass detection limit using a sample of known concentration, the volume of the injection plug must be estimated. The Poiseuille equation can be used to calculate the injection volume of the sample:

$$V = \frac{\Delta P d^4 \pi t}{128 \eta L} \tag{3.1}$$

where $V$ is the volume injected, $\Delta P$ is the pressure applied, $d$ is the capillary internal diameter, $t$ is the duration of pressure application, $\eta$ is the viscosity of the medium, and $L$ is the total capillary length [21]. Optimized fluorescence detection systems can detect $10^{-20}$mol or lower concentrations of fluorescent standard, suggesting that it may be possible to detect fluorescently labeled reporters in a single cell [7, 17, 24]. In single-cell kinase activity studies, $10^{-20}$–$10^{-18}$ mol levels of labeled substrate peptides were detected in HT1080 cells [6, 9], RBL cells [7], and Swiss 3T3 cells [6, 7].

### 3.6.3
### Detection in a Microfluidic Device

Owing to the spatial limitation of microfluidic devices, an epifluorescence optical setup was used for detection [11, 14]. This system was similar to that described earlier, except the exciting laser beam was reflected off a 505 nm long-pass dichroic mirror (Omega Optical, Brattleboro, VT). Fluorescence was detected 3 mM from the point of lysis. A second detection system was added in some cases to detect the fluorescence from intact cells just prior to the point of lysis [14]. In this case, the excitation light was shared by both detection systems. Figure 3.5 illustrates the chemical analysis by microfluidic separation of Ba/F3 cells preloaded with fluorescein and Oregon green [11].

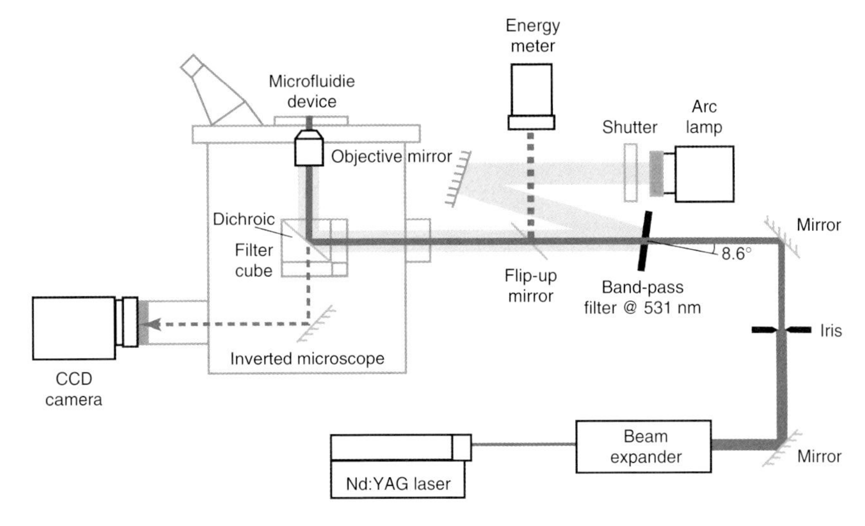


**Figure 3.5** Schematic diagram of the optical delivery system used in microfluidic-based single-cell analysis. The Nd:YAG laser (532 nm, 750 ps) and a beam expander designed for laser-based cell lysis and an epifluorescence optical setup was used for fluorescence detection. (Reproduced with permission from [11].)

## 3.7 Automation and Throughput

### 3.7.1 Increasing Throughput

While results obtained thus far using CE for analysis on the single-cell level have been very impressive, data collection has been time consuming due to the nature of the custom-built CE system. Time consuming steps of positioning the capillary over the cell to be lysed, movement of the capillary from sample site to separation buffer and capillary washing between runs must all be performed by hand. Thus far, traditional CE has allowed for analysis of less than 20 cells per day, which is not practical for measuring large populations [12]. Two strategies exist for increasing the throughput of CE separations: parallel and serial analyses. Parallel analyses consist of multiple capillaries monitored simultaneously, either with concurrent monitoring of all capillaries or with a moving detector that sequentially monitors each capillary in time [40]. While the best known application of parallel CE has been the sequencing of the human genome, parallel analysis may not be feasible for living cells because alignment of all the capillaries over the cells may be difficult to accomplish. The second strategy for increasing throughput is serial analysis, in which a second sample is injected into the same capillary prior to the first sample's exit from the capillary [13]. Given the parameters, serial analysis appears to be the most promising for high-throughput CE analysis of single cells.

### 3.7.2
### Capillary-based Devices

The development of a coaxial flow system in 2007 that allows for the serial injection of cells onto the capillary has increased throughput from less than 20 cells per day to one cell every 2 min [13]. An important caveat is that the cells to be analyzed must remain viable up until the point of lysis. As discussed above, electrophoretic buffer is generally toxic to the cells, so the cells remain in a physiologic buffer prior to lysis and the buffer must be rapidly exchanged for an electrophoretic buffer for the separation. This is accomplished by positioning the cells in a PDMS channel, flowing a physiologic buffer over the viable cells and utilizing a sheath around the capillary to contain the electrophoretic buffer. The sheath controls when the electrophoretic buffer is allowed to flow and can be switched on and off. When the sheath is opened, the electrophoretic buffer runs into the capillary and downstream in the channel, over the cells that have already been sampled, but does not affect viable cells to be sampled. Sequential analysis of the cells occurs by moving the capillary upstream relative to the physiologic buffer flow; hence the next cell has not yet been exposed to the electrophoretic buffer and so remains unperturbed (Figure 3.3a–c). Injection of the next cell occurs prior to the detection of the contents of the first cell; hence more than one assay is present in the capillary at a single moment in time. As long as the contents of the sequential samples do not overlap, samples can be injected and detected serially, increasing the throughput of the analyses.

Throughput of single adherent-cell analysis by CE can be further enhanced through electrical lysis of cells held in individual microwells in combination with the previously discussed coaxial flow system [12]. In studying the feasibility of this method, Marc and coworkers investigated appropriate materials for microwell fabrication, positioning of the electrode in the well, field strengths required for cell lysis, and effects of this field strength on nontarget neighboring cells. The first design of the microwell plates had many flaws, including difficulty in aligning the plates, inability to visualize the cell through an opaque electrode, and the delamination of the SU-8 microwell surface coating from the chips. These problems were addressed and a new design was utilized for the remainder of the study. The newer design incorporated the use of 1002F as the polymer cell trap, which has a greater adhesion to glass and resists delamination. The electrode design was altered so that only one electrode was used in each well, as opposed to two per well, and each electrode addressed more than one cell trap, which aided in the fabrication of the chips. A second electrode was placed in the media surrounding the cell and was used as a common ground for both the lysis and separation steps. Finally, it was determined that an electrical pulse of 10 ms generating a field strength of approximately 10 kV $cm^{-1}$ was both sufficient to lyse cells in 33–66 ms, and showed no effect on cells in neighboring traps positioned 500 μM apart. Of initial concern was the formation of a gas bubble in the well during electrical lysis. If the bubble entered the capillary, there would be a break in the electrical circuit, which would prevent current flow through the capillary and halt the separation. However, it was found that the contents of the lysed cell were rapidly loaded onto the capillary while the gas bubble remained trapped in the microwell. By using cells trapped in a microwell array, it was determined that cells loaded with fluorescein and Oregon green could be analyzed at a rate of one cell in 30 s, giving adequate separation of the two components. This shows much promise in the design of a higher-throughput instrument for CE analysis of single cells.

### 3.7.3 Microfluidic-based Devices

Another way to increase throughput is to use microfluidic devices that incorporate cell-handling, cell-processing and electrophoretic separation on a single chip [14]. This could greatly enhance the throughput while maintaining high separation efficiencies. Similar challenges arise in that the cells must remain viable up until the point of lysis, and thus must not be exposed to toxic electrophoretic buffer prior to lysis. This has been accomplished by chip design as well as application of a pulsed DC voltage for cell lysis immediately prior to separation. A problem encountered was the adhesion of the lipid components of the cellular debris to the microchannel. This was greatly reduced by the addition of an emulsifying agent to the buffer. An additional challenge was the injection of the contents into the separation channel, which was perpendicular to the flow channel, as opposed to the contents continuing straight into the waste channel. This is a function of the fluid velocity toward the waste channel and the electrophoretic velocity of the analyte in the separation channel. It was found that the flow velocity must be slower than the electrophoretic velocity of the slowest moving analyte in order to get an injection into the separation channel. Cells loaded with fluorescein and Oregon green were used on this microchip to demonstrate a 2.2 s separation time of each cell's contents with less than 1% migration time error. This translates to the analysis of 7–12 cells per minute, dramatically increasing the throughput previously seen and showing that microchips may be important in the design for high-throughput electrophoretic analysis of single cells.

## References


1 Sims, C.E. and Allbritton, N.L. (2003) Single-cell kinase assays: opening a window onto cell behavior. *Curr. Opin. Biotechnol.*, **14**, 23–28.

2 Petsko, G.A. and Ringe, D. (2004) *Protein Structure and Function*, New Science Press Ltd, London.

3 Dovichi, N.J. and Hu, S. (2003) Chemical cytometry. *Curr. Opin. Chem. Biol.*, **7**, 603–608.

4 Li, H., Sims, C.E., Wu, H.Y., and Allbritton, N.L. (2001) Spatial control of cellular measurements with the laser micropipet. *Anal. Chem.*, **73**, 4625–4631.

5 Lee, K.J., Mwongela, S.M., Kottegoda, S., Borland, L., Nelson, A.R., Sims, C.E., and Allbritton, N.L. (2008) Determination of sphingosine kinase activity for cellular signaling studies. *Anal. Chem.*, **80**, 1620–1627.

6 Li, H., Sims, C.E., Kaluzova, M., Stanbridge, E.J., and Allbritton, N.L. (2004) A quantitative single-cell assay for protein kinase B reveals important insights into the biochemical behavior

of an intracellular substrate peptide. *Biochemistry*, **43**, 1599–1608.

7 Meredith, G.D., Sims, C.E., Soughayer, J.S., and Allbritton, N.L. (2000) Measurement of kinase activation in single mammalian cells. *Nat. Biotechnol.*, **18**, 309–312.

8 Mwongela, S.M., Lee, K., Sims, C.E., and Allbritton, N.L. (2007) Separation of fluorescent phosphatidyl inositol phosphates by CE. *Electrophoresis*, **28**, 1235–1242.

9 Soughayer, J.S., Wang, Y., Li, H., Cheung, S., Rossi, F.M., Stanbridge, E.J., Sims, C.E., and Allbritton, N.L. (2004) Characterization of TAT-mediated transport of detachable kinase substrates. *Biochemistry*, **43**, 8528–8540.

10 Nelson, A.R., Allbritton, N.L., and Sims, C.E. (2007) Rapid sampling for single-cell analysis by capillary electrophoresis. *Methods in Cell Biology*. Elsevier Academic Press Inc., San Diego.

11 Lai, H., Quinto-Su, P.A., Sims, C.E., Bachman, M., Li, G.P., Venugopalan, V., and Allbritton, N.L. (2008) Characterization and use of laser-based lysis for cell analysis on-chip. *J. R. Soc. Interface*, **5**, S113–S121.

12 Marc, P.J., Sims, C.E., Bachman, M., Li, G.P., and Allbritton, N.L. (2008) Fast-lysis cell traps for chemical cytometry. *Lab Chip*, **8**, 710–716.

13 Marc, P.J., Sims, C.E., and Allbritton, N.L. (2007) Coaxial flow system for chemical cytometry. *Anal. Chem.*, **79**, 9054–9059.

14 McClain, M.A., Culbertson, C.T., Jacobson, S.C., Allbritton, N.L., Sims, C.E., and Ramsey, J.M. (2003) Microfluidic devices for the high-throughput chemical analysis of cells. *Anal. Chem.*, **75**, 5646–5655.

15 Luzzi, V., Lee, C., and Allbritton, N.L. (1997) Localized sampling of cytoplasm from Xenopus oocytes for capillary electrophoresis. *Anal. Chem.*, **69**, 4761–4767.

16 Lee, C., Linton, J., Soughayer, J.S., Sims, C.E., and Allbritton, N.L. (1999) Localized measurement of kinase activation in oocytes of Xenopus laevis. *Nat. Biotechnol.*, **17**, 759–762.

17 Kottegoda, S., Aoto, P.C., Sims, C.E., and Allbritton, N.L. (2008) Biarsenical-tetracysteine motif as a fluorescent tag for detection in capillary electrophoresis. *Anal. Chem.*, **80**, 5358–5366.

18 Griffin, B.A., Adams, S.R., and Tsien, R.Y. (1998) Specific covalent labeling of recombinant protein molecules inside live cells. *Science*, **281**, 269–272.

19 Luzzi, V., Sims, C.E., Soughayer, J.S., and Allbritton, N.L. (1998) The physiologic concentration of inositol 1,4,5-trisphosphate in the oocytes of Xenopus laevis. *J. Biol. Chem.*, **273**, 28657–28662.

20 Luzzi, V., Murtazina, D., and Allbritton, N.L. (2000) Characterization of a biological detector cell for quantitation of inositol 1,4,5-trisphosphate. *Anal. Biochem.*, **277**, 221–227.

21 Sims, C.E., Luzzi, V., and Allbritton, N.L. (2007) Localized sampling, electrophoresis, and biosensor analysis of Xenopus laevis cytoplasm for subcellular biochemical assays, in *Methods in Molecular Biology* (ed. X.J. Liu), Humana Press Inc., Totowa.

22 Wagner, J., Fall, C.P., Hong, F., Sims, C.E., Allbritton, N.L., Fontanilla, R.A., Moraru, I.I., Loew, L.M., and Nuccitelli, R. (2004) A wave of IP3 production accompanies the fertilization $Ca^{2+}$ wave in the egg of the frog, Xenopus laevis: theoretical and experimental support. *Cell Calcium*, **35**, 433–447.

23 Stephens, D.J. and Pepperkok, R. (2001) The many ways to cross the plasma membrane. *Proc. Natl. Acad. Sci. U.S.A.*, **98**, 4295–4298.

24 Sims, C.E., Meredith, G.D., Krasieva, T.B., Berns, M.W., Tromberg, B.J., and Allbritton, N.L. (1998) Laser-micropipet combination for single-cell analysis. *Anal. Chem.*, **70**, 4570–4577.

25 Han, F., Wang, Y., Sims, C.E., Bachman, M., Chang, R., Li, G.P., and Allbritton, N.L. (2003) Fast electrical lysis of cells for capillary electrophoresis. *Anal. Chem.*, **75**, 3688–3696.

26 Nelson, A.R., Borland, L., Allbritton, N.L., and Sims, C.E. (2007)

Myristoyl-based transport of peptides into living cells. *Biochemistry*, **46**, 14771–14781.

27 Borland, L.M., Kottegoda, S., Phillips, K.S., and Allbritton, N.L. (2008) Chemical analysis of single cells. *Annu. Rev. Anal. Chem.*, **1**, 191–227.

28 Brown, R.B. and Audet, J. (2008) Current techniques for single-cell lysis. *J. R. Soc. Interface*, **5**, S131–S138.

29 Rau, K.R., Guerra, A., Vogel, A., and Venugopalan, V. (2004) Investigation of laser-induced cell lysis using time-resolved imaging. *Appl. Phys. Lett.*, **84**, 2940–2942.

30 Rau, K.R., Quinto-Su, P.A., Hellman, A.N., and Venugopalan, V. (2006) Pulsed laser microbeam-induced cell lysis: time-resolved imaging and analysis of hydrodynamic effects. *Biophys. J.*, **91**, 317–329.

31 Li, H., Wu, H.Y., Wang, Y., Sims, C.E., and Allbritton, N.L. (2001) Improved capillary electrophoresis conditions for the separation of kinase substrates by the laser micropipet system. *J. Chromatogr. B*, **757**, 79–88.

32 Jorgenson, J.W. (1986) Electrophoresis. *Anal. Chem.*, **58**, 743A–760A.

33 Weinberger, R. (1993) *Practical Capillary Electrophoresis*, Academic Press, London.

34 Togo, T., Alderton, J.M., Bi, G.Q., and Stienhardt, R.A. (1999) The mechanism of facilitated cell membrane resealing. *J. Cell Sci.*, **112**, 719–731.

35 Bushey, M.M. and Jorgenson, J.W. (1989) Capillary electrophoresis in buffers containing high-concentrations of zwitterionic salts. *J. Chromatogr.*, **480**, 301–310.

36 Phillips, K.S., Kottegoda, S., Kang, K.M., Sims, C.E., and Allbritton, N.L. (2008) Separations in poly(dimethylsiloxane) microchips coated with supported bilayer membranes. *Anal. Chem.*, **80**, 9756–9762.

37 Hu, S.W., Ren, X.Q., Bachman, M., Sims, C.E., Li, G.P., and Allbritton, N.L. (2004) Surface-directed, graft polymerization within microfluidic channels. *Anal. Chem.*, **76**, 1865–1870.

38 Hu, S.W., Ren, X.Q., Bachman, M., Sims, C.E., Li, G.P., and Allbritton, N.L. (2003) Cross-linked coatings for electrophoretic separations in poly(dimethylsiloxane) microchannels. *Electrophoresis*, **24**, 3679–3688.

39 Swinney, K. and Bornhop, D.J. (2000) Detection in capillary electrophoresis. *Electrophoresis*, **21**, 1239–1250.

40 Dovichi, N.J. and Zhang, J. (2000) How capillary electrophoresis sequenced the human genome. *Angew. Chem. Int. Ed.*, **39**, 4463–4468.

# 4
# Ultrasensitive Detection of Low-copy-number Molecules from Single Cells

*Kangning Ren and Hongkai Wu*

## 4.1
## Introduction

Different from conventional methods that are performed with cell populations, single-cell analysis, which has become a highly attractive tool for investigating cellular contents [1], avoids the loss of information associated with ensemble averaging. Methods including integrated fluorescence and single-molecule imaging can be used for quantifying specific proteins inside a single cell [2–4]. However, the need of resolving fluorescence from different probes restricts these analytical techniques to one or a few species at a time. Moreover, their applications are limited to the cases where the cell environment changes the fluorescence of the reporter molecule or endogenous fluorescence interferes with the measurements. In contrast, chemical cytometry based on microfluidics allows the integration of several distinct functions including cell transport, lysis, and separation of cellular contents, enabling the entire process to be performed rapidly on a massively parallel scale. In this chapter, we focus the discussion on the ultrasensitive detection of low-copy-number molecules from single cells.

## 4.2
## Microchip Designs for Single-cell Analysis and/or Cell Manipulation

It is challenging to process the minute, precious samples of single cells in a reproducible, quantitative, and parallel fashion by using conventional methods, while microfluidic device provides powerful and potential tool for that purpose. Some representative microchip designs for single-cell analysis are summarized in this section.

To obtain better reproducibility, it is essential to precisely trap single cells at a fixed position within microchannels, which can be achieved by well designed microfluidic device. Ros' group fabricated a microfluidic device with microobstacles at channel crossings for cell trapping [5]. Sun and Yin fabricated a multidepth microfluidic chip with a weir structure for cell docking and lysing [6]. Gao *et al.*

*Chemical Cytometry*. Edited by Chang Lu

ISBN: 978-3-527-32495-8

developed a microfluidic system that integrated cell sampling, single-cell loading, docking, lysing, and capillary electrophoresis (CE) separation with laser-induced fluorescence (LIF) detection by using a simple glass micro-CE chip [7]. The flow of cells and single-cell loading into the separation channel were controlled by hydrostatic pressure and electrophoretic force, respectively, and the single-cell docking was performed by a sequence of potentials. The 2D electrophoresis (2DE) has also been used for chemical cytometry [8, 9]. In these experiments, individual cells were injected into a capillary; the cell was lysed by contacting with the running buffer contained within the capillary and the cell lysate was subsequently separated by 2DE [10].

The Quake group fabricated micromechanical valves in a microfluidic device for single-cell RNA and DNA analysis. With this approach, they managed to implement all steps including cell capture, cell lysis, mRNA purification, cDNA synthesis, and cDNA purification in the single device [11]. Then they demonstrated a higher throughput microfluidic system for multiplex digital PCR of single cells, in which many parallel reaction chambers (12 samples × 1176 chambers per sample) were formed by micromechanical valves to act as independent PCR reactors [12]. However, the method based on PCR is limited to amplifiable molecules.

Xie *et al.* [13] demonstrated a microfluidics-based assay with single-molecule sensitivity for real-time observation of the expression of $\beta$-galactosidase ($\beta$-gal), a standard reporter for gene expression in living *Escherichia coli* cells. However, because efflux pumps on the cell membrane actively expel foreign organic molecules from the cytoplasm [14], the fluorescent products of $\beta$-gal are not retained in the cell but pumped to the surrounding medium and rapidly diffuse away, therefore the advantage of enzymatic amplification is lost. To recover the fluorescence signal, they trapped cells in closed microfluidic chambers such that the fluorescent product released from the cells accumulated in the small volume of the chambers. The microfluidic device is made of a soft polymer, polydimethylsiloxane (PDMS), and consists of a flow layer containing the cells and a top control layer for valves [15]. Cells were trapped and cultured in an enclosure of 100 pl by actuation of two adjacent valves in the control layer. The microfluidic chip was translated by a motorized stage, allowing multiplexing of data acquisition by repeatedly scanning the chambers with an inverted fluorescence microscope. Typically, scanning of 100 chambers took less than 2 min. To avoid cellular autofluorescence and photodamage to the cell, the excitation laser beam was tightly focused that did not directly illuminate the cell (Figure 4.1).

Zare's group developed a three-state (fully open, half open, and fully closed) value by which a reaction chamber with a volume of 70 pl for the lysis and derivatization of the contents of a single cell was constructed, while the unwanted dilution of the intracellular chemical contents was reduced [16]. Concentration of various amino acids from individual human T cells was determined. The microfluidic device was also used to count low-copy-number proteins in a single cell, which is discussed in detail in later sections [17].

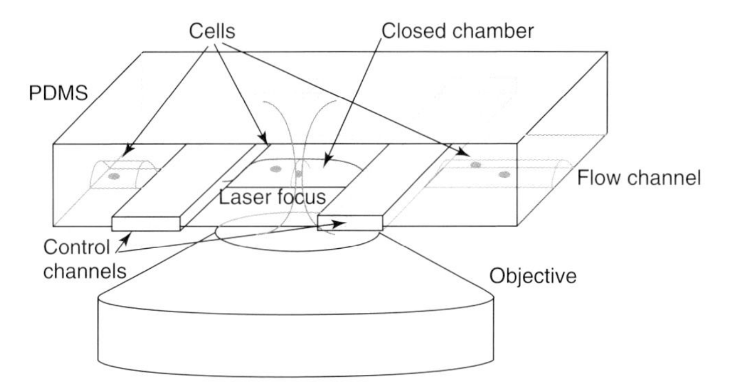


**Figure 4.1** Schematic diagram of the microfluidic chamber used for the enzymatic assay. Cells are trapped inside a volume of 100 pl, formed by compression of a flow channel by two control channels. (Reprint permission to be sought.)

## 4.3 Ultrasensitive Detection Methods for Single-cell Analysis

Low-copy-number proteins (present at fewer than 1000 molecules per cell) are critical in cell functioning. Without amplification procedures, their abundance is far below the detection limits of conventional protein analysis methods, such as enzyme-linked immunosorbent assay (ELISA) and mass spectroscopy. As an ultrasensitive detection method, the single-molecule fluorescence counting technique [18–20] has been used to solve this problem.

### 4.3.1 Fluorescence Detection Method

The power of a miniaturized chemical analysis system is ultimately limited by its sensitivity. Current microanalysis systems mainly rely on four major detection modes: fluorescent, electrochemical, chemiluminescence, and electrochemiluminescence detection. With an extremely high sensitivity, LIF is probably the most widely used among all methods for on-chip chemical cytometry. The spatial coherence characteristic of lasers permits tight focusing that is important for high resolution separation systems, and their narrow emission bandwidth is beneficial for signal to noise ratio. A variety of components including amino acids, peptides, proteins, and nucleic acids from single cells have been detected with LIF at low concentrations. Various LIF detection systems based on different optical arrangements have been successfully and broadly applied in the microfluidic chip-based analysis systems.

In LIF systems, the excitation lasers are often the most bulky component. However, with advances in III–V nitride manufacturing processes, compact and high-power light-emitting diodes (LEDs) in the visual and UV wavelengths are now commercially available at low cost and can be used as excitation sources in optical sensing as alternatives to traditional laser sources. Typically, the width of the channels in microfluidic applications is tens of micrometers; a very tight focal spot is not necessary, otherwise most analytes would not even transit the small focal volume and get excited at the detecting point. For such applications, the spatially incoherent emission from LEDs does not represent a significant drawback. Typically, a $500\,W\,cm^{-2}$ illumination could be achieved with a spot diameter of 0.1 mm. Chiu *et al.* [21] demonstrated the use of these high-power LEDs for sensitive fluorescence detection on chip (Figure 4.2). A 40-mW blue LED was used as the excitation source of a simple chip-based bead sorter for fluorescent beads' enrichment. The blue LED was also used in detection of CE experiments; the results showed a mass detection limit of 200 zmol for fluorescein. In this work, ultrasensitive fluorescence imaging of single rhodamine 123 molecules and individual L-DNA molecules using LED light source was also demonstrated. At a small fraction of the cost of a conventional laser, high-power LEDs provide an effective alternative in many fluorescence applications that demand portability, low cost, and convenience.

Because fixed wavelength of lasers limits the choice of the labeling reagent, the Cheng group constructed a fluorescence detection system for microfluidic CE with a Hg lamp as the excitation source and a photon counter as the detector [22]. Various fluorescent probes with different excitation or emission wavelengths could be adopted owing to the wide spectrum of Hg lamp. The amino acid neurotransmitters in rat pheochromocytoma (PC12) cells were separated and determined, and the intracellular Glu in individual PC12 cells was quantified as $3.5 \pm 3.1$ fmol with this technique.

### 4.3.2
### Fluorescence Labeling

Fluorescent dyes currently used cover the whole visible spectrum, permitting extremely low-concentration detection of samples in femtoliter volume and even

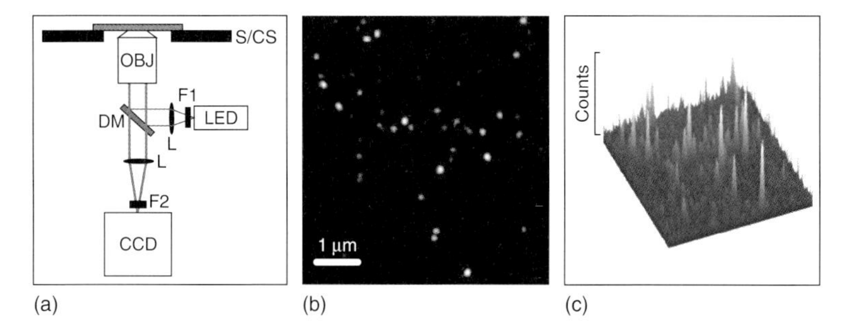


**Figure 4.2** (a) Optical configuration for imaging single molecules with a blue LED. S/CS, stage/coverslip; OBJ, objective; DM, dichroic mirror, L, lens; F1, excitation filter; F2, emission filter. (b) Image of single rhodamine 123 molecules on the surface of a glass coverslip. (c) Three-dimensional intensity plot of the single dye molecules imaged in b. (Reprint permission to be sought.)

single molecules. Many biomolecules have native fluorescence, and laser-induced native fluorescence (LINF) detection can be adopted directly to investigate these molecules. The cellular expression of some proteins and fusion proteins can be analyzed by the natural fluorescence. Label-free LIF detections for proteins based on the UV fluorescence of the three amino acids – tryptophan (Trp), tyrosine (Tyr), and phenylalanine (Phe) have been reported with an ultrahigh sensitivity [23].

CE was commonly used as a very powerful tool to separate the low content components inside a single cell. The Dovichi group separated and detected green fluorescence protein (GFP) expressed in a single bacterium [24]. The background signal was reduced by separating GFP from other native cellular autofluorescent components and the detection limit was improved to 100 ymol (60 copies) for GFP.

The GFP-construct protein (T31N-GFP) in single Sf9 insect cells (*Spodoptera frugiperda*) was separated and determined by native LIF [5, 25]. Black microfluidic chip by incorporating carbon black particles into PDMS could decrease the background fluorescence of the PDMS, which was a major source limiting the fluorescent detection sensitivity; the detection limit of Trp with 25 nM could be achieved, and the single-cell electropherogram with native UV–LIF was demonstrated [26].

Sweedler's group developed a multichannel native fluorescence detection system for CE to obtain more information [27]. In their system, a series of dichroic beam splitters were used to distribute the emitted fluorescence into three wavelength channels: 250–310, 310–400, and >400 nm; three photomultiplier tubes were used to detect the fluorescence in each of the wavelength ranges, respectively. By this system, the analytes could be identified by their multichannel signatures. The instrument was used to detect and identify the neurotransmitters in serotonergic LPeD1 and dopaminergic RPeD1 neurons from *Lymnaea stagnalis*.

However, in most cases, proteins are not autofluorescent and need to be fluorescently labeled in advance. Fluorescent derivatization with a fluorophore significantly improves the detection sensitivity of materials in low concentration and thus is widely used in single-cell- and subcellular analysis. Usually, the derivatization process is nonspecific and incomplete. Progresses are being made on developing more specific and sensitive derivatization agents as well as the labeling techniques, which will facilitate the LIF detection for single-cell analysis.

Specific fluorescence detection has been achieved by combining with immunoreaction. Jin's group used CE with on-capillary immunoreaction for determination of different forms of human interferon-$\gamma$ (IFN-$\gamma$) in single natural killer cells [28]. Immunoreaction between different forms of IFN-$\gamma$ from the lysate of a single cell and their labeled antibody was carried out in a microreactor at the front end of a separation capillary. Then the complexes were separated and detected by CE with LIF detection with a limit of detection (LOD) of zeptomoles.

Intracellular labeling technique was also developed to overcome the diffusion of the analytes during the on-capillary immunoreaction [29]. The immunoreaction was carried out inside a single cell before being injected into the capillary. The cell was then lysed and different forms of IFN-$\gamma$ in the cell were separated and detected with a significant improvement in resolution as compared with those using the on-capillary immunoreaction.

Xiao *et al.* demonstrated an LIF detection method for permeable intact cell without lysis [30]. They used two antibodies, mouse-raised JSB-1, for immunoreaction and LIF detection of the P-glycoprotein (PGP), a transmembrane efflux pump in multidrug resistance research. The labeled cell was injected into a capillary and directly separated without lysis, and the PGP amount on the cell was calculated.

### 4.3.3 Optical Configuration

The confocal LIF system, as an effective arrangement to perform highly sensitive detection, is commonly used on microchip. For example, single-molecule detection of rhodamine 6G was achieved by a confocal microscope [31].

However, the conventional confocal LIF systems are very bulky and complicated with high cost. As alternative, nonconfocal LIF detection systems based on bevel incident laser [32] or orthogonal arrangement [33] (Figure 4.3) were developed for coupling to chip-based systems. By detecting fluorescence light through the sidewall of the chip, an LOD of 1.1 pM fluorescein was obtained [33], which is comparable to that of the confocal LIF systems for chip-based CE.

In principle, single-molecule counting can be used for detection in single-cell analysis with the ultimately high sensitivity. Typically, it was achieved by monitoring the number of fluorescence bursts generated when the molecules flowed through the detection volume. The most common approach to obtain a high $S/N$ is to use confocal fluorescence microscopy, but its detection cross section (approximately 500 nm wide by 1 μm high) is much smaller than the size of an ordinary microfluidic channel, which results in extremely poor detection efficiency. One way to solve this problem is to decrease the dimensions of the channel to the nanometer

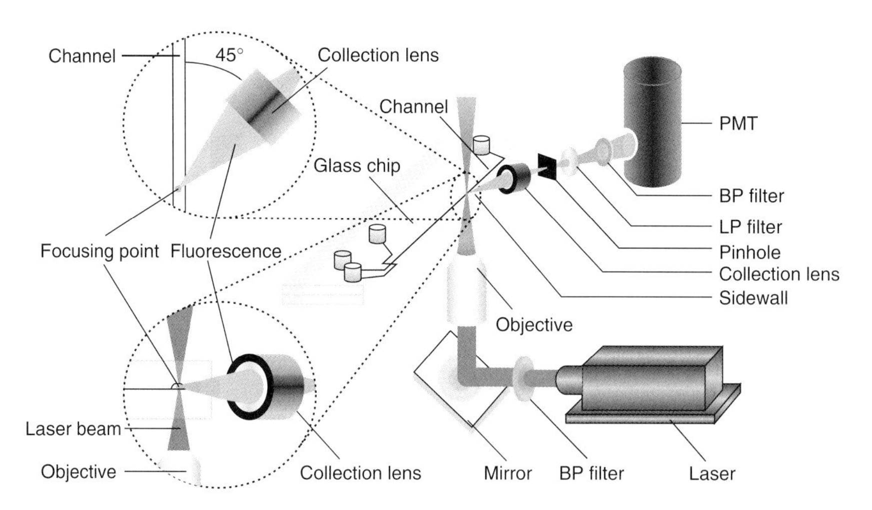


**Figure 4.3** Schematic diagram of the optical setup of an orthogonal LIF detection system. (Reprint permission to be sought.)

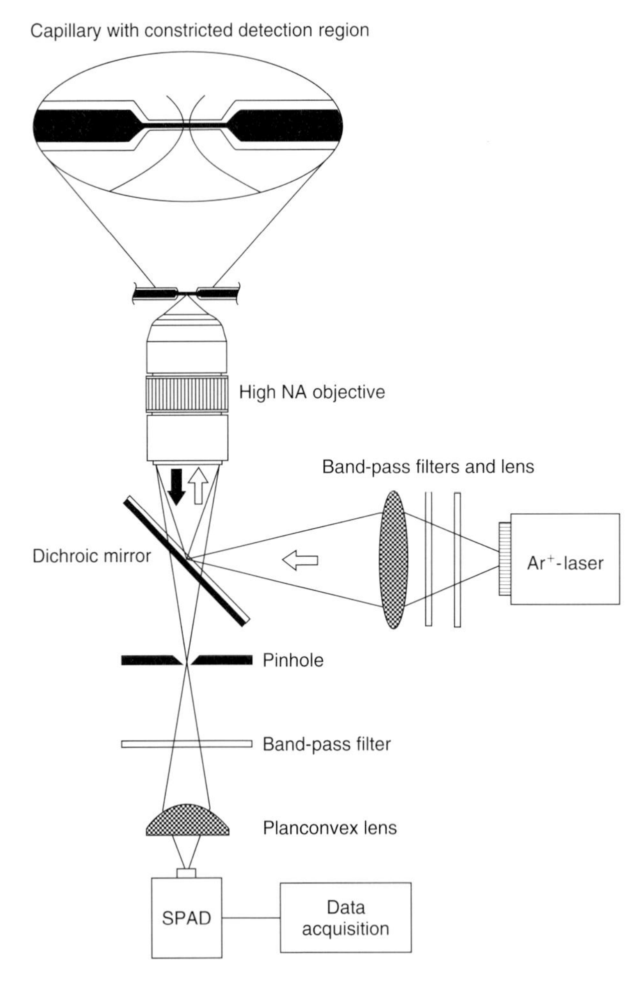


**Figure 4.4** Schematic diagram of the confocal setup for LIF detection with a restrictive channel. (Reprint permission to be sought.)

range (Figure 4.4) so that the entire flow cross section fits into the focus of the confocal microscope [34, 35]. Such small channel dimensions, however, can affect the electrophoretic separation of molecules in cell lysates [36, 37], and can easily cause clogging of the nanochannel with cell debris and other particulate matters.

Zare *et al.* [17] described a more practical way to resolve the counting-efficiency problem. They widened the excitation laser focus in one direction by using a cylindrical optics to illuminate the entire channel width. In their system, the excitation laser beam was focused by a cylindrical lens to form a line at the back focal plane of a high-numerical-aperture objective. The laser beam was collimated in the direction perpendicular to the channel length so that it illuminated a channel width of tens of micrometers. In the other direction, the laser was still focused by the objective to minimize the fluorescence background from out-of-focus excitation. The laser was less tightly focused as compared to confocal systems so that the limit of the channel height by the excitation focus was slightly relaxed (~2 μm). This height was large enough to avoid clogging in their experiments. With this optical configuration, the excitation laser formed a rectangular, curtain-shaped detection region across the channel.

Figure 4.5a shows the formation of a curtain-shaped laser focus in the microchannel by combination of the cylindrical lens and the microscope objective. The laser beam emerging from the optical fiber is collimated with a 100 mm achromatic lens, shaped by a 1 cm × 1 cm square hole, and sent into the microscope through a spherical or cylindrical lens (each having a focal length of 400 mm). The emitted fluorescence is collected by a microscope objective and filtered by a dichroic mirror (400-535-635 TBDR, Omega Optical) and a band-pass filter (HQ675/50m, Chroma). For LIF detection of CE separation, the cylindrical lens is used for excitation, and a photon counting photomultiplier tube module (H6240-01, Hamamatsu) is used for detection, with a 50 μm slit installed at the microscope image plane to reject the out-of-focus emission. For wide-field fluorescence imaging and molecule counting, an intensified charge-coupled device (ICCD) camera (I-Pentamax, Roper Scientific) serves as the detector. In molecule-counting experiments, the power of the laser beam emerging from the objective is about 10 mW, and the line-shaped laser focus at the sample is 50 μm long.

Figure 4.5b shows the z-dependence of the excitation laser strength in three different configuration: (i) wide-field, in which a spherical lens focuses the excitation laser beam to the back focal point of the microscope objective; (ii) cylindrical, in which a cylindrical lens focuses the laser beam to the back focal plane of the objective; and (iii) confocal, in which a parallel laser beam is sent into the objective. The comparison was carried out by imaging the fluorescence from a glass surface coated with Atto 565 labeled streptavidin. When the imaging plane moves away from the focal plane, the confocal configuration has the sharpest drop in excitation

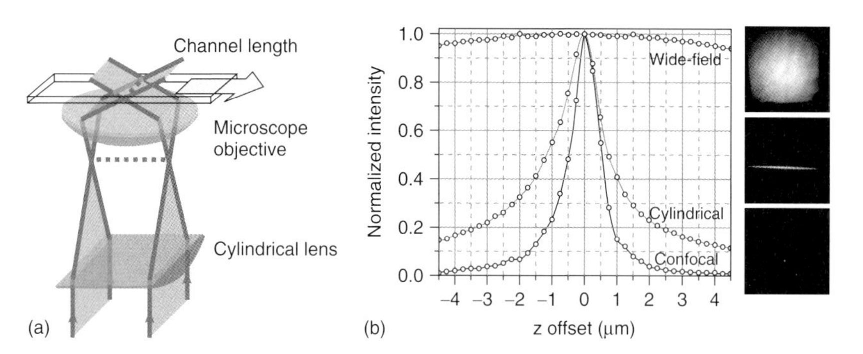


**Figure 4.5** (a) Creation of the detection curtain for molecule counting. (b) Dependence of detected fluorescence from a glass surface coated with Atto 565 labeled streptavidin. (Reprint permission to be sought.)

strength, the cylindrical configuration shows similar but slightly lower z-resolution, and the wide-field configuration has almost constant excitation strength when the z position of the sample changes. A 2-μm channel fits well into the focus of the cylindrical configuration and the out-of-focus background is suppressed.

### 4.3.4
### Molecule-counting Algorithm

The fluorescence from molecules that passed through the curtain can be recorded by an intensified CCD camera. A threshold criterion should be used to identify each molecule as a bright spot in an image frame. To enhance the fluorescence signal, the flow rate can be slowed down when an analyte is expected to pass through the detection curtain.

However, as the excitation laser intensity varies at different positions in the channel cross section, molecules that passed through the periphery of the channel produced lower fluorescence signals, which might be lost in the background noise. As a result, the detection efficiency varied slightly according to the brightness of a specific sample molecule. Zare *et al.* have measured the molecular-counting efficiency of Alexa Fluor 647-labeled streptavidin in a standard "double-T" chip and found that 60% of the injected molecules are counted [17]. A general method was then developed to estimate detection efficiencies directly from counting experiments by varying the threshold of molecule identification. In this way, the actual number of molecules can be derived without additional calibration.

A fluorescent molecule can be recorded by an intensified CCD camera as a bright spot in the image when it travels across the excitation laser focus. Sometimes flashes rather than tracks should be recorded when the motion of the molecules through the detection curtain is faster than the time resolution of the CCD camera. During the time of CCD integration, multiple analyte molecules can pass the detection region. At a relatively low concentration, the resultant fluorescent spots commonly appear at different locations along the detection curtain. Before counting the number of target molecules in a certain frame of the CCD image, a Fourier low-pass filter can be used to reduce the noise in the image. Continuous regions that are above a set threshold can be marked. These regions can be considered to be the signal from a fluorescent molecule if the following two criteria are satisfied: (i) the area of a region is larger than a certain threshold and (ii) the coordinates of the center of mass of a region are within the range of the detection curtain. When the analyte concentration increases, more molecules will be recorded in each image frame, thus increasing the probability of having two or more fluorescent spots very close together. Because each of these spots has a finite size (mainly determined by diffraction and their distance from the focal plane of the objective), when applying the threshold, they will be marked as one continuous region. Therefore, after the threshold is applied, the cross section of the image along the detection curtain should be examined. By identifying local maxima and minima in the cross section, closely spaced molecules can be resolved (Figure 4.6a).

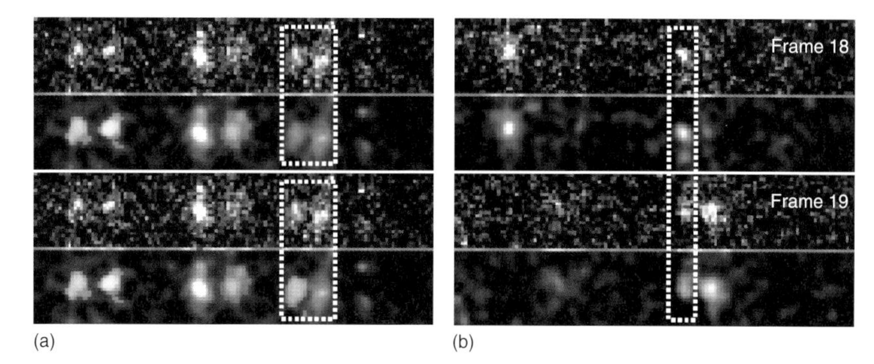


**Figure 4.6** Image analysis procedure for the separation and counting of A647-SA molecules. In each panel, the upper part is the original image recorded by the CCD camera, the lower part is the image after Fourier filtering, and the line between them shows the cross section of the Fourier filtered image along the detection curtain. (a) Note the improvement in identification when overlapped fluorescent spots can be split (lower panel). (b) When one molecule is imaged in two consecutive frames, the fluorescent spot has the same x position in both frames. (Reprint permission to be sought.)

Another source of bias in counting is the possibility that one molecule is imaged in two consecutive frames. In Zare *et al.*'s slow-flow method, the time for a molecule to travel across a 1-μm-wide detection region could be about 2 ms; therefore, if a molecule reaches the detection region at the end of one CCD integration period, it could be recorded in the next integration period as well. Within that short time, the Brownian motion of the molecule is not significant, and hence the molecule could be expected to appear at the same x positions in the two frames. Therefore, the fluorescent spots counted in one image frame could be compared to those in the previous frame with the x positions of their centers of mass. If the difference is within 2 pixels (450 nm), the fluorescent spot in the second frame could be marked as an invalid count (Figure 4.6b). Despite these efforts to compensate for biases in molecule counting, the chance of false negatives increases when the number of molecules in each frame is very high (>10 molecules per frame). By increasing the length of the separation channel, which increases the peak width when the analyte reaches the detection point, the molecules could be spread into more image frames, thereby the number of molecules per frame could be controlled.

The molecule-counting result depends on the threshold chosen for the image analysis. A lower threshold decreases the probability of false negatives in counting but increases that of false positives from background noise. To characterize this effect, an experiment was taken to analyze the total molecule counts from the same experiment (900 frames) with different thresholds [17]. First, a lower limit of threshold was found by calculation of the molecule counts from a blank experiment (no sample is injected). Then, using the optimized threshold, the corresponding overall counting efficiency for streptavidin molecules that have different degrees of labeling was calculated to about 60% (after subtracting the counts from blank experiments).

Two factors can contribute to the incomplete counting of sample molecules: missed molecules in identification (identification efficiency) and loss in transportation from the sample reservoir to the detection point (transportation efficiency). A counting experiment was performed to measure the transportation efficiency on a "double-T" chip that moves the detection point from 5 to 20 mm after the injection junction [17]. The results indicated that the transportation efficiency of a 15 mm channel is nearly 100% and contributes very little to the loss in counting efficiency. Therefore, it can be assumed that the counting efficiency is fully determined by the identification efficiency in the image analysis.

From the threshold analysis, the true sample molecule counts can be estimated directly. Molecules of the same kind at different positions in the channel show different fluorescence intensities because of the different excitation laser intensity. Moreover, molecules distant from the focal plane are dimmer, because their images are blurred by defocusing. These dim molecules are more likely to be rejected by a higher threshold. The same set of images can be analyzed using different thresholds and the true molecule number can be estimated by interpolating the molecule counts to a threshold of zero (which hypothetically should not reject any fluorescence signal). This estimation method was found to be applicable to the major species in single-cell analysis [17]. More sophisticated modeling could provide higher accuracy in estimating the true molecule counts.

## 4.4 Single-cell Analysis with Single-molecule Sensitivity on Integrated Microfluidic Chip

The Zare group further presented some examples to demonstrate the detailed procedures and methods of ultrasensitive detection of low-copy molecules from single cells [17].

For cell manipulation, they used a three-state valve design that has been demonstrated previously [16]. First, the cell was captured in the reaction chamber formed between the three-state valve and a conventional two-state valve, then the cell contents were released by a lysing/labeling buffer, which was injected into the chamber. As the proteins are not naturally fluorescent, fluorescently labeled antibodies were used as a generic method to tag target proteins. In order to preserve the activity of the antibodies, a nonionic detergent, 1 wt% *n*-dodecyl-$\beta$- (DDM), was used as the lysing reagent. The excess labeling reagent was electrophoretically separated from the target proteins after the labeling process. To reduce the loss of the sample during transportation, 0.1 wt% DDM was added to the separation buffer to suppress the protein adsorption on the hydrophobic PDMS channel walls [38]. In addition, a low concentration (~0.005 to ~0.05 wt%) of sodium dodecyl sulfate (SDS) was added to generate sufficient electroosmotic flow by its adsorption to the PDMS surface [39]. Although DDM/SDS separation was used for all of their experiments, they explained that any other electrophoretic/chromatographic separation method that is compatible with a PDMS microfluidic chip could be used.

### 4.4.1
### Microfluidic Chip Fabrication

The PDMS microfluidic devices (Figure 4.7) used in Zare's experiment were fabricated with standard soft photolithography. To produce the silicon masters for the molecule-counting chips, the molecule-counting section was made from a thin layer ($\sim 2\,\mu m$) of negative photoresist (SU-8 2002, MicroChem). The rest of the channels were then fabricated with a $15\,\mu m$ (insect cell analysis chip) or $7\,\mu m$ (cyanobacteria analysis chips) layer of positive photoresist (SPR 220-7). The masters for the channel layer in valve-controlled chips were heated to $115\,^\circ C$ for 30 min to reflow the positive photoresist so that the channels form a smooth, round shape. The masters for the control layer of these chips were made of 40-μm-thick negative photoresist (SU-8 50, MicroChem).

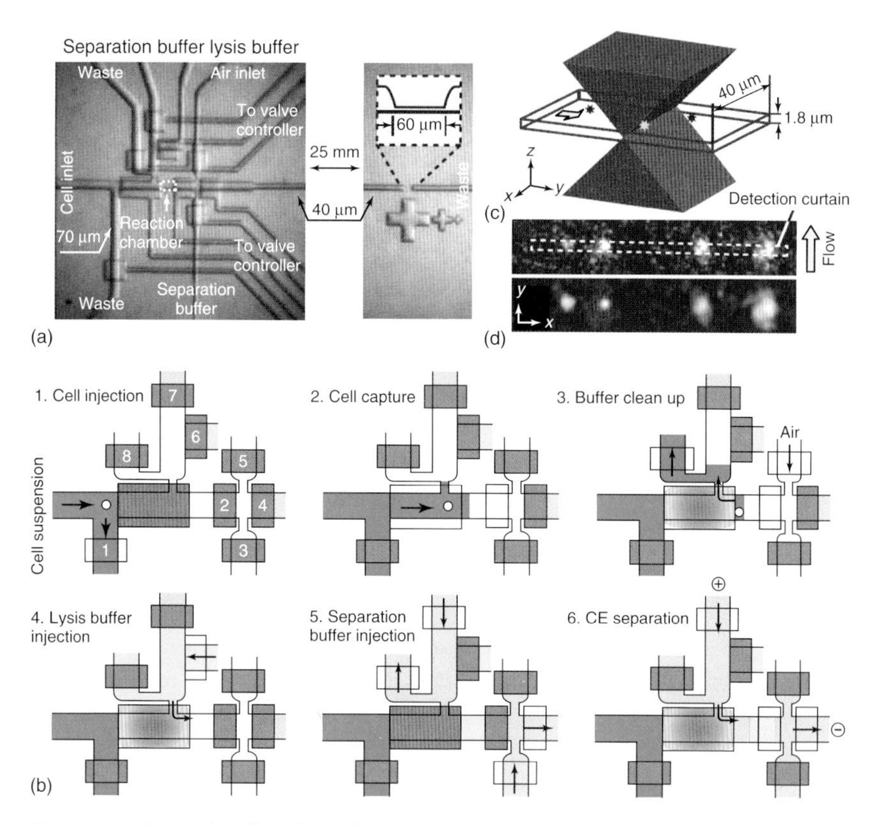


**Figure 4.7** The single-cell analysis chip. (a) Layout of the single-cell chip, showing the cell-manipulation section on the left and the molecule-counting section on the right. (b) Analysis procedure for a mammalian or insect cell. (c) Schematic illustration of the excitation laser focused by the microscope objective and the dimensions of the molecule-counting channel. (d) One frame from the CCD images of fluorescent molecules flowing across the molecule-counting section (upper panel) and the identification results (lower panel). (Reprint permission to be sought.)

### 4.4.2
### Analysis of β2AR in SF9 Cells

Using the single-cell analysis chip, the copy number of a human transmembrane protein, β2 adrenergic receptor (β2AR), which expressed in an insect cell line (SF9), was quantified. β2AR is not naturally fluorescent. Therefore, a short peptide sequence was genetically added to the N terminus, so that it could bind Cy5-labeled monoclonal M1 antibody against FLAG (Cy5-M1) with high affinity (dissociation constant = 2.4 nm) [40].

The separation and imaging experiments were performed on an inverted microscope. The excitation sources were a 532-nm diode-pumped frequency-doubled Nd:YAG laser (Compass 215M, Coherent) and a 638-nm diode laser (RCL-638-25, Crystalaser), which were combined and coupled to the same single-mode optical fiber.

The size of the injection plug could be measured by imaging the injection procedure at the "double-T" junction using 200 nM A647-SA as the sample. An effective plug area could be obtained by dividing the integrated intensity of the injection plug with the intensity in the channels filled with sample solution during the loading step. The injection plug volume could be derived by multiplying this area by the thickness of the channel (7.6 μm).

A647-SA can be separated into multiple peaks using capillary zone electrophoresis and LIF detection. These peaks can be attributed to the charge ladder created when different numbers of negatively charged dyes were labeled on the streptavidin molecule. By inserting a short (10-μm-long) molecule-counting section into the separation channel, this charge ladder was resolved using molecule counting at a low sample concentration. From five different measurements, the effective size of the injection plug was calculated to be 35 ± 4 pl, which corresponds to 1557 ± 174 injected A647-SA molecules when the sample concentration was 73 pM (Figure 4.8).

To measure the average copy number of $\beta_2$AR by anti-FLAG M1 antibody (M1) binding, M1 antibody was labeled with Cy5 succinimidyl ester (GE Health Care) and purified with a gel filtration column. The concentration of M1 was calibrated by measuring the absorption at 280 nm. For single-cell analysis, SF9 cells were harvested 18 h after infection, washed with Dulbecco'sPBS/Ca, and adjusted to a final density of about 1 million cells per ml. Briefly, the cell suspension was injected into the chip using 3 psi of pressure. Valve 1 (Figure 4.7) opened and closed until a cell was close to the three-state valve. The three-state valve was then opened to introduce the cell into the reaction chamber. After the three-state valve was partially closed, a low pressure was added to the air inlet through valve 2 and valve 5 to remove excess DPBS/Ca. The three-state valve was fully closed before filling the channel with lysis/labeling buffer (20 mM HEPES, pH 7.5, 20 nM Cy5-M1, 1 wt% DDM, 1 mM $CaCl_2$) through valve 6. The three-state valve was partially opened to inject the lysis/labeling buffer into the reaction chamber. Valve 2 was closed to confine the volume of injection, and the reaction chamber was filled because of the air permeability of PDMS. Then the three-state valve was fully closed to incubate the

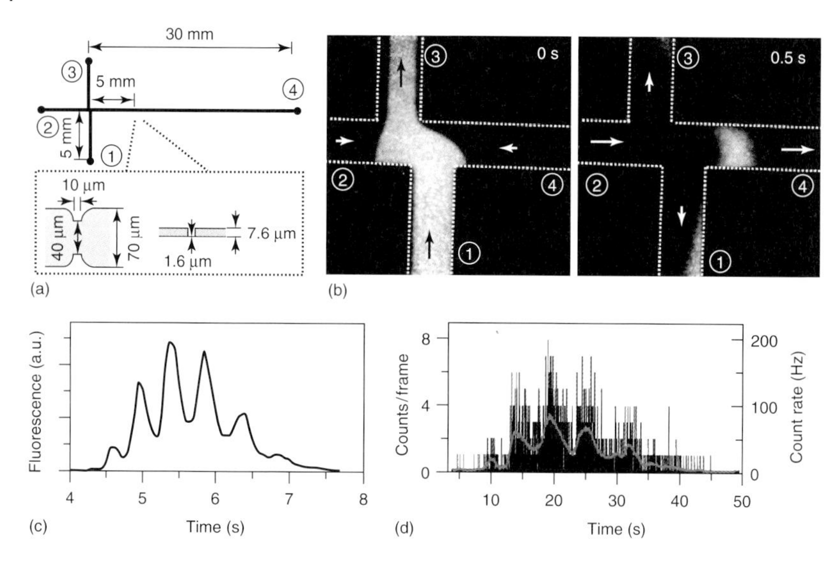


**Figure 4.8** Analysis of A647-SA in a double-T chip. (a) Layout of the "double-T" chip for A647-SA separation. (b) Fluorescence images of the double-T junction when separation starts. Dotted lines show the outline of the channels. Timing starts when the voltage set applied to the chip is switched from loading (1 = 1000 V, 2 = 700 V, 3 = 0 V, and 4 = 1000 V) to separation (1 = 700 V, 2 = 1000 V, 3 = 700 V, and 4 = 0 V). Arrows indicate the flow direction. (c) CE separation of 100 nM A647-SA. (d) Molecule counting of 73 pM A647-SA by lowering the voltage to 1/10 of the ordinary values when the analyte passes the detection curtain, showing the number of identified molecules in each frame of image (black bars) and the average molecule count rate in 1-s time bins (gray line). (Reprint permission to be sought.)

cell with the lysis/labeling buffer for 10 min. At the same time, separation buffer (20 mM HEPES, pH 7.5, 0.1 wt% DDM, 0.02 wt% SDS, 1 mM $CaCl_2$) was injected through valves 3 and 7 to rinse the channels. After the lysis/labeling reaction was complete, a voltage of 1000 V was applied to the chip through valve 7, partially opened three-state valve, valve 2, and valve 4. The image acquisition started 20 s later and an integration time of 20 ms per frame was used. The voltage was lowered to 100 V after the unreacted M1 peak passed the molecule-counting section.

### 4.4.3 Analysis of *Synechococcus*

#### 4.4.3.1 Electrophoretic Separation of *Synechococcus* Lysate

Traditional ways to lyse cyanobacterial cells use strong mechanical forces, such as high pressure or glass bead grinding; both methods were difficult to be integrated into a PDMS microchip design. Zare's group developed a method to lyse *Synechococcus* cells chemically. The cell lysate was diluted at least 10-fold into a sample buffer that contains 20 mM HEPES (pH 7.5), 0.1 wt% DDM, and 0.012 wt% SDS before it was added to the sample reservoir of a "double-T" chip. The other three reservoirs were filled with the separation buffer, which contains

20 mM HEPES (pH 7.5), 0.1 wt% DDM, and 0.045 wt% SDS. The distance between the injection junction and the detection point was 23 mm. No significant change in peak heights was observed through continuous running of the separation, which indicates that the phycobiliprotein complexes were stable in the sample buffer. However, a further increase in the SDS concentration results in gradual dissociation of these protein assemblies.

For facility of the identification of the CE separation peaks, the fluorescence spectra were recorded by an intensified CCD camera on the same microscope. The detection path was modified by inserting a pair of relay lenses and a grating between the microscope and the camera and by placing a 50-μm-wide slit at the image plane of the microscope. With this modification the CCD camera could be used to record wavelength information. By comparing the fluorescence spectra with that in the literature, and by monitoring the change in the electropherogram when adding different antibodies against phycobiliproteins and linker polypeptides, the major peaks in the electropherogram could be identified.

#### 4.4.3.2 *Synechococcus* Analysis Procedure

The lysis and analysis of individual *Synechococcus* cells was performed on a Nikon TE2000-U inverted microscope using a single-cell analysis chip including three reaction chambers. The analysis procedure was carried out in three steps (Figure 4.9a).

1) **Cell capture:** *Synechococcus* cells were treated with lysozyme, washed, diluted into B-PER II, and immediately delivered to the chip from the cell inlet. Then the cells flowed through one of the reaction chambers with a negative pressure applied at the cell outlet by a syringe. The valves of the reaction chamber were opened and closed randomly. At the same time, fluorescence of phycobiliprotein (650–700 nm) was continuously monitored by imaging through a 40× objective using wide-field illumination with a 636 nm laser. When the valve was closed, if no cell or more than one cell was captured, the valve was opened again to let the cell suspension continue to flow; this procedure repeated until an individual cell was trapped, then the next reaction chamber was moved into the view field and the capturing operation was repeated. It took less than 2 min to capture three cells after they were mixed with B-PER II; therefore, no cells were broken during the capture process.
2) **Cell lysis and chip cleaning:** After capture, the lysis was monitored by taking a fluorescent image of each cell every 10 min. The excitation light was controlled by a shutter that was synchronized with the CCD acquisition to minimize the adverse effects (such as photobleaching). During the cell lysing, voltages were applied to wash out the B-PER II solution in the channels (from separation buffer inlet to cell outlet, and then from separation buffer outlets to cell inlet). After all the cells were lysed, the reservoirs were refilled with fresh separation buffer and the chip was washed.

   Figure 4.9b shows a fluorescence image sequence of a *Synechococcus* cell. The fluorescence initially increases, which may be caused by the detachment of

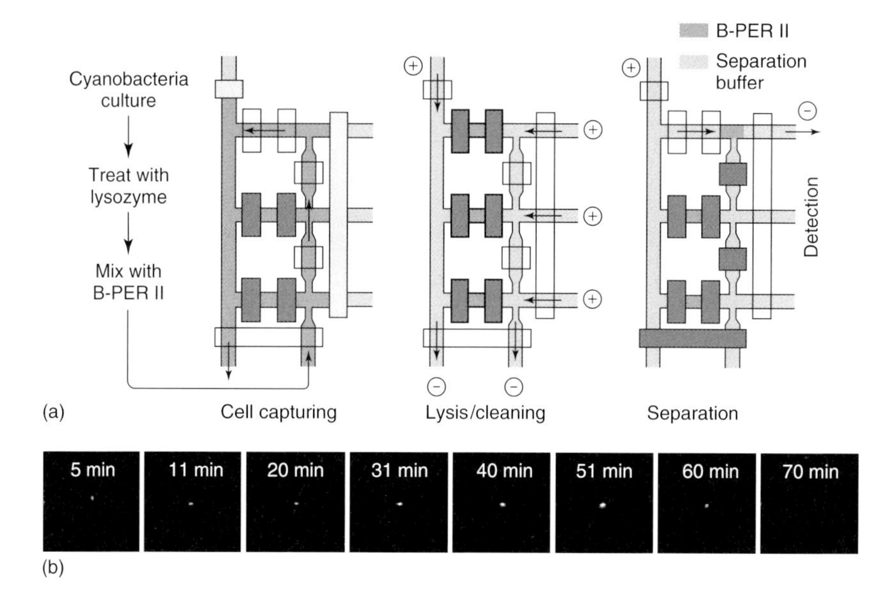


**Figure 4.9** Analysis of individual cyanobacteria cells. (a) Operation procedure of cell capturing, lysis, and analysis. (b) Fluorescence images of a *Synechococcus* cell captured in the reaction chamber at different times during the lysis procedure. (Reprint permission to be sought.)

phycobilisome (PBS) from thylakoid membranes and their partial dissociation. This disruption of the PBS increased fluorescence from membrane-dissociated penicillin – binding protein (PBP) complexes. After 50 min, fluorescence from the cell rapidly decreased, reaching a very low level after 70 min. A comparison of the cell fluorescence intensity at 50 and 70 min following exposure to B-PER II indicated the release of more than 90% of the fluorescent cell contents into the reaction chamber.

3) **Separation:** To start the separation, the excitation path was changed from wide-field configuration to cylindrical configuration, switched from the 40× objective to a 100× 1.4 NA oil immersion objective, and the view field was moved to the detection point in one of the separation channels. The valves of the corresponding reaction chamber were then opened and a 1000 V separation voltage was applied simultaneously. In single-molecule counting, the separation voltage was lowered to 100 V at 18.5 s after the separation starts. The image acquisition started at the same time when the voltage was lowered, and the integration time of the ICCD was 50 ms per frame. Cell lysate in the other two reaction chambers were analyzed sequentially.

After the separation step, the next set of cells can be introduced into the reaction chambers for reinitiation of step (1). Thus, the single-cell analysis chip can be used repeatedly. However, more than 8 h of continuous usage could cause degradation in the resolution of CE separation.

## 4.5 Conclusions

The rapid growth of microfluidics has greatly pushed forward the development of single-cell analysis in the past decade. Various chemical cytometry systems based on microfluidics have been demonstrated for the analysis of different types of molecules in individual cells. However, it is very difficult to quantitatively analyze low-copy-number proteins, which have critical functions in cells; their amount is far below the sensitivity limits of conventional protein analysis methods. Analysis of low-copy-number proteins in individual cells requires the integration of the following functions on chip: (i) careful manipulation of individual cells, (ii) effective suppression of nonspecific adsorption of proteins, and (iii) ultrasensitive detection method.

The integrated microfluidic chip developed by the Zare group satisfied all the requirements for ultrasensitive detection of low-copy-number proteins in single cells. Microvalves are used to control the flow of fluids in microchannels and to isolate and deliver biological cells to specific locations. Furthermore, microvalves can be used to form isolated microchambers with picoliter volume for reaction incubations. The microchambers minimize the dilution of cell contents after lysis and microvalves prevent molecules from diffusing away from the microchambers. A mild neutral surfactant with a hydrophilic sugar group is able to minimize nonspecific adsorption of proteins to single-molecule level. A single-molecule counting technique is used on the microchip for ultrasensitive detection by directly counting the number of minute protein molecules. A cylindrical confocal optical configuration has been adapted to be integrated with the chip by focusing a laser beam on the detection region. With these approaches, the actual number of molecules can be derived without additional calibration, which leads to a convenient way to detect the low-copy-number proteins in single cells. This ultrasensitive method is general to most biomolecules in almost all individual biological cells. Although the throughput of current method needs to be greatly improved, its extremely high sensitivity should be very useful in the study of various biological processes.

## References


1 Cottingham, K. (2004) The single-cell scene. *Anal. Chem.*, **76**, 235A–238A.

2 Wu, J. and Pollard, T. (2005) Counting cytokinesis proteins globally and locally in fission yeast. *Science*, **310**, 310–314.

3 Newman, J.R.S., Ghaemmaghami, S., Ihmels, J., Breslow, D.K., Noble, M., DeRisi, J.L., and Weissman, J.S. (2006) Single-cell proteomic analysis of S-cerevisiae reveals the architecture of biological noise. *Nature*, **441**, 840–846.

4 Yu, J., Xiao, J., Ren, X., Lao, K., and Xie, X.S. (2006) Probing gene expression in live cells, one protein molecule at a time. *Science*, **311**, 1600–1603.

5 Ros, A., Hellmich, W., Regtmeier, J., Duong, T.T., and Anselmetti, D. (2006) Bioanalysis in structured microfluidic systems. *Electrophoresis*, **27**, 2651–2658.

6 Sun, Y. and Yin, X.F. (2006) Novel multi-depth microfluidic chip for single cell analysis. *J. Chromatogr. A*, **1117**, 228–233.

7 Gao, J., Yin, X.F., and Fang, Z.L. (2004) Integration of single cell injection, cell lysis, separation and detection of intracellular constituents on a microfluidic chip. *Lab Chip*, **4**, 47–52.
8 Hu, S., Michels, D.A., Fazal, M.A., Ratisoontorn, C., Cunningham, M.L., and Dovichi, N.J. (2004) Capillary sieving electrophoresis/micellar electrokinetic capillary chromatography for two-dimensional protein fingerprinting of single mammalian cells. *Anal. Chem.*, **76**, 4044–4049.
9 Harwood, M.M., Bleecker, J.V., Rabinovitch, P.S., and Dovichi, N.J. (2007) Cell cycle-dependent characterization of single MCF-7 breast cancer cells by 2-D CE. *Electrophoresis*, **28**, 932–937.
10 Sobhani, K., Fink, S.L., Cookson, B.T., and Dovichi, N.J. (2007) Repeatability of chemical cytometry: 2-DE analysis of single RAW 264.7 macrophage cells. *Electrophoresis*, **28**, 2308–2313.
11 Marcus, J.S., Anderson, W.F., and Quake, S.R. (2006) Microfluidic single-cell mRNA isolation and analysis. *Anal. Chem.*, **78**, 3084–3089.
12 Ottesen, E.A., Hong, J.W., Quake, S.R., and Leadbetter, J.R. (2006) Microfluidic digital PCR enables multigene analysis of individual environmental bacteria. *Science*, **314**, 1464–1467.
13 Cai, L., Friedman, N., and Xie, X.S. (2006) Stochastic protein expression in individual cells at the single molecule level. *Nature*, **440**, 358–362.
14 Nikaido, H. (1994) Prevention of drug access to bacterial targets – permeability barriers and active efflux. *Science*, **264**, 382–388.
15 Unger, M.A., Chou, H.P., Thorsen, T., Scherer, A. and Quake, S.R. (2000) Monolithic microfabricated valves and pumps by multilayer soft lithography. *Science*, **288**, 113–116.
16 Wu, H.K., Wheeler, A., and Zare, R.N. (2004) Chemical cytometry on a picoliter-scale integrated microfluidic chip. *Proc. Natl. Acad. Sci. U.S.A.*, **101**, 12809–12813.
17 Huang, B., Wu, H.K., Bhaya, D., Grossman, A., Granier, S., Kobilka, B.K., and Zare, R.N. (2007) Counting low-copy number proteins in a single cell. *Science*, **315**, 81–84.
18 Castro, A., Fairfield, F.R., and Shera, E.B. (1993) Fluorescence detection and size measurement of single DNA-molecules. *Anal. Chem.*, **65**, 849–852.
19 Chen, D. and Dovichi, N.J. (1996) Single-molecule detection in capillary electrophoresis: molecular shot noise as a fundamental limit to chemical analysis. *Anal. Chem.*, **68**, 690–696.
20 Ma, Y., Shortreed, M.R., Li, H., Huang, W., and Yeung, E.S. (2001) Single-molecule immunoassay and DNA diagnosis. *Electrophoresis*, **22**, 421–426.
21 Kuo, J.S., Kuyper, C.L., Allen, P.B., Fiorini, G.S., and Chiu, D.T. (2004) High-power blue/UV light-emitting diodes as excitation sources for sensitive detection. *Electrophoresis*, **25**, 3796–3804.
22 Shi, B.X., Huang, W.H., and Cheng, J.K. (2007) Determination of neurotransmitters in PC 12 cells by microchip electrophoresis with fluorescence detection. *Electrophoresis*, **28**, 1595–1600.
23 Yan, H., Zhang, B.Y., and Wu, H.K. (2008) Chemical cytometry on microfluidic chips. *Electrophoresis*, **29**, 1775–1786.
24 Turner, E.H., Lauterbach, K., Pugsley, H.R., Palmer, V.R., and Dovichi, N.J. (2007) Detection of green fluorescent protein in a single bacterium by capillary electrophoresis with laser-induced fluorescence. *Anal. Chem.*, **79**, 778–781.
25 Hellmich, W., Pelargus, C., Leffhalm, K., Ros, A., and Anselmetti, D. (2005) Single cell manipulation, analytics, and label-free protein detection in microfluidic devices for systems nanobiology. *Electrophoresis*, **26**, 3689–3696.
26 Hellmich, W., Greif, D., Pelargus, C., Anselmetti, D., and Ros, A. (2006) Improved native UV laser induced fluorescence detection for single cell analysis in poly(dimethylsiloxane) microfluidic devices. *J. Chromatogr. A*, **1130**, 195–200.
27 Lapainis, T., Scanlan, C., Rubakhin, S.S., and Sweedler, J.V. (2007) A multichannel native fluorescence detection system for capillary electrophoretic analysis of

neurotransmitters in single neurons. *Anal. Bioanal. Chem*, **387**, 97–105.

28 Zhang, H. and Jin, W.R. (2004) Determination of different forms of human interferon-gamma in single natural killer cells by capillary electrophoresis with on-capillary immunoreaction and laser-induced fluorescence detection. *Electrophoresis*, **25**, 1090–1095.

29 Zhang, H. and Jin, W. (2006) Single-cell analysis by intracellular immuno-reaction and capillary electrophoresis with laser-induced fluorescence detection. *J. Chromatogr. A*, **1104**, 346–351.

30 Xiao, H., Li, X., Zou, H.F., Yang, L., Yang, Y.Q., Wang, Y.L., Wang, H.L., and Le, X.C. (2006) Immunoassay of P-glycoprotein on single cell by capillary electrophoresis with laser induced fluorescence detection. *Anal. Chim. Acta*, **556**, 340–346.

31 Lyon, W. and Nie, S.M. (1997) Confinement and detection of single molecules in submicrometer channels. *Anal. Chem.*, **69**, 3400–3405.

32 Wang, S.L., Fan, X.F., Xu, Z.R., and Fang, Z.L. (2005) A simple microfluidic system for efficient capillary electrophoretic separation and sensitive fluorimetric detection of DNA fragments using light-emitting diode and liquid-core waveguide techniques. *Electrophoresis*, **26**, 3602–3608.

33 Fu, J.L., Fang, Q., Zhang, T., Jin, X.H., and Fang, Z.L. (2006) Laser-induced fluorescence detection system for microfluidic chips based on an orthogonal optical arrangement. *Anal. Chem.*, **78**, 3827–3834.

34 Foquet, M., Korlach, J., Zipfel, W.R., Webb, W.W., and Craighead, H.G. (2004) Focal volume confinement by submicrometer-sized fluidic channels. *Anal. Chem.*, **76**, 1618–1626.

35 Lundqvist, A., Chiu, D., and Orwar, O. (2003) Electrophoretic separation and confocal laser-induced fluorescence detection at ultralow concentrations in constricted fused-silica capillaries. *Electrophoresis*, **24**, 1737–1744.

36 Pennathur, S. and Santiago, J.G. (2005) Electrokinetic transport in nanochannels. 1. Theory. *Anal. Chem.*, **77**, 6772–6781.

37 Pennathur, S. and Santiago, J.G. (2005) Electrokinetic transport in nanochannels. 2. Experiments. *Anal. Chem.*, **77**, 6782–6789.

38 Huang, B., Wu, H.K., Kim, S., and Zare, R.N. (2005) Coating of poly(dimethylsiloxane) with n-dodecyl-beta-D-maltoside to minimize nonspecific protein adsorption. *Lab Chip*, **5**, 1005–1007.

39 Ocvirk, G., Munroe, M., Tang, T., Oleschuk, R., Westra, K., and Harrison, D.J. (2000) Electrokinetic control of fluid flow in native poly(dimethylsiloxane) capillary electrophoresis devices. *Electrophoresis*, **21**, 107–115.

40 Whelan, R.J., Wohland, T., Neumann, L., Huang, B., Kobilka, B.K., and Zare, R.N. (2002) Analysis of bimolecular interactions using a miniaturized surface plasmon resonance sensor. *Anal. Chem.*, **74**, 4570–4576.

# 5
# Capillary Electrophoresis of Nucleic Acids at the Single-cell Level

*Ni Li and Wenwan Zhong*

## 5.1
## Introduction

Each cell is one unique individual differing significantly even from its close neighbors. For example, bacterial cells have different resistance to antibiotics, which could be attributed to accidental gene mutations that occur in few bacterial cells. Such difference is the basis for the selection of bacterial strain with the capability to survive the antibiotics. Stem cells are another group of cells that show significant variations from cell to cell, each having the ability of renewing themselves through mitotic cell division and differentiating into a diverse range of specialized cell types. What controls the cell development at each growing stage? What determines the final cell phenotype? Only with advanced development in techniques analyzing gene mutation and expression at the single-cell level will all these puzzles be solved. Enabling techniques for single-cell analysis is also beneficial for investigations on cells with extraordinary complexity such as mammalian neurons and tumor cells [1], and for studies dealing with limited availability of homogeneous cell populations such as stem cells. Heterogeneity in tumor cells may indicate disease progress; therefore, single-cell analysis could be helpful in early diagnosis of cancer diagnosis and could enhance the survival rates. Unfortunately, classic analytical methods report only the average composition of a number of cells due to limited sensitivity and sample handling ability. The average does represent the general situation for a population of cells, but it conceals the cell-to-cell differences as well. If we use "S" to stand for the signal from only a few cells of interest and "N" for the background signals from neighboring cells, the classic methods have poor S/N ratio for the few target cells. In order to greatly improve S/N ratio, the most straightforward way is to analyze cells one by one. With the availability of the analytical devices (i.e., capillary electrophoresis (CE), microarray, and microfluidic chips) with small volume and high sensitivity in detection (i.e., laser-induced fluorescence (LIF) and intense CCD camera imaging), researchers have put efforts on analysis of genes and gene expression at the single-cell level in the past 10 years.

*Chemical Cytometry*. Edited by Chang Lu

ISBN: 978-3-527-32495-8

The ideal assays of nucleic acids in single cells should have the following features: sensitive and specific enough to give a "yes" or "no" answer; capable of screening a population of cells within a short time period; and cost-effective. In this chapter, CE-based gene and gene expression analysis is discussed, which demonstrates the basis of constructing affordable and applicable assay platforms. In these platforms, conventional gene analysis assays were performed in miniaturized devices with high degree of automation, and several challenges were addressed, including how to integrate and simplify the processes of cell lysis, extraction of nucleic acids, and gene amplification; how to modify the enzymatic reactions for adaptation to the capillary column reactor; and how to extract as much information as needed from one single cell. At the end of the chapter, envision of the future developments of the CE-based gene analysis is presented.

## 5.2 On-line Cell Analysis

From intact whole cell to detectable nucleic acids, the conventional genetic analysis usually includes breaking up the cells to release nucleic acid materials, extracting and cleaning up the DNA or RNA, and finally detecting the presence of the target of interest. Most likely, amplification strategies are adopted to amplify the target strands, such as polymerase chain reaction (PCR) and reverse transcription polymerase chain reaction (RT-PCR). The overall process is not only time consuming and labor intensive, but also needs sufficient amounts of cells. In addition, multiple reaction steps with tedious liquid handling are susceptible to contaminations and sample loss, which increase the frequency of false negatives/positives and introduce errors in quantification.

If the multistep process is integrated well in an on-line format, starting cell number can be reduced, cell lost can be limited, and analysis time can be reduced greatly. Such improvements are extremely valuable in cases where only a few cells of interest are available and fast analysis is required; for example, in early diagnosis of diseases. Some lead to the development of integrated CE instruments for gene and gene expression analysis in single cells, especially those performed in Yeung's group, which are covered in this section of the chapter.

### 5.2.1 Cell Injection and Lysis

Introducing cells into the capillary is the first step for the on-line assays. CE is quite compatible with single-cell analysis in terms of sample volume. The inner diameters of the capillary are in tens of micrometers, highly suitable to injection of cells with diameters of a few micrometers (5–15 μm) [2]. The on-column detection window also allows *in situ* counting of individual cells inside the capillary when they pass the detection window individually. Thus, both injection of one single cell and continuous flow injection of multiple cells were adopted in the CE-based on-line analysis [2–4]. To inject individual cells, hydrodynamic injection can be used for well suspended cells by creating suction at the outlet end of the capillary [3]. For adherent cells, electrokinetic injection can be employed [5] or laser cavitation and fast electrical lysis developed in the Allbritton group can directly lyse the cells and inject only the cell content into the capillary [6, 7], which has been reviewed elsewhere [8]. In order to monitor the injection process, the inlet capillary tip should be mounted under a microscope by a micromanipulator and point at the cell [3]. In general, single-cell sampling is tedious and yields low throughput. Inspired by the sample introduction procedure in flow cytometry, continuous cell injection by a capillary was developed by Lillard's group as shown in Figure 5.1 [2]. Two capillaries were coupled together with a 5-mm-long Teflon tube, where cell lysing occurred due to mechanical disruption by a dramatic change in the flow property at the gap. The inlet end of the first capillary was dipped into cell suspension buffer to bring in cells continuously via electroosmotic flow (EOF). The second capillary was for separation and detection. In their setup, the capillary inner diameter was 21 μm. The gap between the outlet tip of the injection capillary and the inlet tip of the separation capillary was 5 μm and is smaller than the size of the cell of interest – the red blood cell with a diameter of 7 μm. The Teflon tube junction was immersed in a reservoir containing the lysing buffer. The erythrocyte was deformed in the radial direction as it entered the gap with a smaller dimension while being still driven by

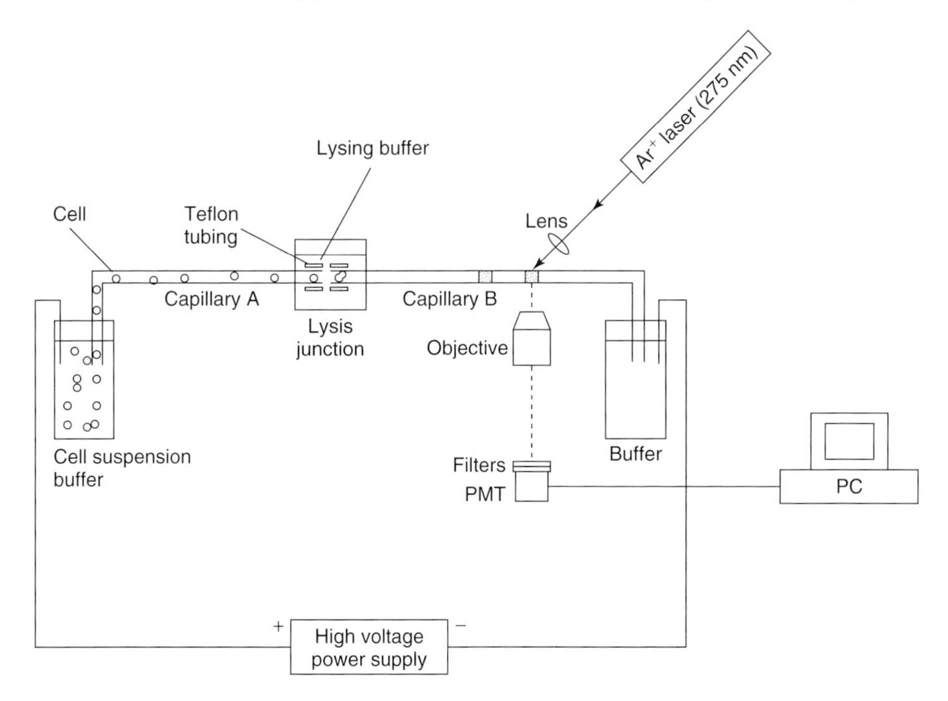


**Figure 5.1** Schematic of instrumental setup for high-throughput CE analysis of single cells with continuous cell introduction [2].

the electric field and the siphoning flow toward the second capillary. Therefore, the cell membrane was ruptured and the intracellular compositions were swept into the secondary capillary by the siphoning action. The average analysis time of one human erythrocyte was less than 4 min including injection, lysis, separation by CE, and detection of the nucleic acid materials with LIF. The continuous cell injection is one big step forward in single-cell analysis because it increased the throughput of cell sampling, which is critical in single-cell gene analysis for the collection of profiling information from cells in different significant cellular stages.

Another simple approach of injecting multiple suspended cells to a defined capillary was to directly flow the highly diluted cell suspension into the capillary. The throughput was greatly improved without sophisticated sample preparation and instrumental setup. It was adopted for selective genotyping of individual cells by capillary PCR [4]. Since the injected cells randomly distributed along the capillary, three on-line procedures for counting the loaded cells and estimating their position inside the capillary were developed to correlate the detected peaks of amplification with the individual cells. The first method was to mount a photomultiplier tube (PMT) downstream of the capillary and count the cells stained with SYTO16 by monitoring the fluorescence from cells flowing by. The number of fluorescence burst indicated the number of cells. However, the number identified by this method was much smaller than that estimated by a hemocytometer. Therefore, a second method was developed by replacing the PMT with a frame-transfer CCD camera and recording the LIF images of the cells (Figure 5.2). The imaging method was more straightforward because the actual cells were seen, but the cell debris from the culture media was one source of deviation of the actual count when the regular round capillary (75 μm i.d.) was adopted. The imaging quality and consistency were then improved by switching to square capillary (75 × 75 μm). Without any staining, cells were counted visually by the partial dark-field microscopy. An occult disk was placed between the light source and the condenser of the microscope to make a partial dark field. Although the dark-field viewing mode did not show the capillary wall as distinct as the bright-field method, it resulted in images with better contrast. This development is valuable to cases where the separation step does not follow immediately after the cell lysis. Normally, amplification reactions are required to increase the detection signal. Then, the sampling capillary can also serve as the reactor and the amplified material from each cell can be delivered to the analysis capillary by pressure and separated by electrophoresis.

Following cell injection, the next step would be cell lysis to release the cellular contents for the subsequent amplification and analysis. The lysis method should be effective enough to achieve a high, ideally 100%, lysing efficiency for single-cell analysis because of the low sample amount available in each cell, and at the same time it should be friendly enough for other integrated parts of the entire assay. To break the cell membrane, conventional methods using both the chemical and physical forces are often applied with adjustments to fit the integration experimental design. The feasibility of disrupting the cell membrane depends greatly on the types of the cells. For chemical lysis, it is very important that the added chemicals should have ideally no inhibition on the following amplification and negligible effect

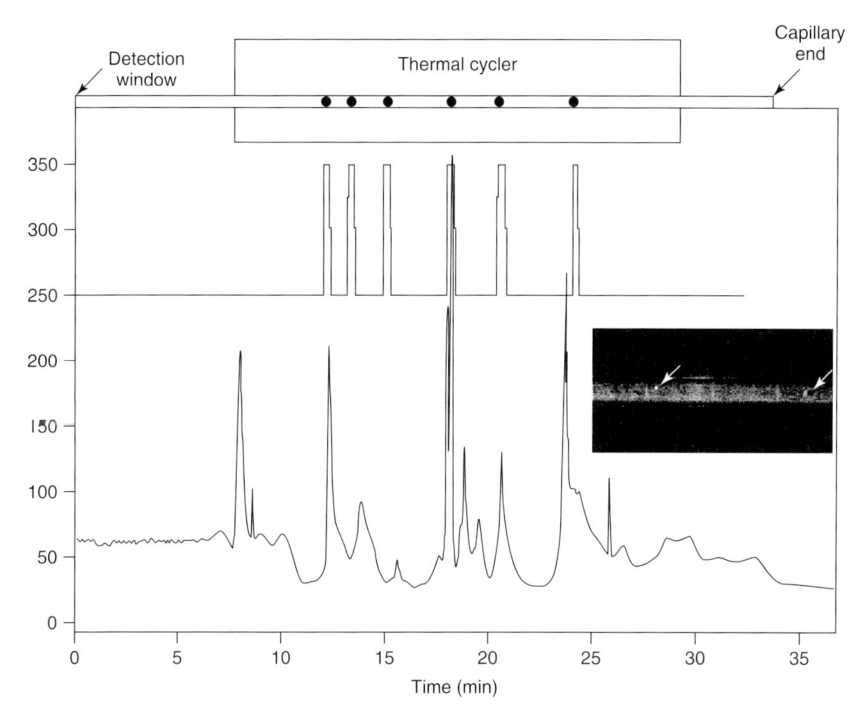


**Figure 5.2** Correlation of starting locations of cells with amplified peaks. The dots were marked positions of cells before reaction, and the bars indicated calculated peak positions based on the migration velocity and the cell locations. The electropherogram below showed the actual amplified peaks with elution times in minutes. Image of cells (without staining) in a 50-μm-i.d. round capillary viewed by partial dark-field microscopy with 10× magnification was inserted [4].

on the electrophoretic separation and detection. Detergents, such as SDS, Triton X-100, and NP40, can be used to break the cell membrane by solubilizing proteins and disrupting lipid–lipid, protein–protein, and protein–lipid interactions. To achieve highly efficient cell lysis by charged surfactants, the concentration of surfactant should be above a certain level. For SDS, the typical effective cell lysing concentration is 0.05–0.1%. However, off-line study showed that a concentration higher than 0.03% in SDS inhibited the subsequent PCR amplification [4]. Neutral surfactants such as NP-9 and Triton X-100 were friendly to PCR but without satisfying lysing efficiency. Enzymatic digestion by Proteinase K is widely adopted as cell lysing reagent for DNA isolation [10, 11]. However, an additional digestion step is needed to remove proteinase K itself before PCR because proteinase K is not compatible with PCR reagent [4]. While chemical lysis may have several limitations, the strategy of combining osmotic pressure and heat worked well for instant cell lysing inside the capillary column. Cells were suspended in hypotonic buffer (50 mM NaCl, 10 mM Tris–HCl, pH 8.0) and immediately mixed with PCR cocktail before loading into the capillary. Before the PCR cycles, a heating step was applied and the temperature was raised to 95 °C for 5 min to achieve efficient on-line cell lysing [4] Other physical cell lysis methods include liquid homogenization, manual grinding, high frequency sound waves, and freeze/thaw cycles. The last two can be easily adopted on-line by automated temperature or ultrasonic wave control. Besides, the capillary with fine diameters can be utilized in constructing junction slightly smaller than cells, where physical squeezing could contribute to cell lysis as shown by Lillard's group in the continuous cell injection [2].

For the studies aiming at obtaining quantitative gene and gene expression information from cells at different growth phases (M, $G_1$, S, and $G_2$), cell synchronization should be carried out before cell injection. It could be an off-line preparation (i.e., shake-off method) [12]. With the development of the microfluidic devices [13], on-line cell synchronization becomes practical as well.

### 5.2.2
### In-column DNA or RNA Amplification with Integrated Devices

In order to monitor specific DNA or RNA in a single cell, either amplifying target of interest or applying ultrasensitive detection system is demanded. Ultrasensitive detection requires highly expensive experimental setup, such as intensified charge-coupled device (ICCD) camera, not a widely acceptable approach because of its high cost. Consequently, nucleic acid amplification is still popular for single-cell analysis because PCR is capable of amplifying tiny amounts of target DNA extracted from even one single cell [14]. Initially, single-cell PCR involved picking up single cells and moving them to individual reaction vials [15, 16], which was time consuming, labor intensive, and with low sample throughput. Thereafter, integrating PCR on-line together with CE separation and LIF detection was developed to increase the analysis throughput. As for RNA analysis, reverse transcriptase should be employed to transcribe the RNA template to the complementary DNA (cDNA) for PCR amplification, which generates several additional challenges in single-cell RNA analysis including the elimination of DNA to reduce false amplification and the thorough removal of the environmental RNase in the reaction and separation capillary. For amplification of both the DNA and RNA materials extracted from single cells, amplification specificity and yield should be noted because the few copies of the target DNA or RNA may coexist with a large amount of interfering background nucleic acid materials. Therefore, successful amplification of DNA or RNA on-line in an integrated device depends greatly on the device design, the connections between different segments for sample preparation, reaction, and separation, and the optimization of the reaction protocols, which would be discussed in the following sections.

#### 5.2.2.1 Stream-lined Instrumental Setup

Since several temperature cycles are required in PCR for denaturing, annealing, and amplification, accurate and fast temperature control is critical for in-column PCR. The commercially available rapid air cycler, which utilizes air flow with high velocity to heat and cool the reaction mixture inside the capillary tubes, was once adopted [4], and one single capillary was used for both amplification and separation with the amplification segment enclosed in the rapid cycler. High convenience in operation and simplicity in the instrumental design are two big advantages of this setup. Cells were introduced into the capillary and counted before cell lysis under the combined action of heat and osmotic pressure as mentioned previously. The cell suspension was diluted to a large extent to ensure that the cells were well apart from each other inside the capillary. Because the two ends of the capillary were sealed with MicroTight unions during PCR to prevent liquid movement, the amplified products in the capillary were concentrated locally at the places where cells with the target DNA fragment located, each zone of the amplified product corresponding to one individual cell. After amplification, the zones containing the amplified products were driven toward the LIF detection window by CE instead of hydrodynamic flow to minimize the band broadening. A low concentration of polyvinylpyrrolidone (PVP;0,4%) was added to the reaction mixture and running buffer to suppress the EOF, and the products were stained with the intercalating dye of SYBR. Green I. Successful PCR on the $\beta$-actin gene in individual cells (human lymphoblast) were achieved with this setup. However, using the same capillary for both amplification and separation imposes limitation on the selection of the separation matrix, which cannot contain polymer concentrations high enough to offer good separation resolution of the amplified products, not to mention DNA sequencing. Moreover, the same capillary was also used for cell sampling as well as cell lysis. Performance of each step could be compromised to some level in order to obtain a good overall result. Owing to the size of the rapid air cycler itself, miniaturization of the entire setup could be problematic as well.

By contrast, minimized temperature control modules, such as the assembled-in-house microthermocycler [17] are more appealing, because every step can be performed in separated sections with elegant design of the capillary connections, enabling simple, independent optimization of individual steps for the enhancement of the overall performance. A simple microthermocycler used for capillary PCR was illustrated in Figure 5.3. A short reference capillary of the same inner and outer diameters as the reaction capillary was used to enclose a thermal couple with water and both ends of it were sealed. Then, both the reaction and reference capillaries were sandwiched between two pieces of the thin brass sheets. Thermal conductive material was applied in the void spaces between the brass sheets, providing effective thermal conduction around the microreactor and the thermal couple probe. The outer surface of one of the brass sheets was attached to a Kapton insulated flexible heater by a double-stick tape [18] or by thermal conductive epoxy [19]. Finally, the brass sheets, the capillaries, and the heater were clamped together by two frames with screws on the four corners to form the microthermocycler. The size of the microthermocycler and the diameter of reaction capillary can be adjusted for the best reaction performance. A proportional-integral-derivative (PID) temperature controller was used to control the temperature profile of the flexible heater and tune the temperature for compensation of temperature overshoot and fluctuation. An aquarium air pump was used alone [18], or together with a microfan [19] to blow air onto the microthermocycler and quickly cool the capillary to the desired temperatures.

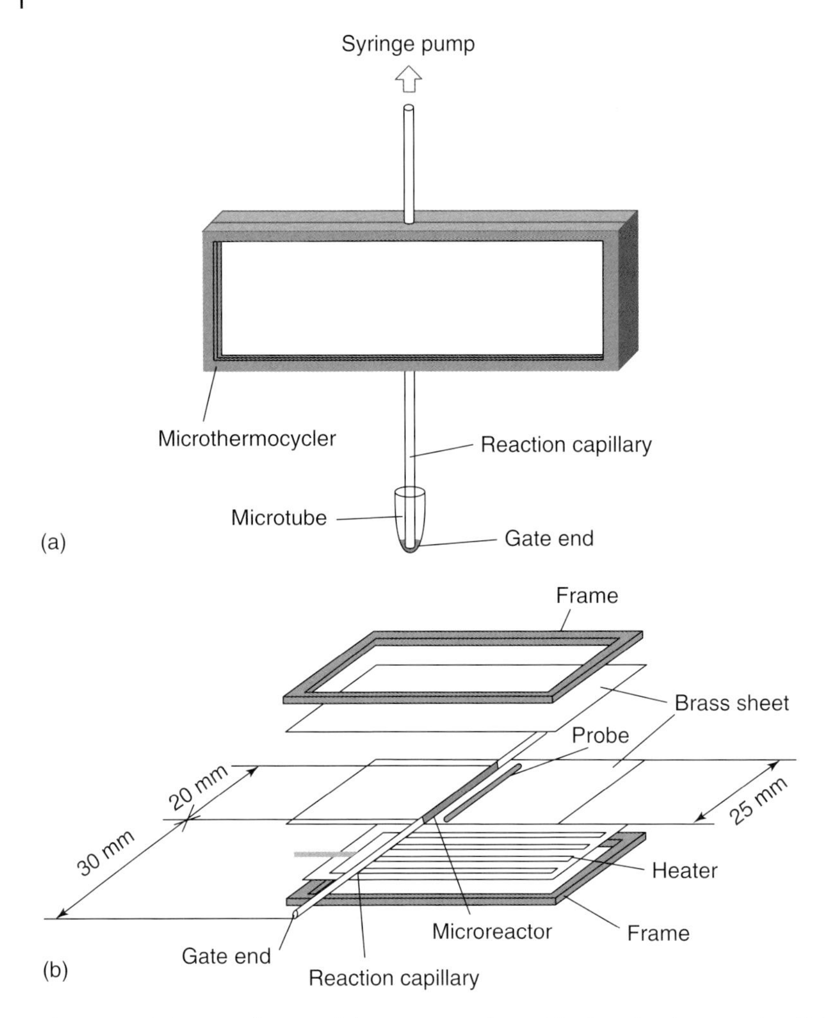


**Figure 5.3** Schematic diagram of the microthermocycler. (a) Sample solution in the microtube is introduced into the reaction capillary from the gate end, and is carried to the microthermocycler by means of the syringe pump. Then, thermal reaction is performed in the microthermocycler. (b) The microreactor (a part of the reaction capillary) and the probe are sandwiched by the brass sheets with the heater. Temperature in the microreactor is controlled by monitoring the temperature in the probe [19].

With this microthermocycler, analysis of the mRNA expression from a few cells was achieved by Matsunaga *et al.* [19] DNase treatment was necessary to remove DNA from the background and extreme care was taken to avoid RNA target degradation. Cell lysis, DNase treatment, and RT-PCR were all performed in the same capillary, a part of which (20 mm long with an overall volume of 1.0 μl) was inserted to the microthermocycler serving as the reactor. One end of the capillary was connected to a syringe pump through a 25-μl gas-tight syringe and a microtight adaptor. The liquid content could then be accurately aspirated or dispensed from or to the microvials in which the other end of the capillary was dipped. To precondition the capillary and inhibit the RNase activity, the reaction capillary was flushed with RNase ZAP®, diethylpyrocarbonate (DEPC)-treated water and the RNase-free 1× Tri-borate EDTA (TBE) buffer sequentially before usage. Cell lysing was realized inside the microthermocycler by heating up the capillary to 95°C for 2 min and keeping it at 65°C for 5 min, and the cell lysate was expelled to the microvial and mixed with DNase cocktail thoroughly by aspiring and dispensing the mixture 12 times with the syringe pump. Then, 1.0 μl reaction mixture was moved into the microthermocycler and incubated at 37°C for 15 min to digest DNA molecules. Similar mixing step was carried out in the following steps of RT-PCR and EtBr staining of the PCR products, followed by the transportation of 2 μl reaction mixture to a second capillary for capillary gel electrophoresis (CGE). Since the separation matrix no longer affected the amplification, a polymer solution of 3% PVP (Mw 1 000 000) with the intercalating dye EtBr was used. The entire analysis took 3 h, significantly shorter than the time taken for the traditional off-line RT-PCR procedure for single-cell analysis. Reliable amplification of the mRNA of $\beta$-actin from approximately 16 human lymphoblast cells was attained. This study well demonstrated that successful enzymatic reactions could be performed in the lab-made microthermocycler, which saves a lot of space, minimizes the sample amount, and is fully compatible with the following CE analysis. The construction of the microthermocycler paved the way for the subsequent, full integration of the PCR and cycle-sequencing reactions with CGE analysis, offering feasibility of high sample throughput and smooth adaptation into a microfabrication system [18].

In such a system, PCR, cycle-sequencing reaction, purification of the sequencing fragments, and CGE separation for the sequencing product were performed in different capillaries as illustrated in Figure 5.4. The capillaries were connected by universal capillary connectors, and programmable syringe pumps were employed to accurately transfer precise volumes of solutions at desired times from the microvials. The whole process was not fully automated yet with some steps still manually operated, for example, preconditioning the capillary and replacing the capillary zone electrophoresis (CZE) buffer with the sequencing buffer in C3, C5, and C6 prior to the CGE separation, which could be operated by robots in the future high-throughput setup. Removing the excess dye-terminators from the cycle-sequencing reaction is necessary to ensure high separation resolution and detection sensitivity. A novel approach was employed by separating the dye-terminators from the long sequencing products with CZE. A freeze/thaw valve was opened to control the material transfer from the cycle-sequencing capillary to the CZE capillary immediately after the reaction by siphoning. Then, CZE was initiated with the application of a positive field at the R2 position. The CZE in normal mode eluted the elongated PCR product earlier than the terminators. When the DNA peak reached the detection window, the positive field was switched to the R3 position to electrokinetically inject the sequencing product into the CGE capillary (C7) (Figure 5.5). Even though complete separation of the sequencing product and dye-terminator was not achieved, the dye peaks were well removed from the sequencing region (>70 bp) with careful injection time adjustment and

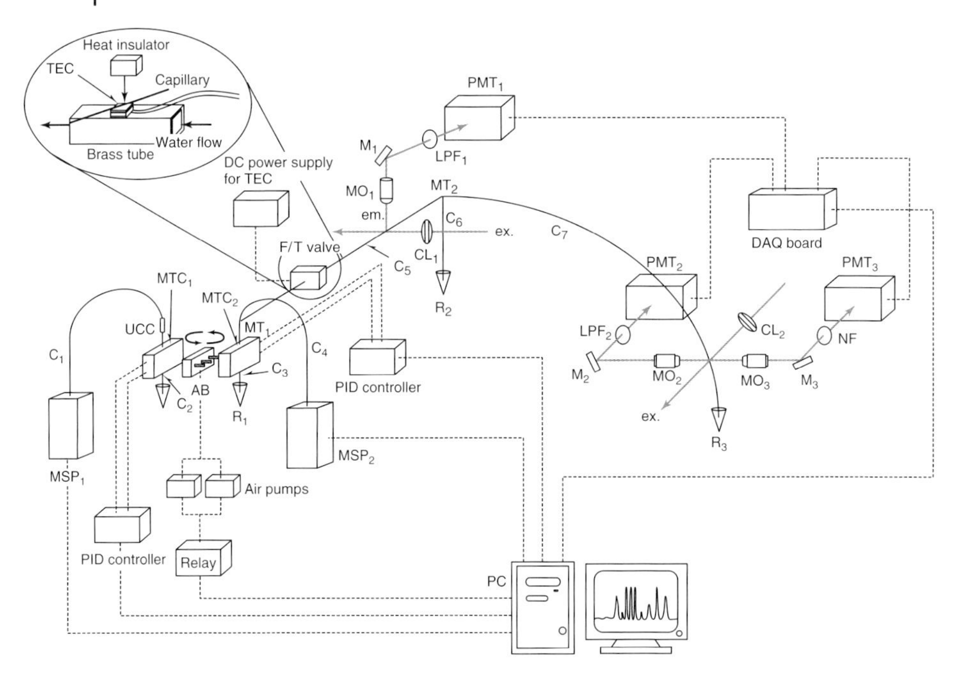


**Figure 5.4** Schematic diagram of instrumental setup for on-line integration of PCR and cycle sequencing in capillaries. AB, air blower; $C_1$ and $C_4$, solution delivery capillaries; $C_2$, reaction capillary for PCR; $C_3$, reaction capillary for cycle-sequencing reaction; $C_5$, CZE purification capillary; $C_6$, waste arm; $C_7$, CGE capillary; $CL_1$ and $CL_2$, convex lenses; F/T valve, freeze/thaw valve; $LPF_1$, 560 nm long pass filter; $LPF_2$, 630 nm long pass filter; $M_1$, $M_2$, and $M_3$, mirrors; $MO_1$, $MO_2$, and $MO_3$, microscope objectives; $MSP_1$ and $MSP_2$, microsyringe pumps; $MT_1$ and $MT_2$, microtees; $MTC_1$ and $MTC_2$, microthermocyclers; NF, 514-nm notch filter; $PMT_1$, $PMT_2$, and $PMT_3$, photomultiplier tubes; $R_1$, $R_2$, and $R_3$, buffer reservoirs; TEC, thermoelectric cooler; UCC, universal capillary connector. The inset shows the F/T valve control system using a TEC [18].

fast electrode switching (~10 s). The overall time from PCR to CGE separation for sequencing only took 4 h. The effectiveness of the setup was demonstrated by the successful sequencing of a 257-bp gene region of the $\beta$-actin in 9.8 ng $\mu l^{-1}$ from human genomic DNA with a volume of 616 nl (total 6 ng, 3 zmol, ~180 cells). This integrated system is compatible with the continuous cell injection and on-line cell lysing discussed previously for genetic analysis starting with a few cells.

#### 5.2.2.2 **Optimization for Reactions**

On-line amplification is quite different from PCR performed in vials with the heating block-based thermocyclers. First of all, the reaction volume of on-line PCR is much smaller than that in the vial format, which significantly shortens the duration for denaturing, annealing, and amplification and thereby the overall PCR time. Then, higher cycle numbers can be used without extending the analysis time. Secondly, the yield for on-line PCR greatly affects the success of the following operations, such as sequencing or detection. For the traditional PCR, if with

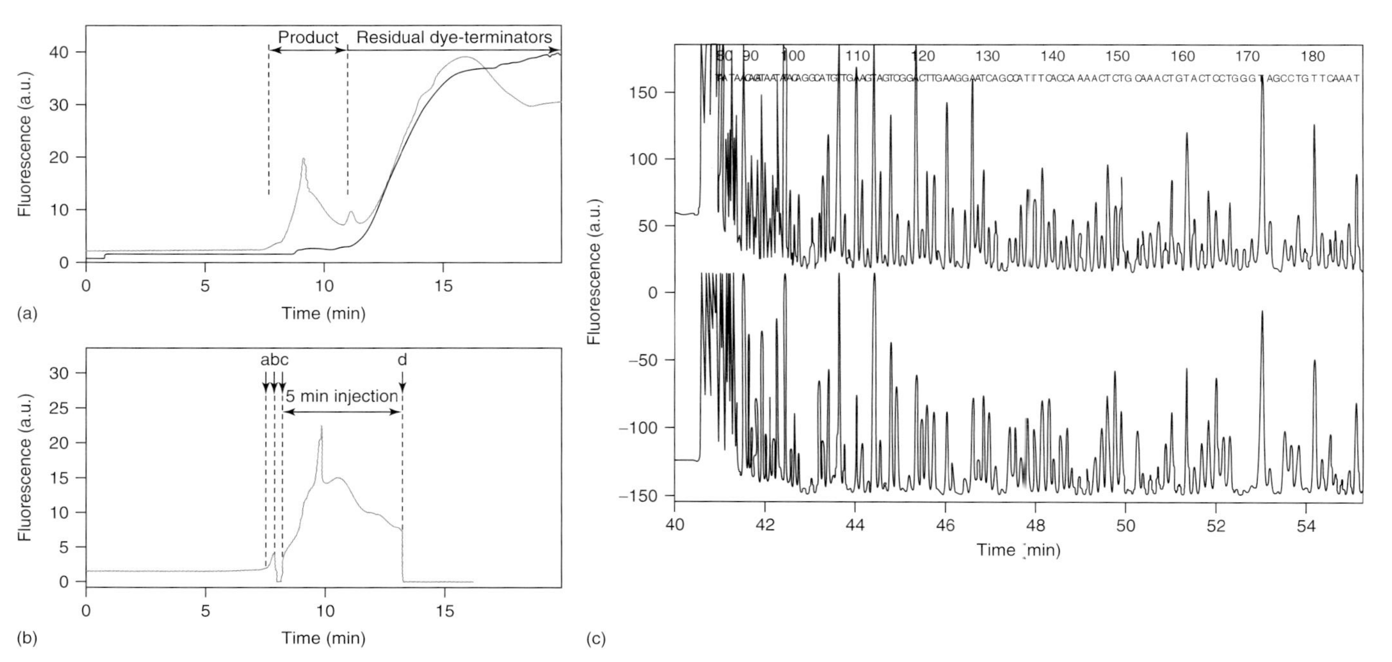


**Figure 5.5** Separation of dye-labeled DNA products and residual dye-terminators (a) and on-line product injection (b). The thin line in (a) represents CZE separation of the product and the residual dye-terminators while the solid line indicates electrophoretic migration of the dye-terminators only. The interval between a and b in (b) represents the time delay before injection. The positive electrode was switched from $R_2$ to $R_3$ between b and c. On-line injection was performed between c and d. Separation electrolyte, 10 mM Tris ± HCl buffer (pH 8.0) contains 2 mM $MgCl_2$, 3 mM KCl, and 0.3% (w/v) PVP. More details refer to the original paper. (c) Part of the CGE separation of on-line purified cycle-sequencing product. Separation matrix, 1.5% PEO (Mr 8 000 000) and 1.4% PEO (Mr 60 000) in 1× TBE with 7 M urea [18].

one-round of reaction, the amount of PCR products are still not enough, more reaction rounds might be performed in the original vials. However, for the integrated on-line system, reactions are performed sequentially without interruption and the sample amount is limited. If the yield of PCR is not enough, it directly suppresses the subsequent sequencing and fails the entire process, which could only be recognized at the end of the analysis. Therefore, it is especially critical to optimize the yield of PCR for on-line assays.

For any enzymatic reaction, enzyme activity, buffer composition, pH value, and reaction temperature are always important factors that can impact the reaction yield. Numerous DNA polymerases and PCR cocktails are available in the market for traditional PCR in vials, facilitating the selection of the most active enzyme and corresponding reaction cocktail [4]. Comparison of the yields can be monitored by slab gel imaging. However, this is only a rough comparison because when the reaction vessels are very small, the diffusion and other factors may influence the yield as well. Further enhancement of PCR yield could be achieved with addition of betaine (trimethylglycine), which protects DNA polymerase from thermal denaturation [20] to improve both the yield and selectivity [21, 22]. To prevent nonspecific adsorption of DNA polymerase to the capillary wall, BSA can be added to the reaction mixture [23], or the enzyme adsorption sites can be saturated by prerinsing the capillary by repeated aspiring and dispensing 1-μl PCR solution not containing the DNA template or dNTPs, which has been found to lead to a threefold yield enhancement [18] Additionally, capillary-based PCR takes much less time for one cycle due to the small volume of the reaction vessel and the fast temperature change provided by microthermocycler. DNA polymerase denaturation by heat should be less after the same cycle number. Herein, larger cycle numbers (i.e., 50 cycles) can be adopted to generate more PCR product when the polymerases are still active. Moreover, since temperature plays an important role on sensitivity and selectivity, several strategies were developed to optimize the temperature cycle. In the early 1990s, the concept of "touchdown" PCR was introduced to increase PCR yield with good specificity by lowering the annealing temperature stepwise [24], because an annealing temperature too low at early cycles could lead to false hybridization of the primers, but a lower annealing temperature at later cycles could stabilize the primer and target ssDNA hybrids, and enhance the yield of amplification. Therefore, "touchdown" PCR was employed in the integrated on-line system that has been discussed here. Similar strategies can be applied for cycle-sequencing reaction, although the cycle-sequencing reaction is generally more robust than PCR and the temperature cycles do not need to be modified.

Another challenge to be overcome for the integration of PCR–cycle-sequencing–CE separation is that PCR product should be purified to remove the excess amount of dNTP and primers before the cycle-sequencing reaction to maintain the correct ratio of dNTP and ddNTP. To simplify the instrumental setup and avoid post-PCR purification, the PCR protocol was modified with lower concentration of primers and dNTPs, increased number of thermal cycles, and smaller PCR volume transferred to the cycle-sequencing reaction cocktail so that the carryover primer and dNTPs could be diluted and reused in the cycle-sequencing reaction.

However, the lower amounts of primer and dNTPs used for PCR was a trade-off. Additionally, threefold excess of the primers for the cycle-sequencing reaction were added to compete with the carryover primers from PCR.

#### 5.2.2.3 Analysis of Amplified Products

LIF is the dominant detection method for single-cell nucleic acid analysis using the integrated systems because of its high sensitivity and simplicity in instrumentation and operation. Intercalating dyes can be added to the running buffer to stain DNA or RNA molecules. Ultrasensitive nucleic stains, such as ethidium bromide [19], SYBR Green I [4], SYBR Green II [25], and SYBR Gold [18] are frequently employed.

In the direct on-line PCR product detection performed in [9], the product peaks were well correlated with the starting location of individual cells inside the capillary [4]. Even though the separation power was limited due to the incompatibility of the sieving matrix with the PCR buffer, the major product peaks were still separated from the primer peak. Different peak heights were detected probably due to the variation in the amplification efficiency, which, on the other hand, indicated the importance of PCR yield optimization. Occasionally, some relatively small peaks showed up at locations without the presence of a cell, and were attributed to the DNA debris from the dead cells in the cell suspension, which were "invisible" in visual cell counting using the partial dark-field method. The good correlation between the cell and the PCR product positions provided a digital readout of the "yes" or "no" answer for the existence of the target DNA without quantitative information.

CGE separation is necessary if the background nucleic acid material would affect the specific target identification. In such cases, using different capillaries for the reaction and CE separation is advantageous. For example, upon transferring the RT-PCR mixture in the reaction capillary to the separation capillary after the reaction, the low-molecular weight RNAs could be separated from the 523-bp RT-PCR product from the breast cancer specific gene (BCSG-1) mRNA and did not interfere with its detection. In the integrated device for PCR and cycle sequencing, the high resolution capability of CE became more demanding for the analysis of the sequencing products [18] Therefore, two separation capillaries were employed in this device besides several others as reactors and transporters. The first one was to remove the excess dye-terminator from the cycle-sequencing products by CZE. The other was for analyzing the sequencing products by CGE. The segmented design allows using different capillaries for different purposes, even though the system complexity is also increased.

## 5.3 Direct Gene and Gene Expression Analysis Without Amplification

Without amplification, detection of specific DNA or RNA targets from one cell is only possible with highly sensitive methods that are able to detect single molecules. High-throughput single-molecule DNA screening based on CE has been demonstrated [26], which opened up the possibility of screening single copy of DNA in an individual cell. The DNA molecules were labeled with the intercalating dye of YOYO-1 at a ratio of 1 dye molecule per 5 bp. LIF imaging using a scientific-grade microscope together with a high-speed ICCD camera determined the electrophoretic mobility of the labeled DNA molecules through the multiframe, streak, and multispot procedures. Each measurement was finished in a few milliseconds. The migration behaviors of many distinct DNA molecules were recorded simultaneously in the images, so that individual DNA molecules were identified based on their specific mobility measured.

If the overall expression profile is to be sorted out, CE equipped with a regular LIF detector can do the work. RNA levels in individual cells at different phases, the pre-DNA-synthesis phase ($G_1$), the DNA synthesis phase (S), the post-DNA synthesis phase ($G_2$), and the mitotic phase (M), were compared with CE without preamplification by Han *et al.* [12]. To differentiate phases, cell synchronization is needed before analysis. Several techniques commonly used by cell biologists can be employed by the single-cell analysis. These techniques include physical shocks, such as temperature shock and X-ray irradiation; chemical shocks, such as isoleucine deprivation, thymidine block, and vinblastin inhibition; and physical selections, such as filtration, centrifugal elution, membrane elution, and mitotic shake-off [12]. Mitotic shake-off is the simplest one and based on the property of cells they become spherical and tenuously attached to the culture flask during the mitosis phase. Gentle agitation of the culture medium by shaking can lead to the dislodging of the mitosis phase cell from the bottom of the flask. These mitosis phase cells then served as the seeds for other phases by growing them for different durations. One single cell (Chinese hamster ovary cell) at a distinct growth phase was injected into the capillary and lysed with a diluted surfactant solution (0.2% SDS in TBE buffer), and the RNA contents released were separated by electrophoresis and detected by LIF. By comparing with RNA standard ladder, the major peaks detected were identified as 18S and 28S rRNA, which showed significantly different expression levels at each phase of the cell cycle (Figure 5.6). This method may be applied to detect diseases closely related to abnormally regulated cell cycles, such as cardiovascular diseases, human neurodegenerative disease, and cancers, or to monitor the effectiveness of drugs that are developed to target specific cell phases [27, 28]. However, more interesting RNA targets, such as the mRNA and small RNA, cannot be detected due to the lack of sensitivity, and the information obtained by this method is limited.

## 5.4 Potential Alternative Techniques for Single-cell Gene and Gene Expression Analysis

The systems discussed above pave the way for sensitive, rapid, and high-throughput analysis of nucleic acids at the single-cell level with or without amplification. However, further developments are demanded to simplify the integrated systems, lower the operation cost, and increase the volume of the information obtained.

To simplify the design of the integrated systems, several strategies can be considered. For example, to improve the effectiveness of the cell lysis process, microfluidic sonicator could be used for real-time disruption of the eukaryotic cells

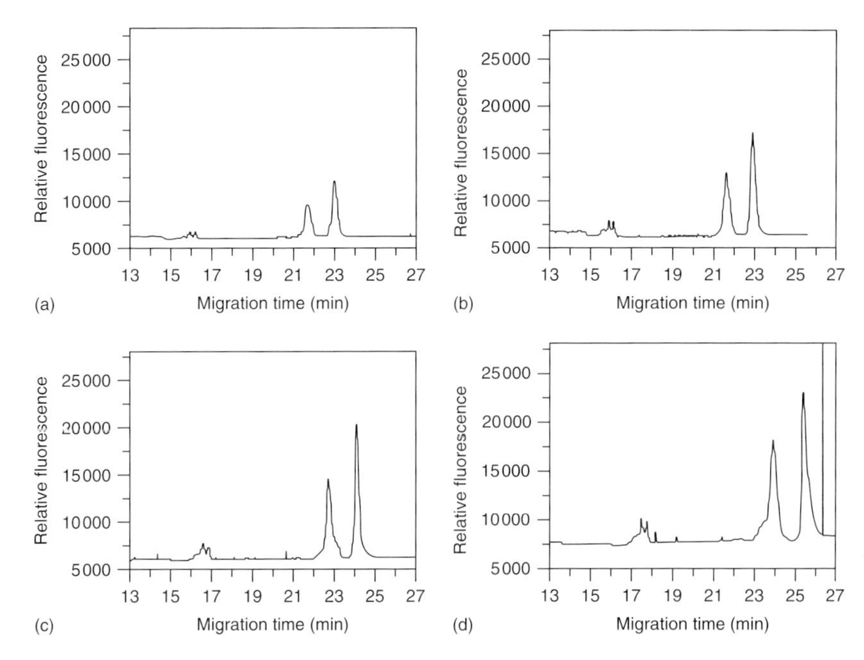


**Figure 5.6** Representative electropherograms indicating the separation of RNA from individual CHO-K1 cells at different phases of the cell cycle: (a) $G_1$, (b) S, (c) $G_2$, (d) M. Eight individual cells from each phase were analyzed for a total of 32 electropherograms. The first group of peaks corresponds to low-molecular mass RNA (including tRNA and 5S rRNA) and the last two peaks (>20 min) correspond to 18S and 28S rRNA [12].

and bacterial spores for DNA analysis [29]. Since the most complicated part of the integrated analysis system is the temperature control module for PCR, isothermal amplification can be employed. Rolling circle amplification is a good replacement for PCR, which can amplify the circular template up to 10 000-fold within a few hours due to the high strand-displacement ability and great processivity of the polymerase [30, 31]. Our group has demonstrated the accurate quantification of the target ssDNA (29 nt) coamplified with an internal standard ssDNA (31 nt) by rolling circle amplification (RCA) using CGE–LIF [25]. This RCA–CE method has adequate sensitivity to detect DNA materials extracted from several hundred cells if the amplification is performed on-line. Since the mature small RNA molecules (17–25 nt) can serve as the primer for RCA and initiate the polymerization from their free 3′-OH groups, we eliminated the ligation step in small RNA detection and improved the reaction efficiency, which further decreased the detection limit by 10-fold. Small RNAs are involved in highly specific regulation of gene expression by the RNA-mediated silencing mechanism in multicellular organisms such as plants and animals [32–34]. Only a few copies of small RNA can effectively silence gene [35]. They play even more critical roles in stem cells by controlling stem cell differentiation [36, 37], cell cycle regulation, apoptosis [38], and maintenance of stemness [39]. MicroRNA is one subgroup of small RNA, the expression patterns of which are found highly related to cancer development and could be potential targets for cancer detection, diagnosis, and prognosis [40, 41]. Our assay can be applied to detect small RNAs in a few cells or even single cells, and therefore is valuable to studies on small RNA expression in stem cells or cancer cells because of the low availability of homogeneous cell population of such cells.

## 5.5 Conclusions

It is well known that the capillary is an appropriate platform for nucleic acids' analysis because of the low sample consumption, low reagent cost, easy dynamic staining, sensitive detection, and potentials for higher sample throughput with the array setups. Facing the challenges in nucleic acids' analysis at the single-cell level, the capillary shows its distinct advantageous features, such as volume compatible with single cells, high heat conductivity, and good optical properties. Therefore, systems consisting of several capillaries were successfully developed to integrate cell introduction, in-column cell lysis, on-line thermal cycle reactions, and CE separation. Such systems are the pioneers in single-cell analysis using electrophoretic techniques and are highly compatible with recent developments such as the CE–RCA assays for continuous improvement in the instrumental design and broader application scope. Future developments in this area can also include sophisticated data analysis tools based on statistic and mathematical modeling to obtain more information from the genetic analysis in a few number of cells, and eventually contribute to deciphering the ultimate difference among individual cells.

## References

1 Toledo-Rodriguez, M. and Markram, H. (2007) *Methods Mol. Biol.*, **403**, 123–139.
2 Chen, S. and Lillard, S.J. (2001) *Anal. Chem.*, **73**, 111–118.
3 Zabzdyr, J.L. and Lillard, S.J. (2001) *Anal. Chem.*, **73**, 5771–5775.
4 Li, H. and Yeung, E.S. (2002) *Electrophoresis*, **23**, 3372–3380.
5 Gilman, S.D. and Ewing, A.G. (1995) *Anal. Chem.*, **67**, 58–64.
6 Han, F., Wang, Y., Sims, C.E., Bachman, M., Chang, R., Li, G.P., and Allbritton, N.L. (2003) *Anal. Chem.*, **75**, 3688–3696.
7 Sims, C.E., Meredith, G.D., Krasieva, T.B., Berns, M.W., Tromberg, B.J., and Allbritton, N.L. (1998) *Anal. Chem.*, **70**, 4570–4577.
8 Arcibal, I.G., Santillo, M.F., and Ewing, A.G. (2007) *Anal. Bioanal. Chem.*, **387**, 51–57.
9 Cheng, T., Shen, H., Giokas, D., Gere, J., Tenen, D.G., and Scadden, D.T. (1996) *Proc. Natl. Acad. Sci. U.S.A.*, **93**, 13158–13163.
10 Taylor, R.W., Taylor, G.A., Durham, S.E., and Turnbull, D.M. (2001) *Nucleic Acids Res.*, **29**, e74/71–e74/78.
11 Roers, A., Montesinos-Rongen, M., Hansmann, M.-L., Rajewsky, K., and Kueppers, R. (1998) *Eur. J. Immunol.*, **28**, 2424–2431.
12 Han, F. and Lillard, S.J. (2002) *Anal. Biochem.*, **302**, 136–143.
13 Kim, U., Shu, C.-W., Dane, K.Y., Daugherty, P.S., Wang, J.Y.J., and Soh,

H.T. (2007) *Proc. Natl. Acad. Sci. U.S.A.*, **104**, 20708–20712.

**14** Li, H., Gyllensten, U.B., Cui, X., Saiki, R.K., Erlich, H.A., and Arnheim, N. (1988) *Nature*, **335**, 414–417.

**15** Von Eggeling, F., Michel, S., Guenther, M., Schimmel, B., and Claussen, U. (1997) *Hum. Genet.*, **99**, 266–270.

**16** Garvin, A.M., Holzgreve, W., and Hahn, S. (1998) *Nucleic Acids Res.*, **26**, 3468–3472.

**17** Pang, H.-M. and Yeung, E.S. (2000) *Nucleic Acids Res.*, **28**, e73, ii–viii.

**18** Hashimoto, M., He, Y., and Yeung, E.S. (2003) *Nucleic Acids Res.*, **31**, e41/41 e41/17.

**19** Matsunaga, H., Anazawa, T., and Yeung, E.S. (2003) *Electrophoresis*, **24**, 458–465.

**20** Santoro, M.M., Liu, Y., Khan, S.M.A., Hou, L.X., and Bolen, D.W. (1992) *Biochemistry*, **31**, 5278–5283.

**21** Weissensteiner, T. and Lanchbury, J.S. (1996) *BioTechniques*, **21**, 1102–1108.

**22** Henke, W., Herdel, K., Jung, K., Schnorr, D., and Loening, S.A. (1997) *Nucleic Acids Res.*, **25**, 3957–3958.

**23** Zhang, N., Tan, H., and Yeung, E.S. (1999) *Anal. Chem.*, **71**, 1138–1145.

**24** Don, R.H., Cox, P.T., Wainwright, B.J., Baker, K., and Mattick, J.S. (1991) *Nucleic Acids Res.*, **19**, 4008.

**25** Li, N., Li, J., and Zhong, W. (2008) *Electrophoresis*, **29**, 424–432.

**26** Shortreed, M.R., Li, H., Huang, W.-H., and Yeung, E.S. (2000) *Anal. Chem.*, **72**, 2879–2885.

**27** Bicknell, K.A. and Brooks, G. (2008) *Curr. Opin. Pharmacol.*, **8**, 193–201.

**28** Copani, A., Guccione, S., Giurato, L., Caraci, F., Calafiore, M., Sortino, M.A., and Nicoletti, F. (2008) *Curr. Med. Chem.*, **15**, 2420–2432.

**29** Marentis Theodore, C., Kusler, B., Yaralioglu Goksen, G., Liu, S., Haeggstrom Edward, O., and Khuri-Yakub, B.T. (2005) *Ultrasound Med Biol*, **31**, 1265–1277.

**30** Dean, F.B., Nelson, J.R., Giesler, T.L., and Lasken, R.S. (2001) *Genome Res.*, **11**, 1095–1099.

**31** Blanco, L., Bernad, A., Lazaro, J.M., Martin, G., Garmendia, C., and Salas, M. (1989) *J. Biol. Chem.*, **264**, 8935–8940.

**32** Rana, T.M. (2007) *Nat. Rev. Mol. Cell Biol.*, **8**, 23–36.

**33** Chapman, E.J. and Carrington, J.C. (2007) *Nat. Rev. Genet.*, **8**, 884–896.

**34** Costa, F.F. (2008) *Gene*, **410**, 9–17.

**35** Reynolds, A., Leake, D., Boese, Q., Scaringe, S., Marshall, W.S., and Khvorova, A. (2004) *Nat. Biotechnol.*, **22**, 326–330.

**36** Zaehres, H., Lensch, M.W., Daheron, L., Stewart, S.A., Itskovitz-Eldor, J., and Daley, G.Q. (2005) *Stem Cells*, **23**, 299–305.

**37** Lakshmipathy, U., Love, B., Goff, L.A., Joernsten, R., Graichen, R., Hart, R.P., and Chesnut, J.D. (2007) *Stem Cells Dev.*, **16**, 1003–1016.

**38** Xu, X.-F., Zhang, Z.-Y., Ge, J.-P., Cheng, W., Zhou, S.-W., Zhang, X., Xu, Q., Wei, Z.-F., and Gao, J.-P. (2007) *J. Gene Med.*, **9**, 1065–1070.

**39** Stadler, B.M. and Ruohola-Baker, H. (2008) *Cell*, **132**, 563–566.

**40** Tricoli, J.V. and Jacobson, J.W. (2007) *Cancer Res.*, **67**, 4553–4555.

**41** Jiang, J., Lee, E.J., Gusev, Y., and Schmittgen, T.D. (2005) *Nucleic Acids Res.*, **33**, 5394–5403.

# 6
# Microfluidic Technology for Single-cell Analysis

*Yan Chenand Jiang F. Zhong*

## 6.1
## Introduction

Molecular biologists increasingly recognized the difficulty of obtaining a population of homogenous cells for molecular analysis, especially for oligo-microarray assays, which may require a few millions of cells. Without intervention procedures such as synchronization, which often changes characteristics of cells, it is very difficult to obtain a homogeneous cell population. Many heterogeneous cell populations were treated as homogenous samples simply due to our inability to obtain homogeneous cell populations for analysis. The human embryonic stem cell (hESC) colony maintained in pluripotent culture condition was regarded as a homogenous pluripotent cell population in many studies. However, recent single-cell analysis of these colonies showed that each colony is a heterogeneous cell population [1].

Homogeneity is a relative concept. In fact, no two cells are identical because they are exposed to different environments. This is similar to the fact that two identical twins are two individuals with different personalities. Perhaps, the only homogenous cell population is a single cell. Single-cell analysis is a critical aspect of molecular biology. However, current experimental tools for molecular biologists are not suitable for analysis of single cells, which are only several picoliters. This is the primary reason why biologists are forced to accept the concept of a "homogenous cell populations" when they know that "homogeneity" is a relative concept.

### 6.1.1
### Limitation of Current Technology

Fluorescence-activated cell sorting (FACS) is the most popular analysis tool for single-cell analysis. It rapidly analyzes the fluorescent profiles of a population of single cells via various labeling techniques. FACS analysis is often limited by applicable fluorescent tags to label biological molecules of a cell. Laser capture microdissection (LCM) is another common method for isolating specific cells from a cell population [2]. Both FACS and LCM are developed to isolate cells with certain characteristics rather than analyze the chemical contents of individual cells. New

*Chemical Cytometry.* Edited by Chang Lu

ISBN: 978-3-527-32495-8

tools are needed for chemical analysis of single cells which are typically in the scale of picoliters.

Despite the difficulty, attempts have been made to explore the unique and critical information that can be obtained from biochemical analysis of single-cell contents. Multiple protocols for single-cell RT-PCR has been reported [3–5]. However, the success of these protocols is often dependent on the experience of the technical personnel. Material lost remains as a major challenge in single-cell mRNA expression analysis. A single mammalian cell contains 20–40 pg of total RNA [6] and only 0.5–1.0 pg of mRNA ($10^5$–$10^6$ mRNA molecules) [7]. This small amount of material presents a major challenge for single-cell mRNA profiling with current tools such as micropipette and microcentrifuge tubes, which are designed for performing biochemical reactions at the microliter scale. The relatively huge dead volumes of micropipette and microcentrifuge tubes lead to significant material loss in single-cell analysis. Performing biochemical reactions in the nanoliter scale is needed to avoid significant material lost in single-cell analysis. Microfluidic devices developed by us can meet the requirement for single-cell analysis by performing biochemical reactions in 10 nl reactors.

### 6.1.2 Microfluidic Devices

Microfluidic devices are formed by integrating functional valves and pumps that can manipulate nanoliters and even picoliters of fluid. Microfluidic devices can be fabricated by multilayer soft lithography (MSL), a micromachining technique that exploits the elasticity and the surface chemistry of silicone elastomers to create monolithic valves [8]. For devices used in our laboratory, polydimethylsiloxane (PDMS) layers with nanofluidic channels were cast from a microfabricated mold, created by standard optical lithography on photoresist. A typical microfluidic device has two layers. One layer consists of fluidic channels where samples and reagents will be introduced and manipulated; the second layer consists of control channels where pressurized air will be introduced to pneumatically actuate the thin membrane between the control and fluidic channel crossing area. The nanoliter dimension of microfluidic device improves efficiency of biochemical reactions. The mRNA-to-cDNA efficiency in our microfluidic device is fivefold higher than those of bulk assay with the same reagents [1].

## 6.2 Biological Significance of Single-cell Analysis

Many gene expression studies with qPCR and microarray have been carried out to dissect the complex human gene regulatory networks. The two major challenges of the field are as follows: (i) Time- and phenotype-averaging expression profiles from bulk assays conducted with total RNA from heterogeneous cell populations are not very informative for gene network study. (ii) Although gene regulation is a continuous event, continuous expression profiles are not available for studying the interactive relationship among genes.

### 6.2.1 Investigate Gene Regulation in Consecutive Developmental Stages

Compiling and clustering single-cell mRNA profiles from consecutive developmental stages is an effective approach to investigate the complex gene regulation networks in mammalian cells. However, current bulk assays can only compare samples from preselected and very distinct cell stages due to sample heterogeneity. The population-averaging effect of bulk assays excludes the possibility to compare mRNA expression profiles from consecutive developmental stages.

Single-cell gene expression profiling has been conducted with LCM and FACS, but obtaining consecutive single-cell expression profiles with these methods is still not possible. The technical barrier for obtaining single-cell mRNA profiles from consecutive developmental stages has forced biologists to investigate gene regulation networks with compromised approaches such as extrapolation of bacterial and yeast data to mammalian cells or reduction of gene networks to include only a few critical members. Consequently, some published data are contradictory. This problem is particularly severe in the investigation of human disease and stem cell regulatory networks. With our microfluidic system for single-cell analysis, we will be able to interrogate individual cells from entire cell populations consisting of cells in a continuous spectrum of differentiation/maturation stages. The continuous single-cell mRNA expression profiles can also reveal stepwise gene–gene interactions within the regulatory networks.

### 6.2.2 Identifying Cancer Stem Cells (CSCs) Molecular Signature

Recent evidence suggests that mammary stem cells and/or progenitors including mammary luminal epithelial and myoepithelial cells may be the targets for oncogenesis by Wnt-1 signaling elements, and that the developmental heterogeneity of different breast cancers is in part a consequence of differential effects of oncogenes on distinct cell types in the breast [9]. Consequently, elegant models have been suggested toward the development of leukemia and different types of breast cancers arising from cancer stem cell populations, but specific markers delineating these lineages have not yet been clearly defined [10]. Animal experiments with limited dilution methods showed that cancer stem cell (CSC) could be enriched to a frequency of 1 CSC per 100–500 cells by cell surface markers which are empirically determined in animal transplantation experiments [11–13]. However, the CSC specific genes (markers) cannot be determined because there is no method to characterize a single cell before transplanting the same cell for tumor initiation. We have developed microfluidic devices that can enable us to identify CSC from acute lymphoblastic leukemia cells (CD34+CD38–CD19+) [14].

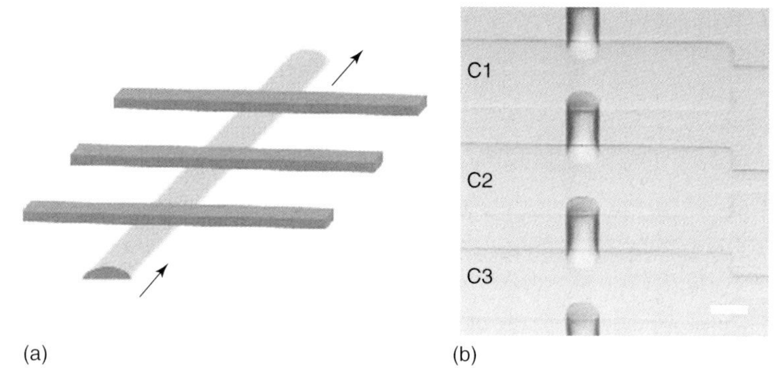


**Figure 6.1** The principle of microfluidic devices. (a) An illustration 3-D diagram of an elastomeric peristaltic pump. The peristaltic pump consists of 3 control channels and 1 flow channel. Peristalsis was typically actuated by the pattern 101, 100, 110, 010, 011, 001, where 0 and 1 indicate "valve open" and "valve closed" respectively. This pattern is named the "120°" pattern, which refers to the phase angle of actuation between the three valves. Other patterns are also possible, including 90° and 60° patterns. The difference in pumping rate at a given frequency of pattern cycling is minimal. (b) A picture of an elastomeric peristaltic pump with three on-off valves. Three control channels (C1, C2 and C3) are placed on top of a flow channel (vertical channel) at 111 configuration (all valves are closed). Scale bar is 100 μm.

Our microfluidic device enables (Figure 6.1) the analysis of several thousands of single cells without expensive equipments and intensive labor. After profiling 500–2500 cells from a CSC-enriched population (5× coverage, at least five CSC will be profiled), we can cluster the gene expression profiles into a dendrogram, a hierarchical organization of the expression profiles based on similarity. These dendrograms will reveal the molecular signature of CSC and its progenitors. Once the CSC molecular signature (its gene expression profile) is identified, systematic discovery of CSC protein markers (especially cell surface markers) is possible by mining the gene expression profiles. The investigation of the molecular developmental pathway of CSC toward tumor cells and the development of CSC targeted therapies will be facilitated by knowledge obtained from single-cell gene expression profiling with our devices.

## 6.3
## Microfluidic Devices in Our Laboratories

Recently, we reported a sophisticated highly integrated microfluidic device, having 26 parallel 10 nl reactors for the study of gene expression in single hESCs [1]. The reaction volume is 10 nl and the micropumps and microvalves have a dead volume of less than 1 pl. All steps of obtaining single-cell cDNA including cell capture, mRNA capture/purification, and cDNA synthesis/purification are performed inside the device. We demonstrated a fivefold higher mRNA-to-cDNA efficiency in reactions performed with microfluidic device as compared to conventional bulk assays.

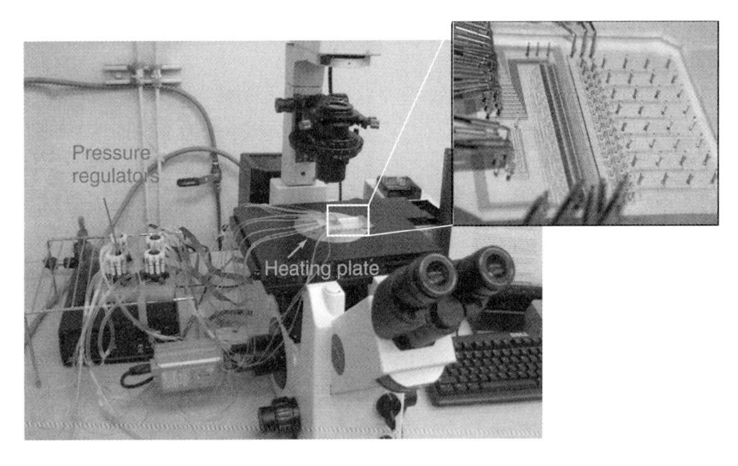


**Figure 6.2** The setup of the microfluidic device for single-cell mRNA extraction. The system includes a microscope, a computer to control air pressure with pressure regulators, a heating stage to heat the microfluidic chip to desired temperatures. The inset showes a typical microfluidic chip.

### 6.3.1 Microfluidic Single-cell mRNA Extraction Device

We have extensively used microfluidic devices in biomedical research [15, 16]. It is in general challenging to perform single-cell gene expression profiling. The difficulty is due to the loss of material during the steps of single-cell capture, lysis, mRNA isolation, and cDNA synthesis. Many successful single-cell gene expression profiling studies have been reported [17–27]. Although these studies showed that valuable information can be obtained from single-cell gene expression profiling, they were limited in their ability to process a large number of cells. However, in order to obtain statistically significant information, large numbers of cells are needed to be profiled.

Figure 6.2 shows the setting of our microfluidic system. The microfluidic valves within the device are controlled by individual pressure regulators (Fluidigm) and are interfaced via 23 gauge pins (New England Small Tube) and tygon tubing (VWR). An NI-DAQ card through a Labview interface (National Instruments) was used to actuate the pressure of the pressure regulators.

### 6.3.2 Functional Components of Single-cell Analysis Devices

The basic components of our device are shown in Figure 6.3, and the device is illustrated in Figure 6.4. The process in brief can be described as follows. Lysis buffer is loaded into the flow channels through lysis buffer inlet until it reaches the waste outlets and leaves no air bubbles. Oligo(dT) beads are then loaded into beads inlet and

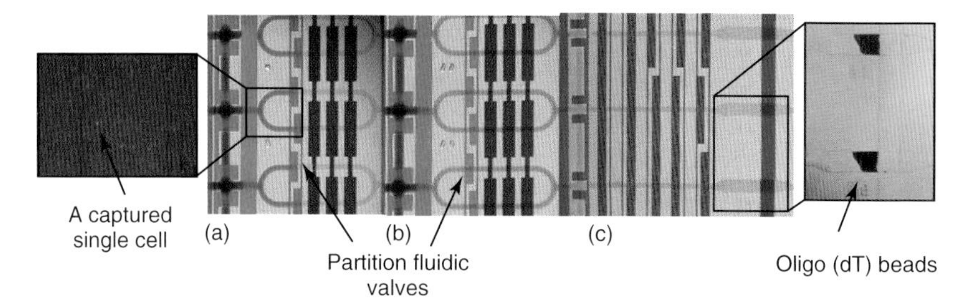


**Figure 6.3** Basic microfluidic components in our microfluidic devices. (a) Microfluidic cell lysis modules constructed by putting control channels (green) under fluidic rings. Individual rings served as cell lysis modules. Yellow fluid represents lysis buffer (right part of the ring). Blue fluid represents cells with PBS. Black blocks are control valves which separate individual rings. An inset showes a fluorescently labeled cell being captured in the cell capture portion (blue) of the ring. (b) After opening the partition valve and mixing lysis buffer with cell by pumping, the cell was disintegrated. This is represented by mixing blue fluid with yellow fluid to produce green fluid. (c) Multiplex control channels (red) were shown to control the open or close of individual fluidic channels. On the left side of the multiplex control net are the cell capture rings, and on the right side are the mRNA capture columns. An inset showes the column loaded with oligo(dT) beads. The clear channel next to the beads creates the leaky sieve valves by partially closing the fluidic channels containing the beads. Sieve valves allow buffers but not the Oligo(dT) beads to pass through.

columns are built serially by addressing flow lines individually with the multiplexer control channels, while keeping the sieve valve (valve present on yellow flow channels Fig. 6.3c) actuated. Once columns are built, excess beads still present in the flow channels are pushed with lysis buffer (introduced from the lysis buffer inlet) to the constructed columns. Cell suspension (blue) is then loaded into cell inlet, and once the suspension flows to its outlet, the reactors are closed off by actuating the control channels. Cells are contained in the cell loading portion and then pushed individually to the left part of the mixing ring. Lysis buffer is then loaded into the right portion of the mixing ring. Cells are then lysed chemically by mixing cells with lysis buffer in the ~10 nl ring. Mixing occurs by executing a peristaltic pump sequence [28, 29] with three control channels over the mixing ring. Cell lysate is then pushed via pneumatic pressure over the affinity columns, followed by washing of the columns with first strand synthesis buffer, dNTPs, and reverse transcriptase (RT).

Once the RT reaction fills the flow channels, first strand synthesis is carried out by heating the device to 40 °C on a thermal microscope stage with the beads utilized as both primers (oligo(dT)$_{25}$ sequences) and a solid phase support. The reaction mixture is flown over the columns for 45 min at a flow rate of ~20 μm $s^{-1}$. Upon completion of the reaction, the waste valves are closed, collection valves opened, and beads sent to output by opening the sieve valves and flowing columns off the device in a serial manner in PCR buffer, by using the fluid multiplexer to address reactors individually. Beads are collected by cutting off the collection wells with beads and centrifugation. Figure 6.4 shows a device we used to obtain cDNAs from 32 single cells simultaneously.

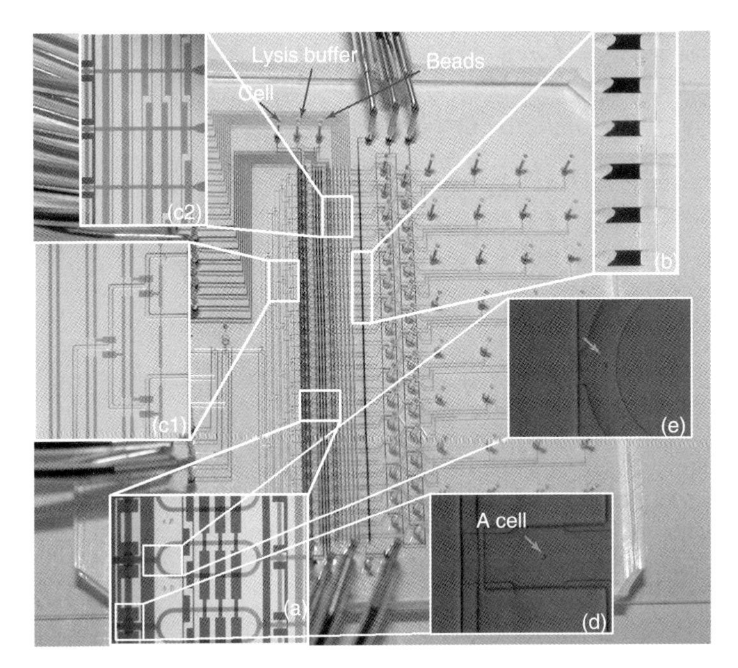


**Figure 6.4** An illustration of manipulating single cells with flow channels in microfluidic devices. A microfluidic single-cell analysis device with 32-nl-scale reactors is filled with food dye. (a) The 10-nl single-cell analysis reactors. Each reactor is a ring with 10 nl volume. The reactor is partitioned into two parts by two microvalves. The right part is filled with the lysis buffer. The left part is filled with the buffer carrying cells (such as phosphate buffered saline). The cell capture compartment is formed by valves around the cell flow channel. (b) Channels stacked with oligo-dT bead columns. (c1, c2) Binary flow networks that allow external pressure to be delivered to individual reactors. The control channels lie on top of the flow channels. (d) An enlarged cell capture compartment shows a captured cell. (e) After opening of valves and applying external pressure via the binary flow network (c1), the cell was delivered into the reactor.

### 6.3.3
### Manipulation of Single Cells

Manipulating the picoliter single cells in microfluidic devices precisely is challenging. We used two methods to manipulate single cells in our nanoliter reactors for single-cell analysis. The first method is a fluidic manipulation. Single cells are manipulated via manipulating the fluid which carries the cells. The second method is optoelectronic tweezers (OETs) technology [30].

As shown in Figure 6.4, a binary network of flow channels is controlled with multiplex control channels. With the multiplex control network, external pressure can be delivered to the 32 individual reactors with computer control. Cells in buffer can be sent to the proximal locations of the reactors by addressing the flow rate and cell concentration. Then external pressure was used to deliver the cells with buffer into individual reactors. Only a minimal amount of buffer was introduced with a cell. This is an important aspect because a large amount of buffer may affect

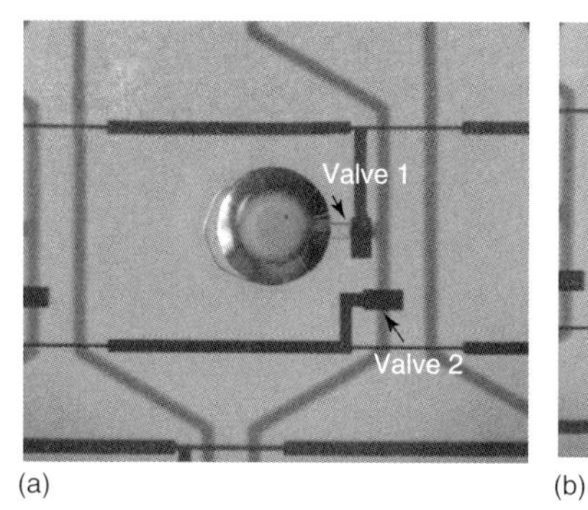


(a)

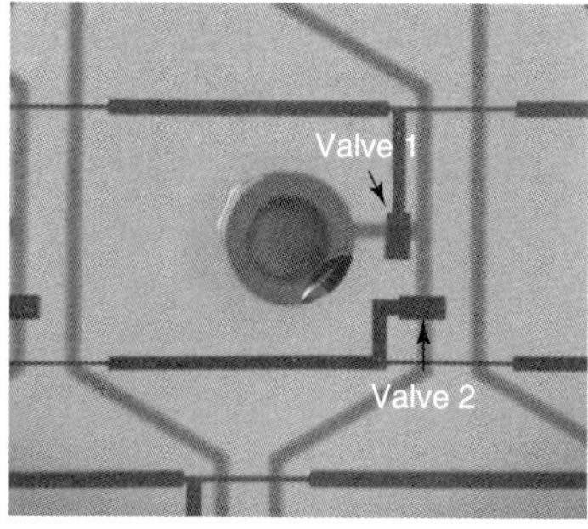


(b)

**Figure 6.5** Collection wells for cDNA collections. Because the poly-$T_{25}$ sequence not only captured the poly-A tail of the mRNA but also served as the primers of the on-chip RT reaction, the newly synthesized cDNA was attached to the beads. These beads were flushed to the collection wells. (a) An empty collection well. The dark dye illustrates two microvalves (valve 1 and 2) that control the input of fluids with beads. Closed valve 1 and open valve 2 direct the fluids to waste outlet. (b) Closed valve 2 and open valve 1 direct the fluid with beads into the collection wells.

the cell lysis and mRNA capture procedure. The major disadvantage of fluidic manipulation is that a complex network of flow and control channels must be designed as shown in Figure 6.4. The microvalves in PDMS devices can be opened and closed precisely, but their ability to control the flow rate is compromised due to the length of the flow channel, especially in these very long flow channels.

Recently, we applied OETs to manipulate single cells in microfluidic devices. OET use light-patterned dielectrophoresis for noncontact and precise cell manipulation. It was first reported by Chiou *et al.* for massively parallel manipulation of single cells in an aqueous media [30]. Taking advantage of the transparent PDMS, a light intensity, 5 orders of magnitude, lower than conventional optical tweezers technology can be used to manipulate cells directly in microfluidic devices (Figure 6.5). Computer automation can be easily applied to manipulate single cells.

## 6.4
## Materials, Methods, and Protocols

### 6.4.1
### Materials

#### Mold Fabrication

1) Three-inch silicon wafer (Silicon Quest International, USA).
2) SU-8 photoresist (Microchem, USA).
3) SPR photoresist (Shipley, USA).
4) Spin processor (Laurell Technology Corporation, USA).
5) Hotplate (VWR, USA).
6) MA6 mask aligner (Karl Suss, USA).

**Chip Fabrication**

1) RTV (General Electric, USA).
2) Hybrid mixer (Keyence, USA).
3) Spincoater (Specialty Coating System, USA).
4) Oven (Fisher Scientific, USA).

**Device Operation**

1) Pressure source (Fluidigm, USA).
2) Twenty-three gauge pins (New England Small Tube, USA).
3) Tygon tubing (VWR, USA).
4) NI-DAQ card (National Instruments, USA).
5) Labview software (National Instruments, USA).

**RNA Capture and First Strand cDNA Synthesis**

1) Dynabeads mRNA Direct Kit (Invitrogen, USA).
2) RNasin Plus RNase Inhibitor (Promega, USA).
3) Sensiscript RT Kit (Qiagen, USA).
4) QuantiTect Multiplex PCR NoRox Kit (Qiagen, USA).

### 6.4.2 Methods

**Control Mold**

1) Spin SU8-2025 at 3000 rpm for 45 s.
2) Soft bake mold for 2 min/5 min at 65 °C/95 °C.
3) Expose mold under a transparency mask with the fluidic design for 30 s on MA6 mask aligner.
4) Bake mold postexposure for 2 min/5 min at 65 °C/95 °C.
5) Develop in SU8 nano developer.
6) Once developed, bake mold at 95 °C for 45 s to evaporate excess solvent.

**Flow Mold**

1) Spin SU8-2015 at 3000 rpm for 45 s.
2) Soft bake mold for 1 min/3 min at 65 °C/95 °C.
3) Expose mold 20 s under a transparency mask with the fluidic design on MA6 mask aligner (7 mW $cm^{-2}$).
4) Bake mold postexposure for 1 min/3 min at 65 °C/95 °C.
5) Develop in SU8 nano developer.
6) Hard bake mold at 150 °C for 2 h for the formation of the 10-μm-high flow channels.
7) Expose mold to HMDS vapor for 90 s.
8) Spin SPR220-7 at 1500 rpm for 1 min.
9) Soft bake mold for 90 s at 105 °C.
10) Expose mold for 100 s on MA6 mask aligner.

11) Develop mold in MF-319 developer and rinse under a stream of $H_2O$.
12) Hard bake 3 h at 200 °C for the formation of the 15-μm-high flow channels.
13) Expose mold to HMDS vapor for 90 s.
14) Spin AZ-50 at 1600 rpm for 1 min.
15) Soft bake mold for 1 min/5 min/1 min at 65 °C/115 °C/65 °C, respectively.
16) Expose mold for 4 min on MJB mask aligner.
17) Develop mold in 3 : 1 of $H_2O$ : 2401 developer. Rinse mold under a stream of $H_2O$.
18) Hard bake 3 h at 200 °C for the formation of the 40-μm-high flow channels.

**Chip Fabrication**

1) Prepare 5 : 1 of GE RTV A : RTV B (mix 1 min, defoam 5 min).
2) Expose flow mold to TMCS vapor for 2 min.
3) Pour 30 g of 5 : 1 of GE RTV A : RTV B on respective flow mold.
4) Degas flow mold under vacuum.
5) Bake flow mold for 45 min at 80 °C.
6) Prepare 20 : 1 of GE RTV A : RTV B.
7) Expose control mold to TMCS vapor for 2 min.
8) Spin 20 : 1 RTV mix at 1800 rpm for 60 s (15 s ramp).
9) Bake control mold for 30 min at 80 °C.
10) Cut devices out of flow mold and punch holes with 650-μm-diameter punch tool.
11) Clean the flow device and align to control mold.
12) Bake the resulting two-layer device for 45 min at 80 °C.
13) Prepare 20 : 1 of GE RTV A : RTV B.
14) Expose blank wafer to TMCS vapor for 2 min.
15) Spin 20 : 1 RTV mix on blank wafer at 1600 rpm for 60 s (15 s ramp).
16) Bake blank wafer for 30 min at 80 °C.
17) Cut out the two-layer device from control mold and punch holes with 650-μm-diameter punch tool.
18) Clean the device and mount on blank wafer. Check for debris and collapsed valves.
19) Bake three-layer RTV device for 3 h at 80 °C.
20) Cut three-layer device out, and mount on glass slide.
21) Bake finished device overnight at 80 °C.

### 6.4.3 Device Operation Protocols

#### 6.4.3.1 Microfluidic Chip Control

The on–off valves in the microfluidic chip are controlled by individual pressure sources via 23 gauge pins and tygon tubing. A NI-DAQ card is utilized through a Labview interface to actuate the pressure sources. A microfluidic station is shown in Figure 6.3.

#### 6.4.3.2 Column Construction and Cell Lysis

The whole process of bead column construction and single-cell lysis is shown in Figure 6.4.

1) Add 1 μl (40 units) RNase inhibitor to 99 μl lysis buffer from Dynabeads mRNA Direct kit. Load the resulting lysis buffer with RNase inhibitor into the flow channel from lysis buffer inlet until it reaches the waste outlets.
2) Pelletize 40 μl of beads from Dynabeads mRNA Direct kit with centrifugation and resuspend beads in 40 μl of lysis buffer. Pelletize the beads again and reduce the lysis buffer to 20 μl. Vortex to resuspend the beads. It is important to resuspend the beads before loading into the microfluidic device from beads' inlet.
3) When it is addressed, the multiplexer (inset c2) opens individual flow lines. This allows the bead columns to be stacked in a serial fashion. The sieve valves are actuated when the beads are flowing in the channels. The sieve valves allow the fluid but not the beads to pass. Once the columns are built, excess beads in the flow channels are pushed into the column with lysis buffer to stack into the column (inset b).
4) Pelletize single cell isolated by FACS with centrifugation, and resuspend cells in 100 μl PBS. Pipette up and down gently before loading cells into microfluidic devices through cell inlet. Cells are contained in cell loading portion (inset d). Then the multiplexer (inset c1) is addressed individually to push the single cell into the left part of the ring (inset e). Lysis buffer is then loaded into the right part of the rings. Cells are lysed chemically by mixing cells with lysis buffer in the ring. Mixing occurs by executing a peristaltic pump (inset a) sequentially.

#### 6.4.3.3 Capturing mRNA, Synthesizing First Strand cDNA, and Recovery of cDNA

1) Prepare 90 μl of RT buffer from Sensiscript RT Kit: 9 μl 10× buffer, 9 μl 5 mM dNTP, 4.5 μl reverse transcriptase, 2.25 μl RNase inhibitor, 65.25 μl $H_2O$.
2) Cell lysate is pushed through the oligo$(dT)_{25}$ columns via pneumatic pressure, the mRNAs are captured by attaching to the oligo$(dT)_{25}$ sequences on the beads. Then the columns are washed by first strand cDNA synthesis buffer. The first strand synthesis is carried out by heating the device to 40 °C on a thermal microscope stage. The reaction mixture is flown over the columns for 45 min. The reaction is completed by heating the device to 70 °C for 15 min.
3) The PDMS microfluidic device is peeled off from the supporting glass slide. Individual collection wells are cut off from the devices and placed in microcentrifuge tube beads with the open end face down. Beads are collected by centrifugation.

#### 6.4.3.4 Analysis of Single-cell cDNA

Beads with attached cDNA from single cells were subjected to multiplex quantitative PCR. Absolute number of mRNA molecules in individual cells can be calculated from a standard curve generated with a known amount of the target DNA, and the efficiency of the cDNA synthesis. The efficiency of the cDNA synthesis in microfluidic devices can be calculated by loading a known amount of standard artificial mRNA (such as GeneChip® Poly-A RNA control from Affymetrix, USA) to the device. Because artificial mRNA does not exist in eukaryotic cells, they can be spiked into the lysis buffer as RNA standard. The volume of each cell lysis module is known, and can be used to determine the amount of spike-in mRNA for calculation of cDNA synthesis efficiency. Besides being used for multiplex quantitative PCR, the cDNA obtained can also be amplified and used for whole genome microarray analysis.

**Notes**

1) It is critical to obtain single-cell suspension by FACS. Clumps of cells or debris could clot the flow channels. Owing to the micrometer size of the channels, the clot is very difficult to be cleared, and often makes the chip useless. Therefore, the FACS procedure is not only for isolating the desired cells but also for obtaining single-cell suspension.
2) Appropriate pressure should be used to control the microfluidic chip. The control channels normally work well under a pressure of 18–22 psi at our laboratory; hence, use the pressure that can completely close the valves, but do not let the pressure go too high.
3) Lysis buffer and beads can be loaded to the flow channels at 2 psi, but cells should be loaded in the channel at a lower pressure of 0.3 psi to control the flow rate.
4) During the mixing of the cells and lysis buffer, the pump valves should be operated at a lower pressure of 16 psi (not fully closing the valve at 18–22 psi) to enable an efficient pumping.
5) In the 45-min first strand cDNA synthesis process, always make sure there is enough cDNA synthesis buffer flowing in the channel so that the bead column does not dry up.

## 6.5
## Conclusions

Understanding the biological events inside a single cell is the key to treat many diseases. Multicellular organisms as sophisticated as human beings depend on individual cells for functioning. As an example, cancer often originates from a single malfunctioning cell. The lack of tool for manipulating nanoliters of reagents is one of the major limitations in current biomedical research. Cell lysate, which is a mixture of a heterogeneous cell population, does not provide sufficient information for studying biological events. Many molecular events are masked by the averaging-effects of cell lysate. Studying and understanding the biochemical events inside a single cell require efficient and affordable technology for manipulating nanoliters of reagents.

Microfluidic devices have great potentials to meet the need of single-cell analysis. The simplicity and inexpensive nature make it possible for routine assays in research. With miniaturization, electronic devices become increasingly efficient and affordable. Inexpensive microfluidic devices have the same potential to be broadly adopted in biochemical research, particularly in single-cell analysis. Once single-cell analysis becomes a routine assay like PCR in laboratories, many biological events will be studied inside individual cells directly rather than being inferred from cell lysate. These microfluidic systems will revolutionize the research of molecular biology.

## References


1 Zhong, J.F., Chen, Y., Marcus, J.S. *et al.* (2008) A microfluidic processor for gene expression profiling of single human embryonic stem cells. *Lab Chip*, **8**, 68–74.

2 Emmert-Buck, M.R., Bonner, R.F., Smith, P.D. *et al.* (1996) Laser capture microdissection.[comment]. *Science*, **274**, 998–1001.

3 Brady, G., Iscove, N.N., Brady, G. *et al.* (1993) Construction of cDNA libraries from single cells. *Methods Enzymol.*, **225**, 611–623.

4 Geigl, J.B., Speicher, M.R., Geigl, J.B. *et al.* (2007) Single-cell isolation from cell suspensions and whole genome amplification from single cells to provide templates for CGH analysis. *Nat. Protoc.*, **2**, 3173–3184.

5 Kurimoto, K., Yabuta, Y., Ohinata, Y. *et al.* (2007) Global single-cell cDNA amplification to provide a template for representative high-density oligonucleotide microarray analysis. *Nat. Protoc.*, **2**, 739–752.

6 Uemura, E. (1980) Age-related changes in neuronal RNA content in rhesus monkeys (Macaca mulatta). *Brain Res. Bull.*, **5**, 117–119.

7 Brady, G. (2000) Expression profiling of single mammalian cells–small is beautiful. *Yeast*, **17**, 211–217.

8 Unger, M.A., Chou, H.P., Thorsen, T. *et al.* (2000) Monolithic microfabricated valves and pumps by multilayer soft lithography. *Science*, **288**, 113–116.

9 Li, Y., Welm, B., Podsypanina, K. *et al.* (2003) Evidence that transgenes encoding components of the Wnt signaling pathway preferentially induce mammary cancers from progenitor cells. *Proc. Natl. Acad. Sci. U.S.A.*, **100**, 15853–15858.

10 Behbod, F. and Rosen, J.M. (2005) Will cancer stem cells provide new therapeutic targets? *Carcinogenesis*, **26**, 703–711.

11 O'Brien, C.A., Pollett, A., Gallinger, S. *et al.* (2007) A human colon cancer cell capable of initiating tumour growth in immunodeficient mice. *Nature*, **445**, 106–110.

12 Al-Hajj, M., Wicha, M.S., Benito-Hernandez, A. *et al.* (2003) Prospective identification of tumorigenic breast cancer cells.[see comment][erratum appears in Proc Natl Acad Sci U S A. 2003 May 27;100(11):6890]. *Proc. Natl. Acad. Sci. U.S.A.*, **100**, 3983–3988.

13 Singh, S.K., Clarke, I.D., Terasaki, M. *et al.* (2003) Identification of a cancer stem cell in human brain tumors. *Cancer Res.*, **63**, 5821–5828.

14 Kong, Y., Yoshida, S., Saito, Y. *et al.* (2008) CD34+CD38+CD19+ as well as CD34+CD38-CD19+ cells are leukemia-initiating cells with self-renewal capacity in human B-precursor ALL. *Leukemia*, **22**, 1207–1213.

15 Kartalov, E.P., Zhong, J.F., Scherer, A. *et al.* (2006) High-throughput multi-antigen microfluidic fluorescence immunoassays. *Biotechniques*, **40**, 85–90.

16 Zhong, J.F., Maltezos, G., Sheriff, Z. *et al.* (2006) Microfluidic systems for studying cell migration regulation. *Lett. Drug Des. Discov.*, **3**, 636–639.

17 Chiang, M.K. and Melton, D.A. (2003) Single-cell transcript analysis of pancreas development. *Dev. Cell*, **4**, 383–393.

18 Kamme, F., Salunga, R., Yu, J. *et al.* (2003) Single-cell microarray analysis in hippocampus CA1: demonstration

and validation of cellular heterogeneity. *J. Neurosci.*, **23**, 3607–3615.
19 Luo, L., Salunga, R.C., Guo, H. *et al.* (1999) Gene expression profiles of laser-captured adjacent neuronal subtypes.[erratum appears in Nat Med 1999 Mar;5(3):355]. *Nat. Med.*, **5**, 117–122.
20 Tietjen, I., Rihel, J.M., Cao, Y. *et al.* (2003) Single-cell transcriptional analysis of neuronal progenitors. *Neuron*, **38**, 161–175.
21 Adjaye, J., Daniels, R., Bolton, V. *et al.* (1997) cDNA libraries from single human preimplantation embryos. *Genomics*, **46**, 337–344.
22 Al-Taher, A., Bashein, A., Nolan, T. *et al.* (2000) Global cDNA amplification combined with real-time RT-PCR: accurate quantification of multiple human potassium channel genes at the single cell level. *Yeast*, **17**, 201–210.
23 Brady, G., Billia, F., Knox, J. *et al.* (1995) Analysis of gene expression in a complex differentiation hierarchy by global amplification of cDNA from single cells.[erratum appears in Curr Biol 1995 Oct 1;5(10):1201]. *Curr. Biol.*, **5**, 909–922.
24 Dixon, A.K., Richardson, P.J., Lee, K. *et al.* (1998) Expression profiling of single cells using 3 prime end amplification (TPEA) PCR. *Nucleic Acids Res.*, **26**, 4426–4431.
25 Eberwine, J., Yeh, H., Miyashiro, K. *et al.* (1992) Analysis of gene expression in single live neurons. *Proc. Natl. Acad. Sci. U.S.A.*, **89**, 3010–3014.
26 Schmidt-Ott, K.M., Tuschick, S., Kirchhoff, F. *et al.* (1998) Single-cell characterization of endothelin system gene expression in the cerebellum in situ. *J. Cardiovasc. Pharmacol.*, **31**, 364–366.
27 Trumper, L.H., Brady, G., Bagg, A. *et al.* (1993) Single-cell analysis of Hodgkin and Reed-Sternberg cells: molecular heterogeneity of gene expression and p53 mutations. *Blood*, **81**, 3097–3115.
28 Chou, H.P., Unger, M.A., and Quake, S.R. (2001) A microfabricated rotary pump. *Biomed. Microdevices*, **3**, 323–330.
29 Marcus, J.S., Anderson, W.F., and Quake, S.R. (2006) Parallel picoliter rt-PCR assays using microfluidics. *Anal. Chem.*, **78**, 956–958.
30 Chiou, P.Y., Ohta, A.T., Wu, M.C. *et al.* (2005) Massively parallel manipulation of single cells and microparticles using optical images. *Nature*, **436**, 370–372.

# 7
# On-chip Electroporation and Electrofusion for Single-cell Engineering

*Ana Valero and Albert van den Berg*

## 7.1
## Introduction

Microfabrication and microfluidic techniques are used for many applications in biology and medicine as tools for molecular biology, biochemistry, and cell biology, or as medical devices and biosensors [1, 2]. Microstructured devices provide significantly enhanced functionality with respect to conventional devices such as lower use of expensive reagents, biomimetic environments, and the ability to manipulate single cells. Moreover, microfabrication enables devices with novel capabilities. These enhancing and enabling qualities are conferred when microfabrication is used appropriately to address the right types of problems [3]. Microfluidics has provided the research community with new methods, especially those based upon laminar flow and diffusion [4, 5], electrokinetic [6–10], and dielectrophoretic effects [11–14]. Utilization of microfluidics and microfabrication in lab-on-a-chip (LOC) applications has increased the potential of high-throughput biochemical assays on individual mammalian cells. Of particular interest is the ability to parallelize up-front assay protocols and still be able to examine and treat every individual cell in the assay separately retrieving single-cell event information [15, 16].

Electroporation as well as electrofusion are broadly used techniques in biology, based on the application of controlled electrical fields to cells; these techniques also benefit from microtechnology. Many adverse effects associated with macroscale electroporation such as local pH variation near electrodes, sample contamination, electric field distortion, and low cell viability can be eliminated by using these techniques in microscale environments. Moreover, *in situ* visualization of the process, real-time monitoring of intracellular processes, low consumption of reagents, and the ability to spatially and temporally control electrical parameters at the single-cell level are several advantages of potential benefits for research areas, also providing rapid optimization of transfection protocols and enhanced cell viability [17–19].

Electroporation is a powerful tool for gene transfection that uses electrical pulses. Currently used methods to introduce foreign DNA into mammalian cells are based

*Chemical Cytometry*. Edited by Chang Lu

ISBN: 978-3-527-32495-8

on bulk procedures in which a large number of cells are simultaneously transfected, electroporated, or virally infected. All of these methods have a number of specific limitations such as limited control over the amount of DNA uptake, the intracellular half-life, and fate of the introduced DNA and site of genomic integration. These limitations represent a serious drawback in situations where genetically modified stem cells have to be produced for therapeutic application including gene therapy and regenerative medicine, especially when these cells are hard to isolate in large enough numbers. Recently, microfluidic devices have shown great benefits for studying a variety of cell processes [20, 21]. Of particular importance is the use of such LOC devices for electroporation, enabling high-efficiency transfer of a variety of macromolecules into cells [19, 22–25].

Cell fusion provides a unique technique to combine the genetic and epigenetic information of two different cells. Cell fusion is used for many different purposes, including generation of hybridomas and reprogramming of somatic cells [26–28]. Standard fusion techniques, however, provide poor and random cell contact, leading to low yields and preventing detailed studies of fusion-mediated reprogramming. The use of LOC devices brings new advantages that are mainly related to the efficiency of the electrofusion process, of which one of the most important advantages is that microsystems allow proper pairing of cells, that is, in close contact, thereby achieving extremely specific fusion.

In this chapter, attention is focused on reviewing current microtechnology-based single-cell engineering devices and enabling LOC technologies, and in particular microfabricated fluidic devices for gene transfer using electroporation, as well as for characterization of fusion-mediated reprogramming of somatic cells.

## 7.2
## Single-cell Electroporation in Microfluidic Devices

Electroporation is the phenomenon that induces breakdown of the cell membrane lipid bilayer, which results in the formation of transient or permanent pores in the membrane; molecules can enter and leave the cell during this permeabilized state [29, 30]. While bulk electroporation appears to be a universal phenomenon, the outcome of an electroporation protocol is cell type–specific and varies among cells in a given population: each cell is affected separately and shows its own characteristic response to the external electric field [31]. Such individuality in response depends on the dimension of the cell, its shape, its relative position to the direction of the electric field, and the structure of particular parts of the membrane [32]. Moreover, there is no effective and real-time feedback on the permeability status of the cell membrane during and shortly after electroporation. The most important drawback of conventional bulk electroporation is, however, that it works with batches of thousands of cells, leading to low cell survival and low cell transfection efficiency.

These and other deficiencies of bulk electroporation can be resolved by performing electroporation in individual cells. Therefore, single-cell electroporation is an interesting and promising approach that opens up a new window of opportunities in manipulating the genetic, metabolic, and synthetic contents of single targeted cells in tissue slices, cell cultures, in microfluidic channels or at specific loci on a chip-based device.

Several methods to target single cells by electroporation have been reported: solid microelectrodes [33], electrolyte-filled capillaries [34], micropipettes (patch-clamp electrodes) [35, 36], and a diversity of chip structures. These techniques make it possible to electroporate a cell without affecting adjacent cells, and thus provide ways to manipulate the biochemical content of single cells and even to subcellular structures (such as organelles), *in vivo* and *in vitro*.

For successful single-cell electroporation, the cell must either be isolated or the electric field must be focused well to target a particular cell. Microfabricated devices are ideally suitable to fulfill both functions of isolating single cells as well as focusing of the electric field. In addition, the use of microfluidic devices for cell electroporation applications offers clear advantages compared to common electroporation setups (Table 7.1). First of all, by applying microelectronic pattern techniques, the distances between the electrodes in the microdevices can be made very small, which means that relatively low potential differences are sufficient to give high electric field strengths in the regions between the electrodes. The electrical design of the pulser is therefore much simpler, which makes it possible to choose from a wider range of pulsers than the common block or exponential decay pulse shapes to be used [37]. Cell handling and manipulation are also easier, as the channels and electrode structures are comparable to the sizes of the cells. As the hydrodynamic regime used in microfluidics is much different than the flow regimes in large-scale equipment, it is possible to make use of specific hydrodynamic effects associated with the laminar regime. The coupling between cell electroporation and separation or detection of the released components is also more direct, as it can be integrated onto the same chip. This makes it possible even to trap single cells and to determine intracellular content or other properties, which is hardly feasible using conventional laboratory-scale equipment. In addition, only small amounts of cells and difficult-to-produce reagents, such as specific plasmids, are needed. On top of this, *in situ* optical inspection and real-time monitoring of the electroporation process (using fluorescent probes, for example) is possible, as the microdevices can be made transparent. Finally, the surface-to-volume ratio in microdevices is relatively large, which results in faster heat dissipation per unit surface area. This makes it possible to distinguish between heat and electric field effects.

In recent years, several publications on microfluidic devices have focused on the process of electroporation, which results in the poration of the biological cell membrane. The devices involved are designed for cell analysis, transfection, or inactivation. Detection is usually achieved through fluorescent labeling or by measuring impedance. So far, most of these devices were merely focused at the electroporation process, but integration with separation and detection processes is expected in the near future. In particular, single-cell content analysis is expected to add further value to the concept of the microfluidic chip. Furthermore, if advanced

**Table 7.1** Electroporation and electrofusion: conventional versus micro.

| Properties | Conventional | Microscale |
|---|---|---|
| Efficiency | | |
| Viability | + | +++ |
| Transfection | + | +++ |
| Electrical | | |
| Electric field form | + (distorted at edge) | +++ (uniform, well focused) |
| Electrode surface | + Large area | ++ smaller area |
| Electrode material | Aluminum | Platinum, gold, Ag/AgCl, n+ polysilicon |
| Chemical | | |
| pH variation | + (often occurs) | +++ (not often) |
| Water dissociation | + (highly) | +++ (little) |
| Metal ions | + ($Al^{3+}$ dissolved) | +++ (no ions dissolved) |
| Optical | | |
| *In situ* visualization | + | +++ |
| Real-time monitoring | + | +++ |
| Others | | |
| Sample volume | Large | Small |
| Cell handling | + | +++ |
| Heat dissipation | + | +++ |
| Throughput | + | +++ (high) |
| Integration | N.D. | +++ |
| Disposability | + | +++ |
| Cost | + | ++ |

+++: high or excellent; ++: medium or good; +: low or poor. N.D.: not determined.

pulse schemes are employed, such microdevices can also enhance research into intracellular electroporation.

### 7.2.1
### Microdevices for Analyzing Cellular Properties or Intracellular Content

Applications of electroporation microdevices vary from those devices which investigate true electroporation properties, like the pore formation process, to applications where electroporation is only an aid to further analysis of the cell content.

Analysis of intracellular materials at the single-cell level presents opportunities for probing the heterogeneity of a cell population. While cell lysis can be accomplished using methods like chemical lysis [38], electroporation has gained popularity as a rapid method for disruption of the cell membrane and release of intracellular contents, since no chemicals need to be added to the system, which could disturb the measurements. Gao *et al.* [39] used electroporation in a microfluidic device to release the cellular content. Their design consists of a simple crossed channel, in which erythrocyte cells were loaded using a pressure gradient. When a cell arrives at the crossing, an electroosmotic flow (EOF) was used to direct the cell into the separation channel, after which the flow is stopped for 15 s to allow the cell to attach to the wall. The cell was then electroporated using a 1-s 1400-V pulse. In this way, the single-cell glutathione content was measured in a reproducible way without the need to use disturbing lysing agents. McClain *et al.* [40] reported on a capillary electrophoresis (CE) chip for single-cell analysis (Figure 7.1a), in which continuous 450 V $m^{-1}$ square wave pulses with a DC offset of 675 V $m^{-1}$ were used. The DC offset provided the necessary potential for separation, while the pulses were used for electroporation. Using this device, cells which were previously loaded with several fluorescent stains were electroporated, which was followed by the separation and measurement of the stains.

Lu *et al.* [41] developed a microelectroporation device for cell lysis based on a sawtooth microelectrode structure to enhance the electric field strength. A straight, 50-μm-high microchannel of the polymer SU-8 was constructed on Pyrex glass, and the side walls consisted of gold sawtooth-shaped electrodes with a tip distance of 30 μm, supported by the polymer SU-8 (Figure 7.1b). Using pressure-driven flow, cells were directed through the channel and electroporated at the place where the electrodes are closest to each other (thus, where the electric field strength is the highest). To avoid electrolysis in the channel, a continuous alternating voltage of 6–8.5 V at 5–10 kHz was applied to electroporate the cells. It was possible with

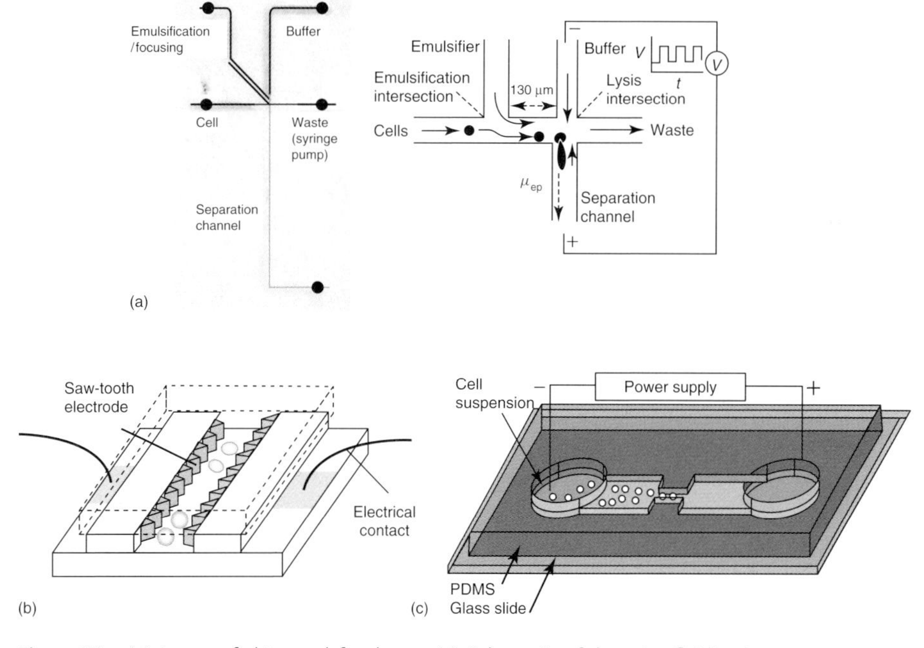


**Figure 7.1** (a) Image of chip used for the cell analysis experiments and schematic of the emulsification and lysis intersections for the microchip design [40]. (b) Schematic of electroporation microdevice for cell lysis [41]. (c) Schematic of the microfluidic electroporative flow cytometry setup. Electroporation occurs in the narrow section of the microfluidic channel when cells flow through [42].

this device to electroporate human carcinoma (HT-29) cells, as assessed using fluorescent acridine orange and propidium iodide staining.

More recently, Bao *et al.* [42] have combined a flow-through electroporation device with flow cytometry (Figure 7.1c) for selective release of intracellular molecules under different electroporation parameters at the single-cell level with a high throughput (~200 cells/s). They examined the release of a small molecule, calcein (MW 600), and a 72-kDa protein kinase, Syk, tagged by enhanced green fluorescent protein (EGFP) from chicken B cells during electroporation at the single-cell level. The effects of the field intensity and the field duration on the release of the two molecules were studied; calcein was released in general at lower field intensities and shorter durations than did SykEGFP. Thus, by tuning the electrical parameters they were able to deplete calcein from the cells before SykEGFP started to release. This approach potentially provides a high-throughput alternative for probing different intracellular molecules at the single-cell level compared to chemical cytometry by eliminating complete disruption of the cell membrane. Using this technique, microfluidic electroporative flow cytometry, single-cell biomechanics [43], and detection of kinase translocation [44] in B cells was also achieved. Owing to the frequent involvement of kinase translocations in disease processes such as oncogenesis, this approach will have utility for kinase-related drug discovery and tumor diagnosis as well as for mechanistic studies of cytoskeleton dynamics in diagnosis and staging of cancers cells in general.

In contrast to the above-discussed microdevices that evidence the on-chip single-cell electroporation principle, other electroporation microdevices have focused on analyzing and understanding the electroporation process itself. Membrane integrity analysis is often performed optically, by measuring the uptake or release of fluorescent markers such as YOYO-1, PI, acridine orange, FLICA, calcein AM or CF. Measuring impedance is another technique that is often used to follow electroporation; this has the advantages of a fast, on-line response, and a noninvasive nature [24, 45–47].

### 7.2.2
### Electroporation Microdevices for Cell Inactivation

In the food processing industry, electroporation is used for the pasteurization of liquid foods in what is known as *pulsed electric field* (PEF) processing [48]. Pasteurization is used to render all spoilage bacteria present in the liquid foods inactive. Therefore, irreversible electroporation of all of the cells needs to take place. The first PEF microreactor was presented by Fox *et al.* [49], which consisted of a 50-μm-deep channel in glass with a 10-μm-deep, 30-μm-long constriction to focus the electric field between the two electrodes (Figure 7.2). It was possible to make a comparison with a preexisting laboratory setup [50], with a typical constriction size of 1 × 2 mm, using artificial vesicles loaded with carboxyfluorescein as a model system. This comparison showed that, despite the difference in length scales, the two devices were comparable when 2-μs square wave pulses of 0–800-V were used. Vesicle electroporation in both devices was studied using the transmembrane

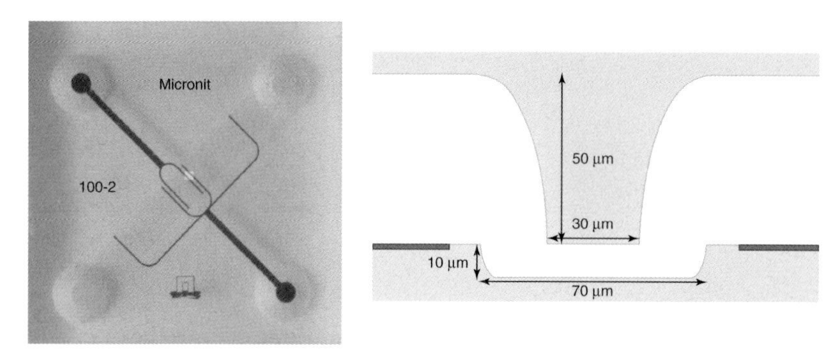


**Figure 7.2** Image and schematic drawing of the PEF microreactor with dimensions.

potential and the total amount of energy added as criteria for comparison. The transmembrane potential turned out to be a good parameter to use when comparing vesicle breakdown as it describes effects happening in the vesicle itself, which eliminates structural effects. Although microtechnology seems a less obvious choice for PEF applications because voluminous flows are processed, the use of microdevices avoids the risks involved with using high voltages and causes any heat generated to be rapidly dissipated. Because of this, microdevices are a tool for process optimization and should aid the exploration of other possibilities in the field of PEF.

### 7.2.3
### Electroporation Microdevices for Gene Transfection

Transfection of cells with foreign DNA is often accomplished by reversible electroporation [51]. However, the treatment protocols are often suboptimal [29] and based upon the application of long-duration pulses (microseconds) with relatively low electric field strengths, which results in an excess amount of inactivated cells. Furthermore, cells exposed to electric fields can be sensitive to substances in the medium such as $Al^{3+}$ ions, which can become solubilized from the electrodes [52]. It is possible to control the circumstances better in microfluidic devices, and hence increase the efficiency of transfection. Besides this advantage, only small amounts of transfection material are needed, and it is possible to make structures where the transfection of more cells in a parallel fashion is possible. Several designs of microdevices for transfection are published, some aiming at single-cell transfection, others at the transfection of larger amounts of cells.

Lin *et al.* [53] constructed a device made of poly(methyl methacrylate) (PMMA) that consisted of a 0.2-mm-high, 5-mm-wide channel with integrated gold electrodes at the top and the bottom of the channel at the electroporation spot (Figure 7.3a). Since the electrode distance was relatively small, only 10-ms pulses of 10 V were required for electroporation. The efficiency of this simple design was proven by transfecting human hepatocellular carcinoma cells (Huh-7) with $\beta$-galactosidase and green fluorescent protein (GFP) genes.

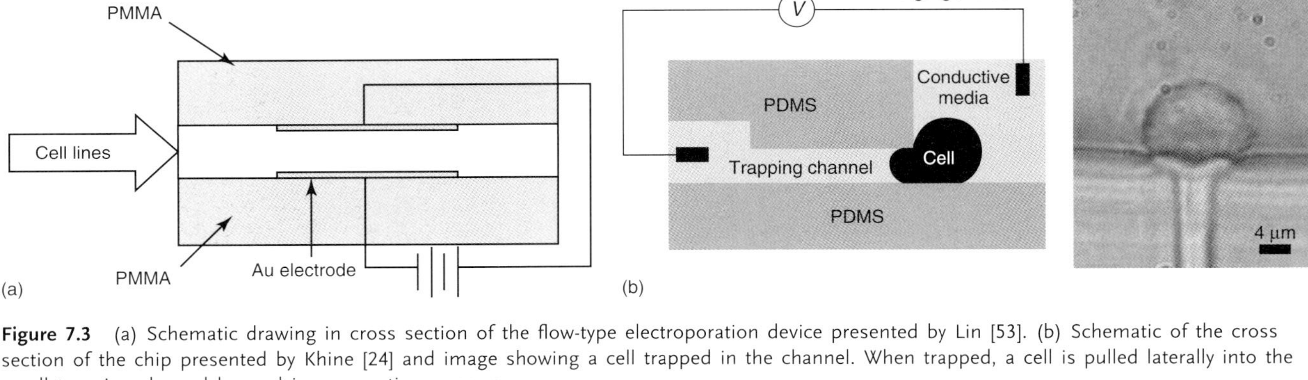


**Figure 7.3** (a) Schematic drawing in cross section of the flow-type electroporation device presented by Lin [53]. (b) Schematic of the cross section of the chip presented by Khine [24] and image showing a cell trapped in the channel. When trapped, a cell is pulled laterally into the small trapping channel by applying a negative pressure.

While Lin *et al.* used a small electrode distance to focus the electric field, it is also possible to accomplish focusing the electric field by introducing a constriction between the two electrodes. Khine *et al.* [18, 24] used this concept in their design (Figure 7.3b), which was originally developed as a multiple patch-clamp array [54]. Although the constriction itself increases the electric field strength, this effect was enhanced in this design by applying a gentle under pressure and sucking a cell partially into the constriction, thereby blocking the constriction completely. The cell could not pass the constriction because the cell diameter (12–17 μm) was approximately four times larger than the constriction (3.1 μm). Low potentials of less than 1 V could be applied using an Ag/AgCl electrode. The release of calcein and the uptake of trypan blue from HELA cells after electroporation were followed visually, but transfection with DNA has not yet been reported with this device.

Huang and Rubinsky, who designed an electroporation analysis device with a microhole in a silicon nitride membrane, also adapted this analysis-oriented design [46] to make it applicable to cell transfection. This was done by creating a flow-through channel on top of a silicon nitride membrane that was approximately 1.5 times the size of a cell (Figure 7.4). The hole in the membrane is situated in the middle of the channel. Once a cell was brought into the microchannel, it was captured in the microhole by a backside pressure, electroporated, uploaded with the desired foreign molecules and then released, exiting the channel on the other side. The microhole in the silicon nitride membrane provided the necessary

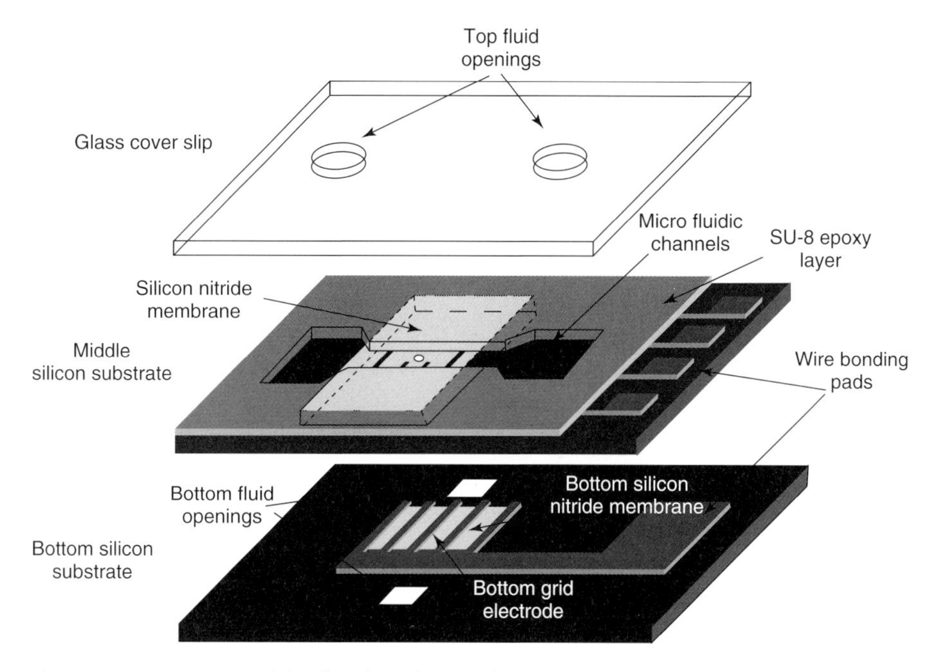


**Figure 7.4** Construction of the flow-through microelectroporation chip with a microhole in a silicon nitride membrane and microfluidic channels for precise cell transport [46].

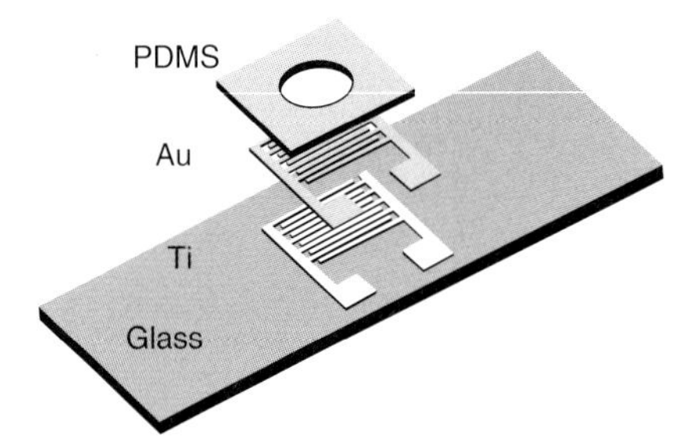


**Figure 7.5** Schematic of the microfluidic chip presented by Lin [55, 56] aimed at the transfection of cells growing on a surface.

enhancement of the electric field. With this microdevice, it was possible to stain cells with fluorescent YOYO-1 using 100-ms, 10-V pulses and transfect them with an enhanced GFP gene.

Whereas the above-mentioned transfection-oriented designs were aimed at cells in solution, Lin *et al.* [53, 55, 56] created a microfluidic chip aimed at the transfection of animal cells growing on a solid surface (Figure 7.5). The chip consisted of a glass wafer with a gold interdigitated electrode structure, which was sealed with a PDMS (polydimethylsiloxane) mold to form a cavity. Cells grew on the glass surface, and by the interdigitated electrode structure could be used to electroporate the surface bound cells. In this way, it was possible to transfect Huh-7 cells, human embryonic kidney cells, and Human Umbilical Vein Endothelial Cells (HUVEC) primary cells with GFP-GFP-DUA DNA. By adding an extra anode electrode above the interdigitated structure, negative DNA plasmids were directed to the cathodes by an electrophoretic potential [56] prior to electroporation, creating a local high concentration of plasmids near the cells at the cathodes. Improved cell transfection was demonstrated by the relatively high concentration of transfected cells near the cathodes as compared to experiments where no electrophoretic forces were used.

Valero *et al.* [19] reported the first microfluidic device capable of transfecting single human mesenchymal stem cells (hMSCs) with high efficiency, while maintaining the viability and the ability of the cell to respond to changes (factors) in its environment. The microfluidic device contained two channels that were connected by microholes, which acted as trapping sites for living cells. The electrodes were positioned such that individual traps could be electrically addressed. Individual cells were successfully electroporated, resulting in expression of GFP-ERK1, in over 75% of the cells. Extracellular signal-regulated kinase, ERK1 is a signaling molecule that is transported from the cytoplasm to the nucleus upon stimulation with external factors like growth factors. In the nucleus, it functions to activate gene transcription. Upon stimulation with fibroblast growth factor (FGF)-2, EGFP-ERK1 was translocated to the nucleus of the hMSCs, while no nuclear translocation was shown in cells electroporated with an EGFP control vector (Figure 7.6). These results demonstrate that the trapped cells survive the electroporation procedure and exhibit

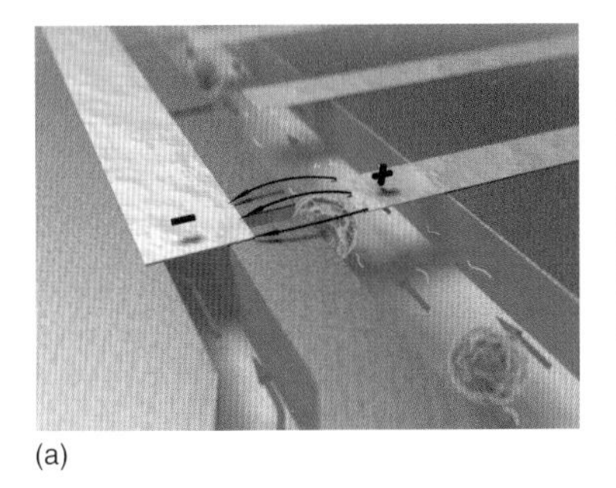

(a)

(b)

**Figure 7.6** (a) Artistic 3D impression of the microfluidic device for single-cell electroporation and gene transfection studies presented by Valero [19]. (b) Fluorescent image of two trapped and transfected hMSCs; EGFP-ERK1 localized along internal structures in the cytoplasm.

biological responses, as shown by others [57, 58]. Moreover, they found indications that protein transport occurs via internal structures in the cytoplasm, possibly actin filaments, indicating that the single-cell experiments can yield information about the mechanism of ERK1 nuclear transport.

## 7.3 Single-cell Electrofusion in Microfluidic Devices

There are a number of methods for carrying out fusion of cells such as chemical (use of polyethylene glycol; PEG) [59, 60], biological (using viruses or receptors) [27, 61], or physical (the use of focused laser beams [62, 63] or applying PEFs [64, 65]). Of these methods cell electrofusion has gained popularity due to its ease of implementation and high efficiency when compared to the other methods. To carry out electrofusion, a suspension of cells (usually at a density of $\sim 10^5$ cells/ml) placed in a fusion chamber are first brought into physical contact by dielectrophoresis using a low-amplitude ($\sim$100–300 V $cm^{-1}$), high-frequency ($\sim$1–3 MHz) AC field. Subsequent application of a short-duration ($\sim$10–50 μs), high-intensity ($\sim$1–10 kV $cm^{-1}$) electric pulse then causes a fraction of the cells that are in close contact to fuse [66]. One of the main drawbacks of this method is therefore the inability of manipulating and fusing cell pairs selectively since it relies in random cell–cell pairing. Moreover, low overall fusion efficiencies are achieved and require antibiotic selection and lengthy subculturing to isolate the desired hybrids. Improving the process of cell fusion not only depends on the mechanism used for initiating membrane fusion but also on controlling how the cells are brought into contact and properly paired. Single cells can be manually immobilized and paired [67] giving precise fusion partners, but this work is tedious and time consuming and results in low throughput of fused cells.

Microfluidic systems provide and attractive and versatile platform for the manipulation, isolation, and transport of selected cells prior their electric field–induced fusion.

There have been previous attempts to take advantage of microfluidics for cell pairing, using either flow-through or immobilization techniques to improve cell contact. Flow-through approaches, in which cells are brought into contact through alternating current fields or biotin–streptavidin coatings, demonstrate that higher membrane fusion efficiencies can be achieved [68–71]. A brief description of these devices is given below.

Tresset *et al.* [69, 70] describe a microdevice comprising heavily doped silicon electrodes sealed between glass substrates, where femtoliter lipid vesicles encapsulating a fluorescent marker could be aligned, porated, and fused (Figure 7.7a). They envision this technology as a flexible alternative to emulsion-based methods of drug and molecule compartmentalizations for complete analysis.

Wang and Lu [71] reported on a microfluidic platform for cell-to-cell electrofusion using a common direct current power supply. In their method, the cells were first conjugated based on biotin–streptavidin interaction, and electrofusion was then conducted by flowing the linked cells through a simple microfluidic channel with geometric variation (Figure 7.7b–c) under continuous direct current voltage.

Techaumnat *et al.* [72] presented the use of electric field constriction created by a microfabricated structure made of SU-8 and PDMS to realize high-yield electrofusion of biological cells (Figure 7.7d–e). The method used an orifice on an electrically insulating wall (orifice plate) whose diameter is as small as that of the cells. Owing to the field constriction created by the orifice they could induce the controlled magnitude of membrane voltage selectively around the contact point, regardless of the cell size. The field constriction also ensured 1 : 1 fusion even when more than two cells were forming a chain at the orifice. Experiments using plant protoplasts or mammalian cells show that the process was highly reproducible, and a yield higher than 90% was achieved.

However, these approaches lack the ability to properly pair and fuse unmodified cells, and the overall yield of desired fusions remains low. Recently, Skelley and Kirak [73] have introduced a microfluidic device containing a dense array of weir-based passive hydrodynamic cell traps for cell electrofusion. By combining these cell-trapping cups and a three-step loading protocol, thousands of cells could be immobilized and paired at once. The device provided insight into the cell fusion process and allowed the researchers to decouple fluorescence exchange and membrane reorganization. Since the device is also compatible with both chemical and electrical fusion protocols, comparison of the fusion efficiencies was also possible. The utility of this microdevice for pairing and fusing different cell types was demonstrated for several cell lines as, for example, NIH3T3 fibroblasts (3T3s), myeloma cells, B cells, mouse embryonic stem cells (mESCs), and mouse embryonic fibroblasts (mEFs), improving fusion efficiencies by up to >50%. Moreover, fused cells maintained their viability and morphology off-chip, and reprogramming of mEFs was observed. Therefore, this device is foreseen as innovative tool to characterize fusion-mediated reprogramming of somatic cells (Figure 7.8).

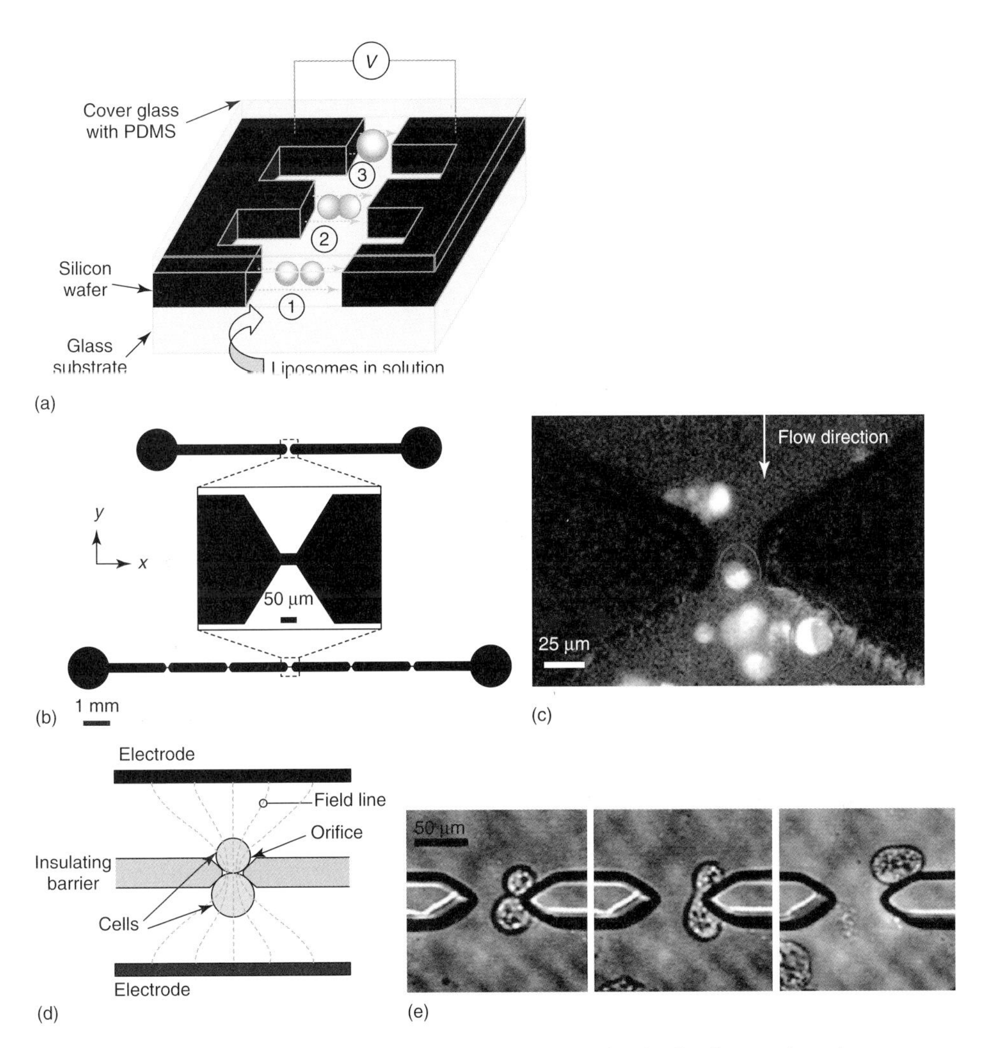


**Figure 7.7** (a) Schematic view of the microfluidic device for electrical manipulation of lipid vesicles [69, 70]; (1) liposomes alignment (AC voltage), (2) membrane breakdown (DC pulses), and (3) hybrid vesicle formation. (b) Layout of single- and five-pulsed microfluidic devices for cell electrofusion with a narrow section of 50 × 40 µm [71]. (c) Fluorescent image of fused cells, between one cell labeled by calcein AM and one unlabeled, immediately after flowing through the narrow section. (d) Schematic of the microfluidic device for cell electrofusion based on field constriction [72]. (e) Fusion between two cells smaller than the orifice; cells attracted to the orifice surface by dielectrophoresis, breakdown of membrane at the contact point due to electric pulse and cell released from the orifice.

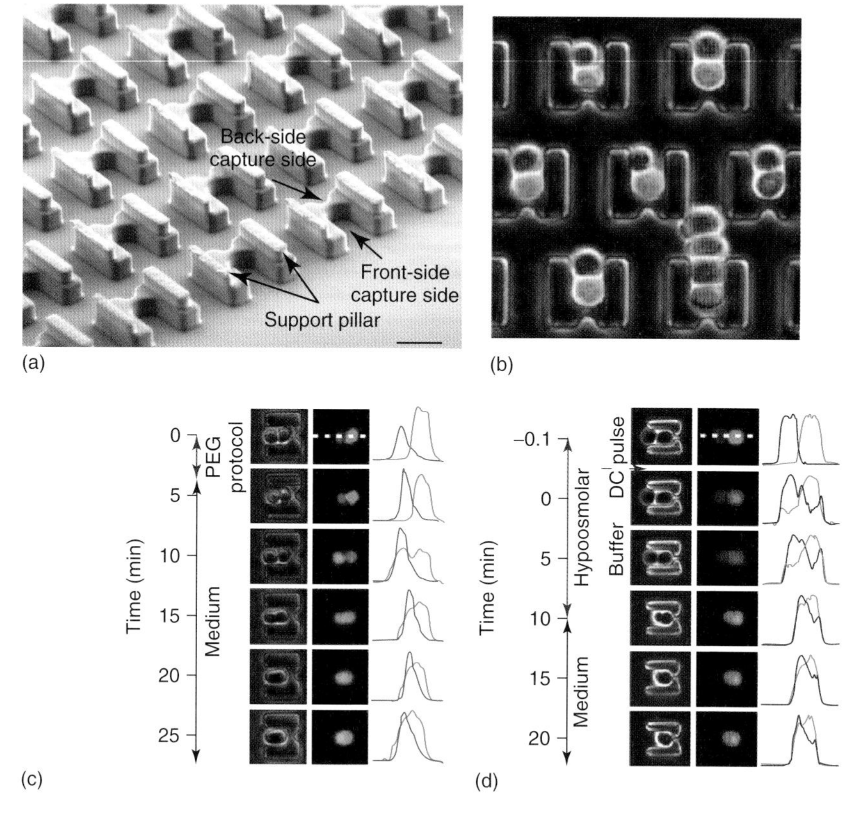


**Figure 7.8** (a) Scanning electron micrograph of the microfluidic device for cell capture and pairing. (b) Red and green fluorescent image overlay of 3T3s cells loaded and capture. (c) PEG fusion of GFP-expressing mESCs and Hoechst-stained mEFs. Phase-contrast images show the status of membrane reorganization and fluorescence overlay images and line scans through the cells (dashed line) demonstrate the exchange of fluorescence. (d) Electrofusion of DsRed- and eGFP-expressing mouse 3T3s. Immediately after the fusion pulse, exchange of fluorescence was observed, outlining the nuclei of the cells [73].

## 7.4 Conclusions

The combination of microfluidic devices/chips and biology is of potential benefit in research areas such as biotechnology, life sciences, drug delivery, and so on. In the past few years, various approaches have been developed to advance these fields in a synergistic manner, of which microdevices for single-cell electroporation and electrofusion are discussed here.

Owing to the successful and promising results, we predict that microfluidic devices can be used for highly efficient small-scale "genetic modification" of cells and biological experimentation, offering possibilities to study cellular processes at the single-cell level. Future applications might be small-scale production of cells for therapeutic application under controlled conditions. Since the experimental conditions for manipulation of cells in chips can be tightly controlled while maintaining normal physiological responses, microfluidic devices offer prospects to study dynamic processes at the single-cell level.

It is expected that integrated devices where combinations of electroporation, separation, and analysis occur will emerge, such as devices with integrated chromatography, electrophoresis or isoelectric focusing steps for separation, and mass spectroscopy, electrochemical, and fluorescent methods for analysis. Present designs usually require multiple manual steps in order to insert the cells, electroporate them, and measure the effects. Until now, no devices have been reported where all of these steps have been integrated in an automated way, preferably with multiple samples in parallel, which could greatly enhance the application of microtechnological analysis.

One advantage in combining microfluidics with cell fusion is the possibility for creating rapidly and efficiently a combinatorial library of fused (or porated) cells. With the increasing use of monoclonal antibodies, intense efforts in proteomics and the pressing need for a high-throughput cell-based format to screen for drug leads, engineered cell lines with diverse yet well-controlled characteristics will become ever more critical in these applications. Therefore, by providing a convenient platform on which different techniques for cell fusion and poration can be integrated easily with powerful fluidic methods for cell manipulation and analyses, microfluidics-based cell fusion is well poised to address these emerging challenges [66].

## References


1. Andersson, H. and van den Berg, A. (2004) *Lab on a Chips for Cellomics*, Kluwer Academic Publishers.
2. Li, P.C.H. (2006) *Microfluidic Lab-on-a-Chip for Chemical and Biological Analysis and Discovery*, vol. **94**, CRC Press.
3. Voldman, J., Gray, M., and Schmidt, M. (1999) Microfabrication in biology and medicine. *Annu. Rev. Biomed. Eng.*, **1**, 401–425.
4. Helton, K.L. and Yager, P. (2007) Interfacial instabilities affect microfluidic extraction of small molecules from non-Newtonian fluids. *Lab Chip*, **7**, 1581–1588.
5. Munson, M.S., Cabrera, C.R., and Yager, P. (2002) Passive electrophoresis in microchannels using liquid junction potentials. *Electrophoresis*, **23**, 2642–2652.
6. Dittrich, P.S., Tachikawa, K., and Manz, A. (2006) Micro total analysis systems. Latest advancements and trends. *Anal. Chem.*, **78**, 3887–3907.
7. Ohno, K., Tachikawa, K., and Manz, A. (2008) Microfluidics: applications for analytical purposes in chemistry and biochemistry. *Electrophoresis*, **29**, 4443–4453.
8. Pennathur, S. and Santiago, J.G. (2005) Electrokinetic transport in nanochannels. 1. Theory. *Anal. Chem.*, **77**, 6772–6781.
9. Vilkner, T., Janasek, D., and Manz, A. (2004) Micro total analysis systems. Recent developments. *Anal. Chem.*, **76**, 3373–3385.

10 Vrouwe, E.X., Luttge, R., and van den Berg, A. (2004) Direct measurement of lithium in whole blood using microchip capillary electrophoresis with integrated conductivity detection. *Electrophoresis*, **25**, 1660–1667.

11 Braschler, T., Demierre, N., Nascimento, E., Silva, T., Oliva, A.G., and Renaud, P. (2008) Continuous separation of cells by balanced dielectrophoretic forces at multiple frequencies. *Lab Chip*, **8**, 280–286.

12 Demierre, N., Braschler, T., Linderholm, P., Seger, U., van Lintel, H., and Renaud, P. (2007) Characterization and optimization of liquid electrodes for lateral dielectrophoresis. *Lab Chip*, **7**, 355–365.

13 Fiedler, S., Shirley, S.G., Schnelle, T., and Fuhr, G. (1998) Dielectrophoretic sorting of particles and cells in a microsystem. *Anal. Chem.*, **70**, 1909–1915.

14 Seger, U., Gawad, S., Johann, R., Bertsch, A., and Renaud, P. (2004) Cell immersion and cell dipping in microfluidic devices. *Lab Chip*, **4**, 148–151.

15 Dovichi, N.J. and Pinkel, D. (2003) Analytical biotechnology – tools to characterize cells and their contents – editorial overview. *Curr. Opin. Biotechnol.*, **14**, 3–4.

16 Tixier-Mita, A., Jun, J., Ostrovidov, S., Chiral, M., Frenea, B., LePioflue, B., and Fujita, H. (2004) Proceedings of MicroTotal Analysis System, microTAS, p. 180.

17 Ionescu-Zanetti, C., Blatz, A., and Khine, M. (2008) Electrophoresis-assisted single-cell electroporation for efficient intracellular delivery. *Biomed. Microdevices*, **10**, 113–116.

18 Khine, M., Ionescu-Zanetti, C., Blatz, A., Wang, L.P., and Lee, L.P. (2007) Single-cell electroporation arrays with real-time monitoring and feedback control. *Lab Chip*, **7**, 457–462.

19 Valero, A., Post, J.N., van Nieuwkasteele, J.W., ter Braak, P.M., Kruijer, W., and van den Berg, A. (2008) Gene transfer and protein dynamics in stem cells using single cell electroporation in a microfluidic device. *Lab Chip*, **8**, 62–67.

20 Andersson, H. and van den Berg, A. (2003) Microfluidic devices for cellomics: a review. *Sens. Actuators B Chem.*, **92**, 315–325.

21 Valero, A., Merino, F., Wolbers, F., Luttge, R., Vermes, I., Andersson, H., and van den Berg, A. (2005) Apoptotic cell death dynamics of HL60 cells studied using a microfluidic cell trap device. *Lab Chip*, **5**, 49–55.

22 Fox, M.B., Esveld, D.C., Valero, A., Luttge, R., Mastwijk, H.C., Bartels, P.V., van den Berg, A., and Boom, R.M. (2006) Electroporation of cells in microfluidic devices: a review. *Anal. Bioanal. Chem.*, **385**, 474–485.

23 Huang, Y. and Rubinsky, B. (2002) Flow-through micro-electroporation chip for high efficiency single-cell genetic manipulation, *Workshop on Solit-State Sensors, Actuators and Microsystems*, Hilton Head Island, South Carolina, Jun 02–06, 2002, Elsevier Science Sa, Hilton Head Island, SC, pp. 205–212.

24 Khine, M., Lau, A., Ionescu-Zanetti, C., Seo, J., and Lee, L.P. (2005) A single cell electroporation chip. *Lab Chip*, **5**, 38–43.

25 Olofsson, J., Levin, M., Stromberg, A., Weber, S.G., Ryttsen, F., and Orwar, O. (2007) Scanning electroporation of selected areas of adherent cell cultures. *Anal. Chem.*, **79**, 4410–4418.

26 Blau, H.M., Pavlath, G.K., Hardeman, E.C., Chiu, C.P., Silberstein, L., Webster, S.G., Miller, S.C., and Webster, C. (1985) Plasticity of the differentiated state. *Science*, **230**, 758–766.

27 Kohler, G. and Milstein, C. (1975) Continuous cultures of fused cells secreting antibody of predefined specificity. *Nature*, **256**, 495–497.

28 Miller, R.A. and Ruddle, F.H. (1976) Pluripotent teratocarcinoma-thymus somatic-cell hybrids. *Cell*, **9**, 45–55.

29 Chang, D.C., Chassy, B.M., and Saunders, J.A. (1992) *Guide to Electroporation and Electrofusion*, Academic Press, San Diego.

30 Weaver, J.C. and Chizmadzhev, Y.A. (1996) Theory of electroporation: a review. *Bioelectrochem. Bioenerg.*, **41**, 135–160.

31 Huang, Y. and Rubinsky, B. (1999) Micro-electroporation: improving the

efficiency and understanding of electrical permeabilization of cells. *Biomed. Microdevices*, **2**, 145–150.

32 Ferret, E., Evrard, C., Foucal, A., and Gervais, P. (2000) Volume changes of isolated human K562 leukemia cells induced by electric field pulses. *Biotechnol. Bioeng.*, **67**, 520–528.

33 Lundqvist, J.A., Sahlin, F., Aberg, M.A.I., Stromberg, A., Eriksson, P.S., and Orwar, O. (1998) Altering the biochemical state of individual cultured cells and organelles with ultramicroelectrodes. *Proc. Natl. Acad. Sci. U.S.A.*, **95**, 10356–10360.

34 Nolkrantz, K., Farre, C., Brederlau, A., Karlsson, R.I.D., Brennan, C., Eriksson, P.S., Weber, S.G., Sandberg, M., and Orwar, O. (2001) Electroporation of single cells and tissues with an electrolyte-filled capillary. *Anal. Chem.*, **73**, 4469–4477.

35 Haas, K., Sin, W.C., Javaherian, A., Li, Z., and Cline, H.T. (2001) Single-cell electroporation for gene transfer in vivo. *Neuron*, **29**, 583–591.

36 Rae, J.L. and Levis, R.A. (2002) Single-cell electroporation. *Pflugers Arch. Eur. J. Physiol.*, **443**, 664–670.

37 Schoenbach, K.H., Beebe, S.J., and Buescher, E.S. (2001) Intracellular effect of ultrashort electrical pulses. *Bioelectromagnetics*, **22**, 440–448.

38 Wheeler, A.R., Morishima, K., Arnold, D.W., Rossi, A.B., Zare, R.N., van den Berg, A., Olthuis, W., and Bergveld, P. (2000) *Micro Total Analysis Systems, 2000*, Kluwer, Dordrecht, p. 623.

39 Gao, J., Yin, X.F., and Fang, Z.L. (2004) Integration of single cell injection, cell lysis, separation and detection of intracellular constituents on a microfluidic chip. *Lab Chip*, **4**, 47–52.

40 McClain, M.A., Culbertson, C.T., Jacobson, S.C., Allbritton, N.L., Sims, C.E., and Ramsey, J.M. (2003) Microfluidic devices for the high-throughput chemical analysis of cells. *Anal. Chem.*, **75**, 5646–5655.

41 Lu, H., Schmidt, M.A., and Jensen, K.F. (2005) A microfluidic electroporation device for cell lysis. *Lab Chip*, **5**, 23–29.

42 Bao, N., Wang, J., and Lu, C. (2008) Microfluidic electroporation for selective release of intracellular molecules at the single-cell level. *Electrophoresis*, **29**, 2939–2944.

43 Bao, N., Zhan, Y.H., and Lu, C. (2008) Microfluidic electroporative flow cytometry for studying single-cell biomechanics. *Anal. Chem.*, **80**, 7714–7719.

44 Wang, J., Bao, N., Paris, L.L., Wang, H.Y., Geahlen, R.L., and Lu, C. (2008) Detection of kinase translocation using microfluidic electroporative flow cytometry. *Anal. Chem.*, **80**, 1087–1093.

45 He, H.Q., Chang, D.C., and Lee, Y.K. (2008) Nonlinear current response of micro electroporation and resealing dynamics for human cancer cells. *Bioelectrochemistry*, **72**, 161–168.

46 Huang, Y. and Rubinsky, B. (2003) Flow-through micro-electroporation chip for high efficiency single-cell genetic manipulation. *Sens. Actuators A Phys.*, **104**, 205–212.

47 Huang, Y., Sekhon, N.S., Borninski, J., Chen, N., and Rubinsky, B. (2003) Instantaneous, quantitative single-cell viability assessment by electrical evaluation of cell membrane integrity with microfabricated devices. *Sens. Actuators A Phys.*, **105**, 31–39.

48 Knorr, D., Angersbach, A., Eshtiaghi, M.N., Heinz, V., and Lee, D.U. (2001) Processing concepts based on high intensity electric field pulses. *Trends Food Sci. Technol.*, **12**, 129–135.

49 Fox, M., Esveld, E., Luttge, R., and Boom, R. (2005) A new pulsed electric field microreactor: comparison between the laboratory and microtechnology scale. *Lab Chip*, **5**, 943–948.

50 Pol, I.E., Mastwijk, H.C., Bartels, P.V., and Smid, E.J. (2000) Pulsed-electric field treatment enhances the bactericidal action of nisin against Bacillus cereus. *Appl. Environ. Microbiol.*, **66**, 428–430.

51 Neumann, E., Sowers, A.E., and Jordan, C.A. (1989) *Electroporation and Electrofusion in Cell Biology*, Plenum Press, New York.

52 Loomishusselbee, J.W., Cullen, P.J., Irvine, R.F., and Dawson, A.P. (1991) Electroporation can cause artifacts due to solubilization of cations from the

electrode plates – aluminum ions enhance conversion of inositol 1, 3, 4, 5-tetrakisphosphate into inositol 1, 4, 5-trisphosphate in electroporated L1210 cells. *Biochem. J.*, **277**, 883–885.

53 Lin, Y.C., Jen, C.M., Huang, M.Y., Wu, C.Y., and Lin, X.Z. (2001) Electroporation microchips for continuous gene transfection. *Sens. Actuators B Chem.*, **79**, 137–143.

54 Seo, J., Ionescu-Zanetti, C., Diamond, J., Lal, R., and Lee, L.P. (2004) Integrated multiple patch-clamp array chip via lateral cell trapping junctions. *Appl. Phys. Lett.*, **84**, 1973–1975.

55 Lin, Y.C., Li, M., Fan, C.S., and Wu, L.W. (2003) A microchip for electroporation of primary endothelial cells. *Sens. Actuators A Phys.*, **108**, 12–19.

56 Lin, Y.C., Li, M., and Wu, C.C. (2004) Simulation and experimental demonstration of the electric field assisted electroporation microchip for in vitro gene delivery enhancement. *Lab Chip*, **4**, 104–108.

57 Ando, R., Mizuno, H., and Miyawaki, A. (2004) Regulated fast nucleocytoplasmic shuttling observed by reversible protein highlighting. *Science*, **306**, 1370–1373.

58 Costa, M., Marchi, M., Cardarelli, F., Roy, A., Beltram, F., Maffei, L., and Ratto, G.M. (2006) Dynamic regulation of ERK2 nuclear translocation and mobility in living cells. *J. Cell Sci.*, **119**, 4952–4963.

59 Davidson, R.L. and Gerald, P.S. (1976) Improved techniques for induction of mammalian-cell hybridization by polyethylene-glycol. *Somatic Cell Genet.*, **2**, 165–176.

60 Pontecorvo, G. (1975) Production of mammalian somatic-cell hybrids by means of polyethylene-glycol treatment. *Somatic Cell Genet.*, **1**, 397–400.

61 Jahn, R., Lang, T., and Sudhof, T.C. (2003) Membrane fusion. *Cell*, **112**, 519–533.

62 Steubing, R.W., Cheng, S., Wright, W.H., Numajiri, Y., and Berns, M.W. (1991) Laser-induced cell-fusion in combination with optical tweezers – the laser cell-fusion trap. *Cytometry*, **12**, 505–510.

63 Wiegand, R., Weber, G., Zimmermann, K., Monajembashi, S., Wolfrum, J., and Greulich, K.O. (1987) Laser-induced fusion of mammalian-cells and plant-protoplasts. *J. Cell Sci.*, **88**, 145–149.

64 Vienken, J. and Zimmermann, U. (1982) Electric field-induced fusion – electrohydraulic procedure for production of heterokaryon cells in high-yield. *FEBS Lett.*, **137**, 11–13.

65 Zimmermann, U. and Vienken, J. (1982) Electric field-induced cell-to-cell fusion. *J. Membr. Biol.*, **67**, 165–182.

66 Chiu, D.T. (2001) A microfluidics platform for cell fusion – commentary. *Curr. Opin. Chem. Biol.*, **5**, 609–612.

67 Stromberg, A., Ryttsen, F., Chiu, D.T., Davidson, M., Eriksson, P.S., Wilson, C.F., Orwar, O., and Zare, R.N. (2000) Manipulating the genetic identity and biochemical surface properties of individual cells with electric-field-induced fusion. *Proc. Natl. Acad. Sci. U.S.A.*, **97**, 7–11.

68 Bakker Schut, T.C., Kraan, Y., Barlag, W., de Leij, L., de Grooth, B.G., and Greve, J. (1993) Selective electrofusion of conjugated cells in flow. *Biophys. J.*, **65**, 568–572.

69 Tresset, G. and Iliescu, C. (2007) Electrical control of loaded biomimetic femtoliter vesicles in microfluidic system. *Appl. Phys. Lett.*, **90**, 173091–173093.

70 Tresset, G. and Takeuchi, S. (2004) A microfluidic device for electrofusion of biological vesicles. *Biomed. Microdevices*, **6**, 213–218.

71 Wang, J. and Lu, C. (2006) Microfluidic cell fusion under continuous direct current voltage. *Appl. Phys. Lett.*, **89**, 234102–234103.

72 Techaumnat, B., Tsuda, K., Kurosawa, O., Murat, G., Oana, H., and Washizu, M. (2008) High-yield electrofusion of biological cells based on field tailoring by microfabricated structures. *IET Nanobiotechnol.*, **2**, 93–99.

73 Skelley, A., Kirak, O., Suh, H., Jaenisch, R., and Voldman, J. Microfluidic control of cell pairing and fusion. (2009) *Nat. Methods*. **6**, 147–152.

# 8
# Electroporative Flow Cytometry for Single-cell Analysis

*Chang Lu, Jun Wang, Ning Bao, and Hsiang-Yu Wang*

## 8.1
## Introduction

Electroporative flow cytometry (EFC) combines electroporation with flow cytometric detection. Electroporation is applied in EFC as a tool to physically process or perturb cells at the single-cell level. Electroporation occurs when cells experience an electrical field with the intensity beyond a certain threshold [1–3]. During electroporation the electrical field opens up pores in the cell membrane. Such pores allow the delivery of foreign molecules and the release of intracellular materials from/into the surrounding solution. When the intensity and the duration of the field applied for electroporation exceed certain thresholds, irreversible disruption of the cell membrane, or cell lysis, occurs. Conventionally, pulsed voltage is applied to generate the field necessary for electroporation. The most common application of electroporation is to deliver exogenous macromolecules such as DNA into cells *in vitro* and *in vivo* [4–6]. However, electroporation (or electric lysis) is also applied to release intracellular materials [7–9]. The latter application of electroporation is most relevant to the EFC assays developed so far. In EFC, cells are detected individually during or after electroporation in the manner of flow cytometry. Flow cytometric approach allows a definitive discrimination, cell by cell, by sending the cells through a detection point in a carrier flow [10, 11]. In this case, the flow cytometric detection ensures that EFC has a high throughput that allows a cell population of reasonable size to be studied. This is critical for ensuring that the data are reflective of the population distribution and best simulate the *in vivo* results when primary materials including different cell subsets are analyzed.

EFC is closely related to earlier work on chemical cytometry in which the breaching of cell membrane serves as an essential step for cellular analysis [7, 8, 12–15]. There are also important differences: first, EFC does not require (as a matter of fact, tries to avoid) complete lysis of the cells at one physical location of the device because such a requirement often limits the throughput of the method. Owing to this reason, in principle, EFC can have a throughput matching that of flow cytometry ($\sim 10^4$ cells/s). Second, chemical cytometry emphasizes on the

*Chemical Cytometry*. Edited by Chang Lu

ISBN: 978-3-527-32495-8

detection of intracellular molecules after lysis. In contrast, most EFC applications discovered so far generate the results by examining the cells instead of molecules during or after electroporation. Third, chemical cytometry is typically designed for quantifying the expression levels of intracellular molecules at the single-cell level. As detailed below, EFC has been applied to a variety of different problems in molecular and cell biology that are not limited to detection of protein expression level.

We have discovered several applications for EFC so far. One application is to use EFC to detect intracellular protein translocation by detecting the release of the target molecule after electroporation [16]. The basic assumption is that the difficulty in the release of an intracellular molecule is determined by its subcellular location and its binding to other subcellular structures. When the molecule is deep inside cells (e.g., in the nucleus) or binds to some subcellular compartment (e.g., plasma membrane), these situations make the release of the molecule by electroporation, difficult. By examining how much the molecule is removed by electroporation from an individual cell, we can deduce the subcellular location of the molecule. The other application we found for EFC was to use it to study the deformability of cells [17]. Visible cell expansion can be easily spotted and quantified during flow-through electroporation. Our hypothesis is that with the same electric parameters applied, more deformable cells would expand more rapidly and more substantially during such processes. By quantifying the cell expansion due to electroporation, we can measure the biomechanical properties for each cell. These two applications are focused on molecular and cell biology, respectively, with no apparent connections with each other. Other than the above applications, which are fairly distinct from typical chemical cytometry methods, another application for EFC that is very closely related to chemical cytometry is to selectively release certain intracellular molecules for analysis without complete lysis of cells in order to boost the throughput [18]. We fully expect that more applications of EFC will be discovered in the future.

## 8.2
## Flow-through Electroporation under Constant Voltage

We demonstrated flow-through electroporation of bacterial and mammalian cells based on constant voltage [19–22]. This technique allows flowing cells to be electroporated in one or multiple sections of a microfluidic channel with reduced cross-sectional area when a constant voltage is established across the channel. This technique deviates from traditional electroporation technique by eliminating the application of electric pulses. The use of constant DC voltage allows electroporation to be conducted using the same electric source for electrophoresis. This dramatically lowers the cost and logistics burden for integrated devices. Furthermore, constant DC voltage permits uniform treatment of all cells in the electroporation field. Such traits will be difficult when discrete electric pulses are applied. Flow-through electroporation provides a perfect platform for coupling cell screening with electroporation.

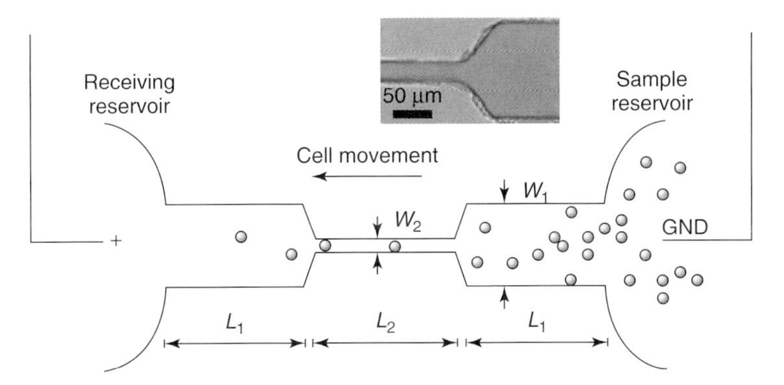


**Figure 8.1** The schematic of a flow-through electroporation device. Cells were loaded into the sample reservoir and flowed to the receiving reservoir in a DC field. Electroporation was confined in the narrow section of the channel due to the amplified field inside. The inset shows a microscope image of a part of a fabricated device. The devices used in this study have the following dimensions: $L_1 = 2.5$ mm, $L_2 = 2.0$ mm, $W_1 = 213$ µm, $W_2 = 33$ µm.

In flow-through electroporation, we applied geometric variation to a microfluidic channel to create a local high field in a geometrically defined section. We controlled the overall voltage across the channel so that only the field intensity in the defined section would produce electroporation. We were able to adjust the duration for the cells to be electroporated in the high field by controlling the velocity of the cells and the length of the electroporation section.

A flow-through electroporation device in its simplest form is shown in Figure 8.1. The devices had the following dimensions: $L_1 = 2.5$ mm, $L_2 = 2.0$ mm, $W_1 = 213$ µm, $W_2 = 33$ µm. When a constant DC voltage is applied at a conductor (in this case, a buffer-filled channel) the potential drop at individual sections of the conductor is proportional to its resistance within the section. Like any conductor, the resistance within a certain section of a microfluidic channel is determined by the conductivity, the length, and the cross-sectional area. For a channel with a uniform depth and a varying width as shown in Figure 8.1a, the field strength $E$ is different in various sections. The field strength in the wide section ($E_1$) and in the narrow section ($E_2$) can be calculated using the below equations:

$$E_1 = \frac{V}{2L_1 + L_2\left(\frac{W_1}{W_2}\right)} \tag{8.1}$$

$$E_2 = \frac{V}{2L_1\left(\frac{W_2}{W_1}\right) + L_2} \tag{8.2}$$

$$\frac{E_2}{E_1} = \frac{W_1}{W_2} \tag{8.3}$$

The width $W_2$ was always much smaller than $W_1$ in our design. This resulted in much higher field strength in the narrow section compared to those in the other

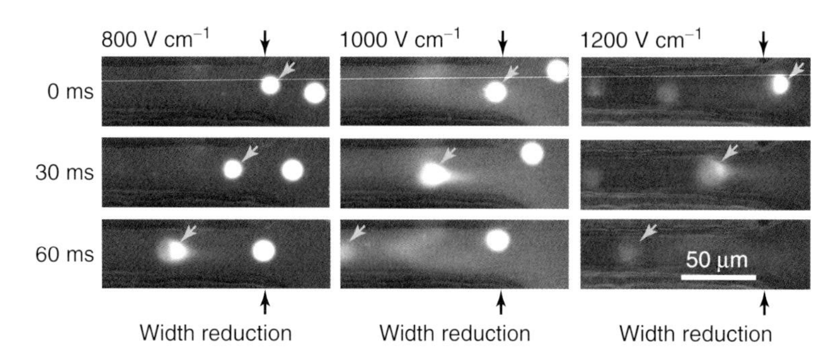


**Figure 8.2** The image series showing the rapid release of intracellular materials when $E_2$ was between 800 and 1200 V cm$^{-1}$. The CHO cells were loaded with calcein AM. The images were taken at a frame rate of 33 Hz. The arrows indicate where the width reduction occurs in the channel.

two sections when a DC field was applied across the whole length of the device. Similar geometric variations have been shown to create local electrical fields as high as 0.1 MV cm$^{-1}$ [23, 24]. Figure 8.2 shows that electroporation (indicated by the release of fluorescent calcein loaded in CHO cells) occurred immediately when the cells entered the narrow section.

Flow-through electroporation devices can also have multiple narrow sections. In that case, the treatment of cells is equivalent to applying multiple electric pulses. Such configuration is superior to devices with single narrow (electroporation) section when gene delivery into cells is the intended application [22]. Similar flow-through devices can also be used for cell electrofusion [25].

## 8.3 Electroporative Flow Cytometry for Detecting Protein Translocation

Translocation of a protein between different subcellular compartments is a common event during signal transduction in living cells. Integrated signaling cascades often lead to the relocalization of protein constituents such as translocations between the cytosol and the plasma membrane or nucleus. Such events can be essential for the activation/deactivation and biological function of the protein. Determination of protein translocation within cells has been traditionally carried out using methods such as subcellular fractionation/Western blotting or imaging of a few cells. These techniques either obtain only the bulk average information of the population or lack the high throughput for studying each cell in a large cell population. Conventional flow cytometry is intrinsically insensitive to the subcellular location of the probed protein. Laser scanning cytometry (LSC) has been used for quantifying nuclear/cytoplasmic distribution of a fluorescently labeled protein based on solid-phase cytometry technique [26–28]. However, LSC is not particularly effective for observing cytosol/membrane translocation. The algorithm of quantification based on image analysis is complex and lacks robustness. More importantly, the throughput of LSC is typically less than 100 cells/s, compared to $10^4$ cells/s offered by flow cytometry [29].

One important application established for EFC is to detect protein translocation in cells [16]. Compared to the above techniques, EFC has both the high throughput and the single-cell resolution to study translocation in a cell population. EFC detects protein translocation based on the principle that the release of an intracellular protein from cells by flow-through electroporation is sensitive to the subcellular location of the protein. When the protein is deep in the cell (e.g., in the nucleus) or binds to certain subcellular compartments, its release by electroporation is slowed down or prohibited. We studied the case of the kinase Syk translocating from the cytosol to the plasma membrane using EFC. The protein-tyrosine kinase, Syk, is a prime example of a protein that translocates to the plasma membrane as part of its role in signal transduction. Syk is essential for the survival, proliferation, and differentiation of B lymphocytes, processes regulated by signals sent from the cell surface receptor for antigen (B cell antigen receptor (BCR)) [30, 31]. Syk is the prototype kinase of the Syk/Zap-70 family [32, 33]. In mature B cells, clustering of the BCR by interactions with antigens (or artificially through interactions with anti-IgM antibodies) leads to recruitment of Syk to the aggregated BCR, which binds to the phosphorylated receptor through a tandem pair of N-terminal SH2 domains and couples the receptor to multiple intracellular signaling networks including the Ras/Erk, phospholipase C$\gamma$/NF-AT, and PI3K/Akt pathways [34, 35]. In the case of Syk translocation to the plasma membrane, because a fraction of Syk moves to bind to the surface receptor, the release of the kinase due to electroporation is less from cells with translocation than from those without translocation.

We studied the release of enhanced green fluorescent protein (EGFP)-tagged Syk from chicken DT40 cells with different activation states using microfluidic EFC [16]. As shown in Figure 8.3, we applied the flow-through electroporation technique

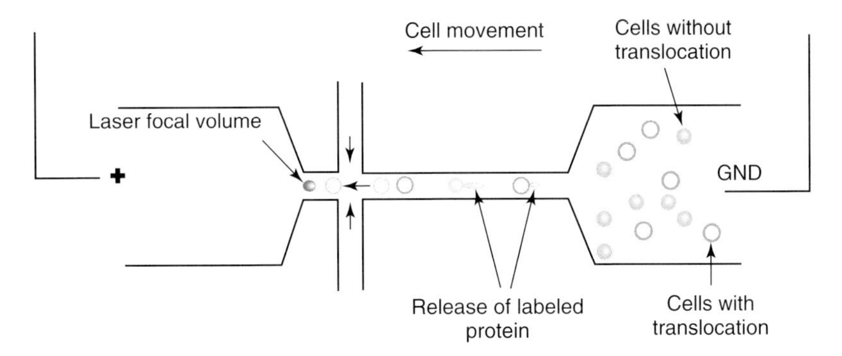


**Figure 8.3** The schematic of microfluidic EFC for studying kinase translocation from cytosol to plasma membrane. Cells are electroporated and intracellular materials including fluorescently labeled kinase are released in the narrow section of the channel before laser-induced fluorescence intensity from each cell is measured. The protein distribution in the cells with and without translocation is exaggerated in the drawing.

that electroporated single cells when they flowed through the narrow section of the channel. The horizontal channel was composed of two wide sections and one narrow section with the depth of the whole channel being uniform. When a constant DC voltage was established across the channel, the field strength in the narrow section was approximated to be 10 times higher than in the wide sections due to the difference in the cross-sectional area ($E_2/E_1 = W_1/W_2$) [20, 21]. Electroporation exclusively occurred in the narrow section because the field intensity in the wide sections was too weak to compromise the cell membrane [20, 21]. The cells flowed through the horizontal channel carried by a pressure-driven flow generated by a syringe pump. In order to screen cells at the single-cell level, hydrodynamic focusing was applied by having the buffer flow into the horizontal channel from the two vertical channels at equal flow rates (supported by a second syringe pump). The duration for cells to stay in the narrow section (the electroporation section) of the channel was determined by their velocity and the length of the section. The velocity of cells here was determined by the infusion rate of the syringe pump with little influence from the electric field. We had one detection point at the exit of the electroporation section where we measured the fluorescence from single cells after the release of EGFP-tagged Syk due to electroporation.

Syk translocation from the cytosol to the plasma membrane was well-established by biologists using cellular fractionation/Western blotting and fluorescence imaging [36, 37]. Syk coupled to EGFP (SykEGFP) expressed in Syk-deficient chicken DT40 cells has been shown to respond to anti-IgM antibody stimulation by translocating from cytoplasmic and nuclear compartments to the cross-linked BCR at the plasma membrane [37]. In the absence of Lyn, the receptor–Syk complexes can persist at the inner side of the plasma membrane without being internalized for more than 1 h at 37 °C [37]. In our experiment, we examined SykEGFP-expressing chicken DT40 cells lacking both Syk and Lyn (SykEGFP-DT40-$Syk^-$-$Lyn^-$) to ensure that the localization of Syk at the plasma membrane lasted long enough for us to finish the tests. The cells were stimulated at room temperature (22 °C) by anti-IgM antibody. The translocation was confirmed by cellular fractionation/Western blotting. A small fraction (~17%) of SykEGFP moved from the cytosol to the plasma membrane after the stimulation as shown by Western blotting. This represents a delicate, but significant change in the state of the cells.

Using the device in Figure 8.3, we were able to establish histograms of the fluorescent intensities from single cells after their electroporation and release of intracellular molecules, for cell populations with or without anti-IgM stimulation and translocation. Such histograms were taken with various field intensities and durations in the electroporation section. Figure 8.4 shows histograms of the fluorescence intensity from cell populations treated under different conditions and electroporated under different electrical parameters. In each histogram, the $y$ axis shows the percentile frequency of detection and the $x$ axis represents the fluorescence intensity (in channels). In Figure 8.4a, we show the histograms of the fluorescence intensity generated by the cell samples stimulated by anti-IgM (black) and those that were not stimulated (grey), in the microfluidic EFC device. The cells stayed in the narrow section for around 120 ms. When we increased

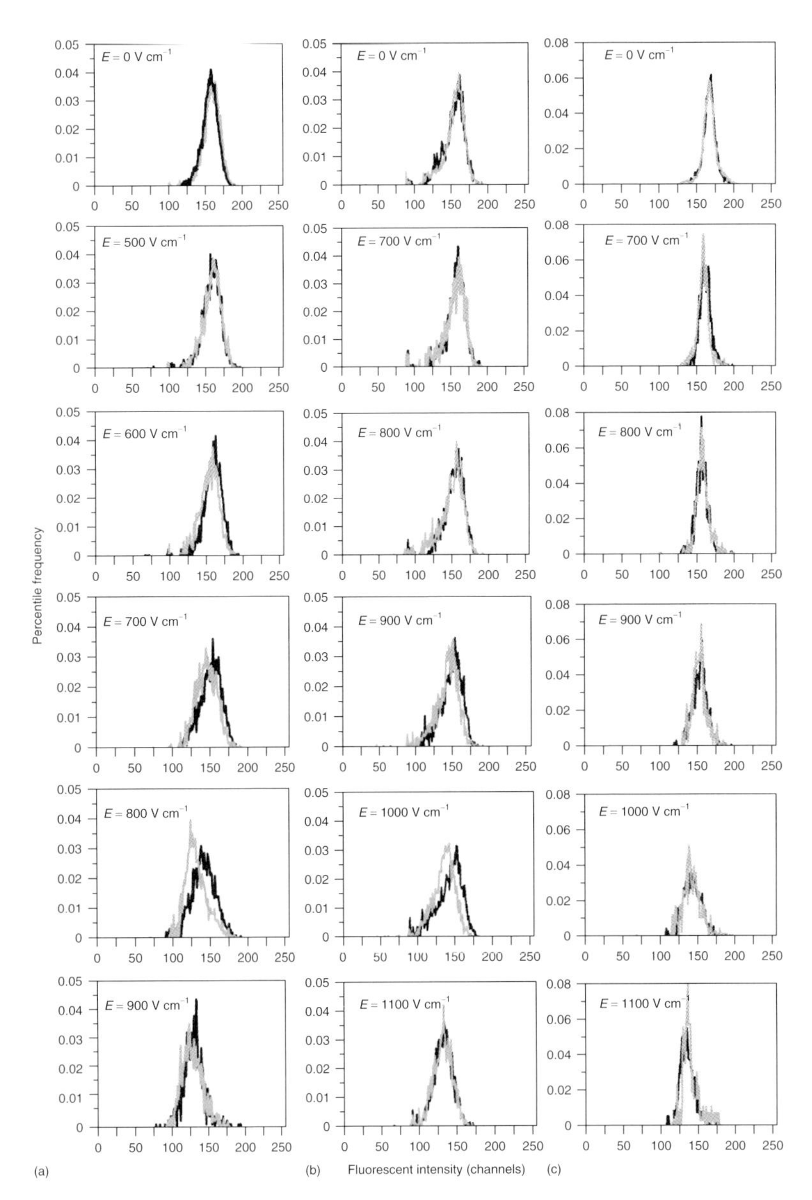


**Figure 8.4** The histograms of the fluorescent intensity of DT40 cells detected by the microfluidic EFC under different electric field intensities and durations. SykEGFP-DT40-Syk$^-$-Lyn$^-$ cells were applied in (a) and (b), while calcein AM stained DT40-Syk$^-$-Lyn$^-$ cells were used in (c). The black curves were generated by cells stimulated by anti-IgM (with translocation) and the grey curves were obtained from cells without stimulation/translocation. The data in (a) and (b) were obtained with different electroporation durations of 120 and 60 ms, respectively. The duration in (c) was 60 ms. The field intensity in the narrow section is indicated for each histogram.

the field intensity in the narrow section from 0 to 900 V $cm^{-1}$, the fluorescence intensity of the cell population (stimulated with anti-IgM or not) shifted to the lower end. The translocation did not make any difference in the histogram until the field intensity increased to 600 V $cm^{-1}$. At this field intensity the two histograms did not totally overlap and the stimulated cell population had a slightly higher fluorescence intensity compared to that of the other population without stimulation and translocation. This difference increased further when we increased the field intensity to 700 and 800 V $cm^{-1}$. At 900 V $cm^{-1}$, the two histograms overlapped again. To observe the effects of the duration of time spent in the electroporation field on the release of SykEGFP, we applied a microfluidic EFC device with a shorter narrow section. With the flow rates kept the same as in the previous experiment, the electroporation duration before detection was only half of that in the first experiment under the same field intensity. Figure 8.4b shows that the two histograms from the cell populations with and without stimulation overlapped very well up to 700 V $cm^{-1}$. Difference between the two histograms started to show up at 800 V $cm^{-1}$ and reached a maximum at around 1000 V $cm^{-1}$ before the two histograms were not distinguishable again at 1100 V $cm^{-1}$. Compared to the data in Figure 8.4a, with the electroporation duration at one half, a higher field intensity was needed to achieve a similar level of differentiation between the stimulated cell population and the unstimulated one. In Figure 8.4a and b, the difference between the two cell populations was the most pronounced in the medium range of the field intensity. Such a difference was not present without the electric field and it diminished at very high field intensity. To confirm that such differentiation was not caused by the interaction between the plasma membrane and the antibody, we also performed the same experiment with DT40-$Syk^{-}$-$Lyn^{-}$ cells that were not labeled by expression of SykEGFP, but were instead stained with calcein AM. In live cells, the nonfluorescent calcein AM is converted to green fluorescent calcein, after acetoxymethyl ester hydrolysis by intracellular esterases. As shown in Figure 8.4c, we did not observe a significant difference between the cell population with added anti-IgM and the population without the antibody at any field intensity. This confirms that the differentiation was closely related to the translocation of SykEGFP.

Figure 8.5 shows the mean fluorescence intensity of the cell populations plotted against the field intensity for all the three experiments in Figure 8.4. It was found that the optimal field intensities for detecting translocation to the plasma membrane in a cell population was around 700 and 800 V $cm^{-1}$ with a duration of 120 ms or 1000 V $cm^{-1}$ with a duration of 60 ms (significantly different at $P < 0.01$).

Phase contrast images after the tests confirmed that the plasma membrane of the cells had no significant fragmentation. The throughput we achieved was ~150 cells/s. Our data indicate that the electroporative release of an intracellular kinase is closely related to the activation state of the cells and whether translocation occurs. Although there is only a small fraction of the kinase engaging in such translocation (~17% based on Western blotting), such a change can be readily recognized by EFC. The diminishing of the differentiation at the high field

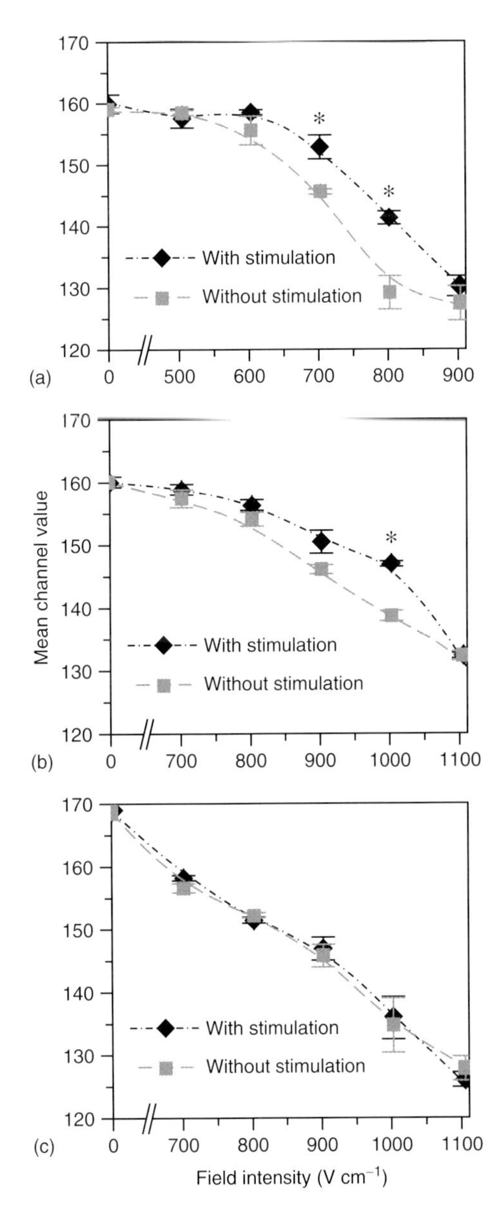


**Figure 8.5** The variation of the mean fluorescence intensity value of the cell population at different field intensities with and without stimulation by anti-IgM. SykEGFP-DT40-Syk$^{-}$-Lyn$^{-}$ cells were applied in (a) and (b), while calcein AM stained DT40-Syk$^{-}$-Lyn$^{-}$ cells were used in (c). The data in (a) and (b) were obtained with different electroporation durations of 120 and 60 ms, respectively. The duration in (c) was 60 ms. The error bars were generated by carrying out the experiments in triplicate. (*) indicates significant difference at $P < 0.01$, calculated using unpaired t test with equal variance.

intensities is possibly due to the dissociation of bound Syk from the plasma membrane due to Joule heating or excessive poration.

Detecting kinase translocation using EFC may find important applications not only in basic molecular biology studies but also in drug discovery. Owing to the frequent involvement of kinase activation in cancer development and tumorigenesis, kinase inhibitors are an important category of cancer drugs (e.g., Gleevec by Novartis). Depending on the kinase being targeted, the response of a cell population to these kinase inhibitors can be evaluated using EFC when translocation is a requirement for the activation of the kinase.

## 8.4
## Electroporative Flow Cytometry for Measuring Single-cell Biomechanics

Differences in biomechanical properties have important implications for cell signaling, cytoadherence, migration, invasion, and metastatic potential. The cytoskeleton, the internal scaffolding comprising a complex network of biopolymeric molecules, provides the framework of cells that determines the shape and mechanical deformation properties. Mammalian cells typically contain three distinct types of polymeric biomolecules in the cytoskeleton: actin microfilaments, intermediate filaments, and microtubules [38]. Disease development can often dramatically affect cell biomechanics. For example, the structures of the cytoskeleton and the extracellular matrix are often transformed by cancer [38, 39]. During the cell's progression from a fully mature, postmitotic state to an immortal cancerous cell, the cytoskeleton experiences a reduction in the amount of constituent polymers and accessory proteins and a restructuring of the biopolymeric network [38–42]. The altered cytoskeleton also changes the ability of cancer cells to contract or stretch, which determines the mechanics of deformation. In general, malignant cells exhibit lower resistance to deformation than normal cells. Metastatic cancer cells are even more deformable than nonmetastatic cells.

There have been a number of tools developed over the years for measuring the mechanical properties of cells. Early techniques such as filtration [43] or micropipette aspiration [44–46] were used to study cell deformation based on suction of cells into capillaries or pores. It was found that a direct correlation existed between an increase in deformability and progression from a nontumorigenic cell to a tumorigenic and metastatic cell [43, 46]. Atomic force microscopy (AFM) can be used to probe an attached cell by applying a local force and measuring local structural properties using a hard indentor [47–49]. AFM was used to study both cell lines and primary cells and the results showed that a normal cell has a Young's modulus 1 order of magnitude higher than a cancerous one [47, 50]. Magnetic tweezers can also be applied to study the viscoelastic properties of a cell by attaching magnetic beads onto the cell surface and applying magnetic forces while tracking the bead location [51–53]. In microplate manipulation, a cell can be seized between two microplates with the more flexible one serving as a sensor of the applied force while unidirectional compression and traction is applied [54–56]. Optical tweezers can also be applied to the studies of cell elasticity and mechanotransduction by manipulating beads attached to the cell surface [57–59]. Arrays of microneedles were fabricated to map cell forces [60–62]. Microfabricated devices and sensors were used to apply forces and study cell biomechanics [63]. Several methods have recently been developed for high-throughput studies. A microfluidic optical stretcher was used to examine the elasticity of cells in a continuous flow at a throughput of 1 cell/min [64, 65]. The behavior of cells squeezing through microfluidic channels with a cross-sectional area smaller than that of the cells can be monitored for characterizing cell deformability [66, 67]. On the basis of this mechanism, a throughput of 50–100 cells/min was recently demonstrated [67].

In our recent work, we applied EFC to study cell biomechanics at the single-cell level with a throughput of 5 cells/s. This represents a high-throughput approach to study cell biomechanics in a large cell population. The throughput can be further improved by using a faster camera. We noticed from our earlier work on flow-through electroporation that cells expanded rapidly in their size once they flowed into the electroporation section of a microfluidic device [20, 21]. This is generally believed to be due to the influx of the surrounding solution into cells after the cell membrane is compromised by electroporation [68]. We hypothesized that how rapidly a cell expanded due to electroporation correlated with the deformability of the cell. Thus, by observing the swelling of flowing cells during electroporation, we would be able to measure the cell biomechanics at the single-cell level.

To test the hypothesis, three cell types with increasing metastatic potential and deformability (MCF-10A, MCF-7, and 12-*O*-tetradecanoylphorbol-13-acetate (TPA)-treated MCF-7) were screened by microfluidic EFC with a throughput of ~5 cells/s. MCF-10A is a nontumorigenic epithelial cell line derived from benign breast tissue. These cells are immortal, but otherwise normal, noncancerous mammary epithelial cells. MCF-7 is a corresponding line of human breast cancer cells (adenocarcinoma) obtained from the pleural effusion. These cells are nonmotile, nonmetastatic epithelial cancer cells. TPA-treated MCF-7 cells were generated by treating MCF-7 cells with 100 nM TPA for 18 h. The treatment of MCF-7 cells using TPA introduces a dramatic increase (18-fold) in the invasiveness and the metastatic potential of these cells according to the literature [69]. The swelling of single cells was monitored in real time during their flow-through electroporation using a CCD camera with a frame rate of 16 frames/s. As shown in Figure 8.6, in EFC experiments, cells flowed through a microfluidic channel with a narrow section at the center, while constant voltage was established across the channel. Electroporation and swelling occurred when cells flowed into the narrow section. We used a CCD camera to record time series of images of cells after they flowed into the entry of the narrow section. In most cases, visible expansion in the cell size was observed immediately after a cell entered the narrow section (as shown in the inset image of Figure 8.6).

We studied the swelling of these three cell types with field intensities of 200, 400, and 600 V $cm^{-1}$ in the narrow section and recorded the percentile cell size change during the first 200 ms after the cells flowed into the narrow section. With the field strength of 200 V $cm^{-1}$, there was essentially no expansion in the cell

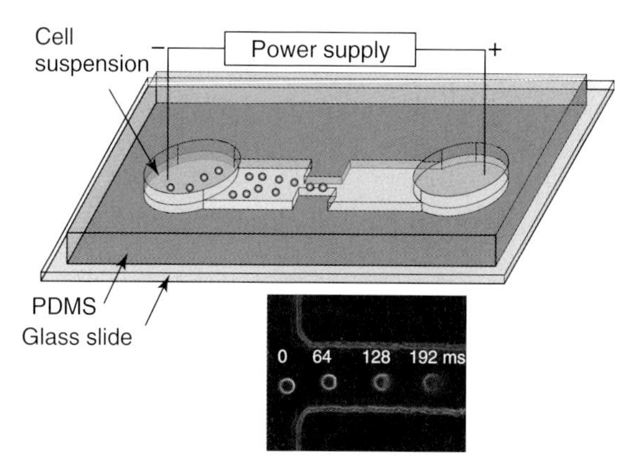


**Figure 8.6** The schematic of microfluidic EFC for single-cell biomechanics studies. Electroporation occurs in the narrow section of the microfluidic channel when cells flow through. The depth of the channel was 32 μm and the widths of the narrow and wide section(s) were 58 and 392 μm, respectively. A constant voltage was established across the channel. A CCD camera monitored a part of the narrow section including the entry. The inset image shows the cell size change of the same cell (MCF-7) at different time points flowing in the channel with the field intensity of $400\,\mathrm{V\,cm^{-1}}$ in the narrow section.

size during the 200-ms period for all three cell types, due to the fact that the field intensity was lower than the electroporation threshold. With the increase of the field strength to $400\,\mathrm{V\,cm^{-1}}$, all three cell types were observed to swell substantially with TPA-treated MCF-7 swelling the most and MCF-10A cells the least. At 192 ms after entering the narrow section, the average size of MCF-10A cells increased to 126% of their original size, while MCF-7 and TPA-treated MCF-7 cells increased to 148 and 170%, respectively. With a further increase of the field strength to $600\,\mathrm{V\,cm^{-1}}$, the difference among the cell lines diminished despite the fact that each cell type swelled more rapidly than at $400\,\mathrm{V\,cm^{-1}}$.

Our approach allowed us to obtain data on the swelling of each cell in a population and put together a histogram describing the cell population. In Figure 8.7, we show that, based on the percentile swelling of a small population of cells under a set of optimized conditions (with $400\,\mathrm{V\,cm^{-1}}$ in the narrow section and at 192 ms after electroporation), we were able to establish histograms that differentiated the cell types. We fit the histograms using normal distributions and the data yielded statistically significant differences between any two cell types. The results indicated that the more malignant and metastatic cell type had higher deformability. Although the average cell size increased substantially during the electroporation for all three cell types, about 15% of MCF-10A cells and 4% of MCF-7 cells actually decreased in their size (smaller than 100%) as shown in the histograms. In this case, the release of intracellular materials decreased the cell size when the swelling due to the influx was not pronounced for some cells. Such a feature would not have been visible if the cells were not examined at the single-cell level.

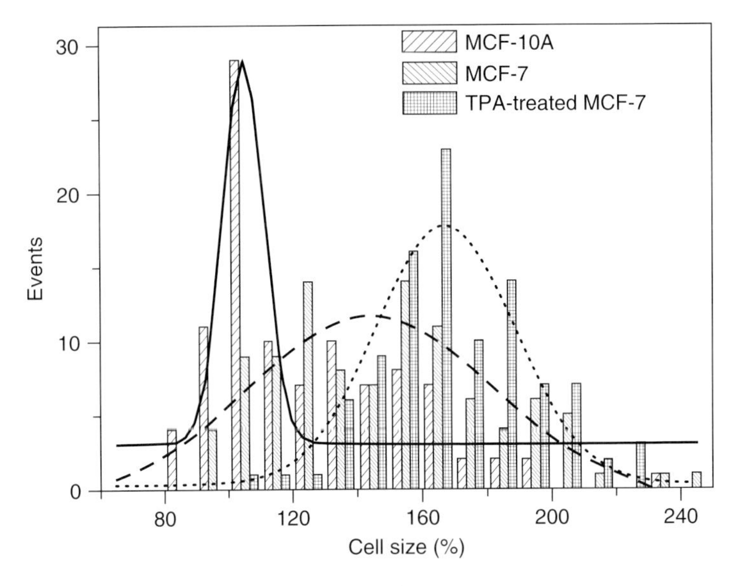


**Figure 8.7** Histograms on the swelling of MCF-10A, MCF-7, and TPA-treated MCF-7 cells under the field intensity of 400 V cm $^{-1}$ (in the narrow section) and at the time point of 192 ms. Each histogram includes data from ~100 cells. The curves were added based on the assumption of normal distribution for each population.

We also found that the treatment of cells with cytoskeleton-disrupting reagents (e.g., colchicine) substantially varied the EFC data [17]. This confirmed that the EFC data were reflective of the cytoskeletal dynamics and properties.

EFC offers the unique advantage of studying a cell population by collecting data on the biomechanical properties of single cells given its high throughput. Owing to the relevance of cell biomechanics to disease processes such as tumorigenesis, EFC has the potential to become a powerful tool for studying the cell biology involved in these processes.

## 8.5 Electroporative Flow Cytometry for Selectively Releasing and Analyzing Specific Intracellular Molecules

Single-cell analysis based on released intracellular materials or chemical cytometry, typically requires disruption of the cell membrane and release of the intracellular contents before the intracellular molecules are analyzed by tools such as electrophoresis. Chemical cytometry represents a very powerful approach that provides the molecular signature of the intracellular contents at the single-cell level. However, chemical cytometry is largely limited by its throughput (<100 cells/min), due to the fairly low speed for both cell lysis and electrophoresis [9].

We recently showed that EFC might offer rapid analysis of specific intracellular molecules at the single-cell level by selectively releasing one or a group of intracellular molecules under particular flow-through electroporation conditions [18]. Different intracellular molecules can be released under different sets of electroporation conditions (e.g., different field intensities and durations) and get analyzed. By having different release "thresholds" for various molecules, flow-through electroporation serves as a tool to separate intracellular molecules from one another in EFC, playing the role of electrophoresis in chemical cytometry. Such analysis of intracellular molecules (still occurring at the single-cell level) can have a very high throughput (~200 cells/s) as we demonstrated.

As a proof-of-principle, we examined the release of calcein (MW ~600) and a protein kinase, Syk (72 kDa, tagged by EGFP) from B cells during EFC using a device similar to the one in Figure 8.3. We found that the electroporation-based release occurred quite differently for the two molecules, calcein and SykEGFP. SykEGFP required higher field intensity for its electroporative release than calcein, presumably due to the fact that the molecular size of SkyEGFP is significantly larger than that of calcein, and the higher field intensity creates larger pores in the membrane [70].

When cells contained both calcein and SykEGFP (calcein AM loaded SykEGFP-DT40-Syk$^-$ cells), we were able to selectively release calcein first, without releasing SykEGFP, by tuning the electrical parameters. As shown in Figure 8.8, the histogram generated by the cell population containing both SykEGFP and calcein was located at substantially higher fluorescent intensity than that of the

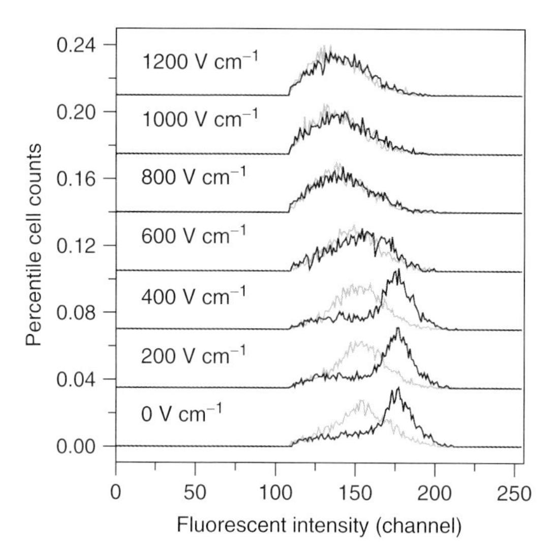


**Figure 8.8** Comparison between the histograms of fluorescent intensity generated by calcein AM loaded SykEGFP-DT40-Syk cells (black) and those generated by SykEGFP-DT40-Syk cells (gray) under different electric field intensities with the duration of 100 ms.

cell population containing only SykEGFP, when the examination was taken at 0, 200, and 400 V $cm^{-1}$. However, the two histograms overlapped when the field intensity was increased to 600 V $cm^{-1}$ and beyond, since calcein was depleted from the cells at this field intensity and duration before the release of SykEGFP started at 800 V $cm^{-1}$. In this case, the selective release of intracellular molecules during electroporation allowed us to single out one type of molecule each time at the single-cell level by tuning the electrical parameters. When combined with a proper detection method, the released molecule can be potentially analyzed at the single-cell level as demonstrated in previous work [71, 72] and different molecules can be analyzed under different operational conditions.

## 8.6 Conclusions

EFC is a newly emerging technique for examining cells by combining flow-through electroporation with flow cytometry. It has been used to detect protein translocation, measure cell biomechanics, and selectively analyze specific intracellular molecules with high throughput. In EFC, electroporation provides a powerful means to break the membrane barrier for access of intracellular molecules or rapid perturbation of the cell state. More applications of EFC remain to be discovered. EFC will find important applications in basic molecular and cell biology research, diagnosis and staging of diseases, and drug discovery.

## Acknowledgments

The authors acknowledge Wallace H. Coulter Foundation, NSF grant CBET-0747105, USDA CSREES NRI grant 2008-01370, and a cooperative agreement with the ARS-USDA, project number 1935 42000-035, through the Center for Food Safety Engineering at Purdue University for research funding.

## References

1 Weaver, J.C. (1993) Electroporation – a general phenomenon for manipulating cells and tissues. *J. Cell. Biochem.*, **51**, 426–435.

2 Weaver, J.C. and Chizmadzhev, Y.A. (1996) Theory of electroporation: a review. *Bioelectrochem. Bioenerg.*, **41**, 135–160.

3 Teissie, J., Golzio, M., and Rols, M.P. (2005) Mechanisms of cell membrane electropermeabilization: a minireview of our present (lack of?) knowledge. *Biochim. Biophys. Acta*, **1724**, 270–280.

4 Neumann, E., Schaefer-Ridder, M., Wand, Y., and Hofschneider, P. (1982) Gene transfer into mouse lyoma cells by electroporation in high electric fields. *EMBO J.*, **1**, 841–845.

5 Rols, M.P. *et al.* (1998) In vivo electrically mediated protein and gene transfer in murine melanoma. *Nat. Biotechnol.*, **16**, 168–171.

6 Aihara, H. and Miyazaki, J. (1998) Gene transfer into muscle by electroporation in vivo. *Nat. Biotechnol.*, **16**, 867–870.

7 Han, F.T. *et al.* (2003) Fast electrical lysis of cells for capillary electrophoresis. *Anal. Chem.*, **75**, 3688–3696.
8 McClain, M.A. *et al.* (2003) Microfluidic devices for the high-throughput chemical analysis of cells. *Anal. Chem.*, **75**, 5646–5655.
9 Wang, H.Y. and Lu, C. (2006) Microfluidic chemical cytometry based on modulation of local field strength. *Chem. Commun. (Camb)*, 3528–3530.
10 Givan, A.L. (2001) *Flow Cytometry: First Principles*, Wiley-Liss, New York.
11 Ormerod, M.G. (ed.) (2000) *Flow Cytometry*, Oxford University Press, Oxford.
12 Dovichi, N.J. and Hu, S. (2003) Chemical cytometry. *Curr. Opin. Chem. Biol.*, **7**, 603–608.
13 Hogan, B.L. and Yeung, E.S. (1992) Determination of intracellular species at the level of a single erythrocyte via capillary electrophoresis with direct and indirect fluorescence detection. *Anal. Chem.*, **64**, 2841–2845.
14 Meredith, G.D., Sims, C.E., Soughayer, J.S., and Allbritton, N.L. (2000) Measurement of kinase activation in single mammalian cells. *Nat. Biotechnol.*, **18**, 309–312.
15 Wu, H.K., Wheeler, A., and Zare, R.N. (2004) Chemical cytometry on a picoliter-scale integrated microfluidic chip. *Proc. Natl. Acad. Sci. U.S.A.*, **101**, 12809–12813.
16 Wang, J. *et al.* (2008) Detection of kinase translocation using microfluidic electroporative flow cytometry. *Anal. Chem.*, **80**, 1087–1093.
17 Bao, N., Zhan, Y., and Lu, C. (2008) Microfluidic electroporative flow cytometry for studying single-cell biomechanics. *Anal. Chem.*, **80**, 7714–7719.
18 Bao, N., Wang, J., and Lu, C. (2008) Microfluidic electroporation for selective release of intracellular molecules at the single-cell level. *Electrophoresis*, **29**, 2939–2944.
19 Wang, H.Y., Bhunia, A.K., and Lu, C. (2006) A microfluidic flow-through device for high throughput electrical lysis of bacterial cells based on continuous dc voltage. *Biosens. Bioelectron.*, **22**, 582–588.
20 Wang, H.Y. and Lu, C. (2006) Electroporation of mammalian cells in a microfluidic channel with geometric variation. *Anal. Chem.*, **78**, 5158–5164.
21 Wang, H.Y. and Lu, C. (2006) High-throughput and real-time study of single cell electroporation using microfluidics: effects of medium osmolarity. *Biotechnol. Bioeng.*, **95**, 1116–1125.
22 Wang, H.Y. and Lu, C. (2008) Microfluidic electroporation for delivery of small molecules and genes into cells using a common DC power supply. *Biotechnol. Bioeng.*, **100**, 579–586.
23 Jacobson, S.C., Culbertson, C.T., Daler, J.E., and Ramsey, J.M. (1998) Microchip structures for submillisecond electrophoresis. *Anal. Chem.*, **70**, 3476–3480.
24 Plenert, M.L. and Shear, J.B. (2003) Microsecond electrophoresis. *Proc. Natl. Acad. Sci. U.S.A.*, **100**, 3853–3857.
25 Wang, J. and Lu, C. (2006) Microfluidic cell fusion under continuous direct current voltage. *Appl. Phys. Lett.*, **89**, 234102.
26 Deptala, A., Bedner, E., Gorczyca, W., and Darzynkiewicz, Z. (1998) Activation of nuclear factor kappa B (NF-kappaB) assayed by laser scanning cytometry (LSC). *Cytometry*, **33**, 376–382.
27 Bedner, E., Li, X., Kunicki, J., and Darzynkiewicz, Z. (2000) Translocation of Bax to mitochondria during apoptosis measured by laser scanning cytometry. *Cytometry*, **41**, 83–88.
28 Ozawa, K., Hudson, C.C., Wille, K.R., Karaki, S., and Oakley, R.H. (2005) Development and validation of algorithms for measuring G-protein coupled receptor activation in cells using the LSC-based imaging cytometer platform. *Cytometry A*, **65A**, 69–76.
29 Pozarowski, P., Holden, E., and Darzynkiewicz, Z. (2005) in *Cell Imaging Techniques: Methods and Protocols* (eds D.J. Taatjes and B.T. Mossman), Humana Press, Totowa, NJ, pp. 165–192.
30 Takata, M. *et al.* (1994) Tyrosine kinases Lyn and Syk regulate B cell receptor-coupled $Ca2+$ mobilization through distinct pathways. *EMBO J.*, **13**, 1341–1349.

31 Turner, M. *et al.* (1995) Perinatal lethality and blocked B-cell development in mice lacking the tyrosine kinase Syk. *Nature*, **378**, 298–302.

32 Zioncheck, T.F., Harrison, M.L., and Geahlen, R.L. (1986) Purification and characterization of a protein-tyrosine kinase from bovine thymus. *J. Biol. Chem.*, **261**, 15637–15643.

33 Zioncheck, T.F., Harrison, M.L., Isaacson, C.C., and Geahlen, R.L. (1988) Generation of an active protein-tyrosine kinase from lymphocytes by proteolysis. *J. Biol. Chem.*, **263**, 19195–19202.

34 Campbell, K.S. (1999) Signal transduction from the B cell antigen-receptor. *Curr. Opin. Immunol.*, **11**, 256–264.

35 DeFranco, A.L. (1997) The complexity of signaling pathways activated by the BCR. *Curr. Opin. Immunol.*, **9**, 296–308.

36 Zhou, F., Hu, J., Ma, H., Harrison, M.L., and Geahlen, R.L. (2006) Nucleocytoplasmic trafficking of the Syk protein tyrosine kinase. *Mol. Cell. Biol.*, **26**, 3478–3491.

37 Ma, H. *et al.* (2001) Visualization of Syk-antigen receptor interactions using green fluorescent protein: differential roles for Syk and Lyn in the regulation of receptor capping and internalization. *J. Immunol.*, **166**, 1507–1516.

38 Ben-Ze'ev, A. (1985) The cytoskeleton in cancer cells. *Biochim. Biophys. Acta*, **780**, 197–212.

39 Rao, K.M. and Cohen, H.J. (1991) Actin cytoskeletal network in aging and cancer. *Mutat. Res.*, **256**, 139–148.

40 Cunningham, C.C. *et al.* (1992) Actin-binding protein requirement for cortical stability and efficient locomotion. *Science*, **255**, 325–327.

41 Katsantonis, J. *et al.* (1994) Differences in the G/total actin ratio and microfilament stability between normal and malignant human keratinocytes. *Cell Biochem. Funct.*, **12**, 267–274.

42 Moustakas, A. and Stournaras, C. (1999) Regulation of actin organisation by TGF-beta in H-ras-transformed fibroblasts. *J. Cell Sci.*, **112** (Pt 8), 1169–1179.

43 Ochalek, T., Nordt, F.J., Tullberg, K., and Burger, M.M. (1988) Correlation between cell deformability and metastatic potential in B16-F1 melanoma cell variants. *Cancer Res.*, **48**, 5124–5128.

44 Evans, E. and Yeung, A. (1989) Apparent viscosity and cortical tension of blood granulocytes determined by micropipet aspiration. *Biophys. J.*, **56**, 151–160.

45 Hochmuth, R.M. (2000) Micropipette aspiration of living cells. *J. Biomech.*, **33**, 15–22.

46 Ward, K.A., Li, W.I., Zimmer, S., and Davis, T. (1991) Viscoelastic properties of transformed cells: role in tumor cell progression and metastasis formation. *Biorheology*, **28**, 301–313.

47 Lekka, M. *et al.* (1999) Elasticity of normal and cancerous human bladder cells studied by scanning force microscopy. *Eur. Biophys. J.*, **28**, 312–316.

48 Mahaffy, R.E., Shih, C.K., MacKintosh, F.C., and Kas, J. (2000) Scanning probe-based frequency-dependent microrheology of polymer gels and biological cells. *Phys. Rev. Lett.*, **85**, 880–883.

49 Rotsch, C., Jacobson, K., and Radmacher, M. (1999) Dimensional and mechanical dynamics of active and stable edges in motile fibroblasts investigated by using atomic force microscopy. *Proc. Natl. Acad. Sci. U.S.A.*, **96**, 921–926.

50 Cross, S.E., Jin, Y.S., Rao, J., and Gimzewski, J.K. (2007) Nanomechanical analysis of cells from cancer patients. *Nat. Nanotechnol.*, **2**, 780–783.

51 Wang, N., Butler, J.P., and Ingber, D.E. (1993) Mechanotransduction across the cell surface and through the cytoskeleton. *Science*, **260**, 1124–1127.

52 Bausch, A.R., Moller, W., and Sackmann, E. (1999) Measurement of local viscoelasticity and forces in living cells by magnetic tweezers. *Biophys. J.*, **76**, 573–579.

53 Bausch, A.R., Ziemann, F., Boulbitch, A.A., Jacobson, K., and Sackmann, E. (1998) Local measurements of viscoelastic parameters of adherent cell surfaces by magnetic bead microrheometry. *Biophys. J.*, **75**, 2038–2049.

54 Beil, M. *et al.* (2003) Sphingosylphosphorylcholine regulates keratin network architecture and visco-elastic properties

of human cancer cells. *Nat. Cell. Biol.*, **5**, 803–811.

55 Desprat, N., Richert, A., Simeon, J., and Asnacios, A. (2005) Creep function of a single living cell. *Biophys. J.*, **88**, 2224–2233.

56 Thoumine, O. and Ott, A. (1997) Time scale dependent viscoelastic and contractile regimes in fibroblasts probed by microplate manipulation. *J. Cell Sci.*, **110** (Pt 17), 2109–2116.

57 Laurent, V.M. *et al.* (2002) Assessment of mechanical properties of adherent living cells by bead micromanipulation: comparison of magnetic twisting cytometry vs optical tweezers. *J. Biomech. Eng.*, **124**, 408–421.

58 Sleep, J., Wilson, D., Simmons, R., and Gratzer, W. (1999) Elasticity of the red cell membrane and its relation to hemolytic disorders: an optical tweezers study. *Biophys. J.*, **77**, 3085–3095.

59 Wang, Y. *et al.* (2005) Visualizing the mechanical activation of Src. *Nature*, **434**, 1040–1045.

60 Tan, J.L. *et al.* (2003) Cells lying on a bed of microneedles: an approach to isolate mechanical force. *Proc. Natl. Acad. Sci. U.S.A.*, **100**, 1484–1489.

61 du Roure, O. *et al.* (2005) Force mapping in epithelial cell migration. *Proc. Natl. Acad. Sci. U.S.A.*, **102**, 2390–2395.

62 Sniadecki, N.J. *et al.* (2007) Magnetic microposts as an approach to apply forces to living cells. *Proc. Natl. Acad. Sci. U.S.A.*, **104**, 14553–14558.

63 Galbraith, C.G. and Sheetz, M.P. (1997) A micromachined device provides a new bend on fibroblast traction forces. *Proc. Natl. Acad. Sci. U.S.A.*, **94**, 9114–9118.

64 Guck, J. *et al.* (2001) The optical stretcher: a novel laser tool to micromanipulate cells. *Biophys. J.*, **81**, 767–784.

65 Guck, J. *et al.* (2005) Optical deformability as an inherent cell marker for testing malignant transformation and metastatic competence. *Biophys. J.*, **88**, 3689–3698.

66 Shelby, J.P., White, J., Ganesan, K., Rathod, P.K., and Chiu, D.T. (2003) A microfluidic model for single-cell capillary obstruction by Plasmodium falciparum-infected erythrocytes. *Proc. Natl. Acad. Sci. U.S.A.*, **100**, 14618–14622.

67 Rosenbluth, M.J., Lam, W.A., and Fletcher, D.A. (2008) Analyzing cell mechanics in hematologic diseases with microfluidic biophysical flow cytometry. *Lab Chip*, **8**, 1062–1070.

68 Golzio, M. *et al.* (1998) Control by osmotic pressure of voltage-induced permeabilization and gene transfer in mammalian cells. *Biophys. J.*, **74**, 3015–3022.

69 Johnson, M.D., Torri, J.A., Lippman, M.E., and Dickson, R.B. (1999) Regulation of motility and protease expression in PKC-mediated induction of MCF-7 breast cancer cell invasiveness. *Exp. Cell Res.*, **247**, 105–113.

70 Benz, R., Beckers, F., and Zimmermann, U. (1979) Reversible electrical breakdown of lipid bilayer membranes: a charge-pulse relaxation study. *J. Membr. Biol.*, **48**, 181–204.

71 Gao, N. *et al.* (2006) High-throughput single-cell analysis for enzyme activity without cytolysis. *Anal. Chem.*, **78**, 3213–3220.

72 Gao, N., Zhao, M., Zhang, X., and Jin, W. (2006) Measurement of enzyme activity in single cells by voltammetry using a microcell with a positionable dual electrode. *Anal. Chem.*, **78**, 231–238.

# 9
# Ultrasensitive Analysis of Individual Cells via Droplet Microfluidics

*Robert M. Lorenz and Daniel T. Chiu*

## 9.1
## Introduction

The cell is the fundamental unit of biology, which spans all levels of life from simple single-cell to complex multicellular organisms, and accordingly serves the dual purposes of structure and function. There exists a variety of techniques capable of examining cells on an individual level, such as capillary electrophoresis (CE) [1], flow cytometry [2], fluorescence microscopy [3], electrochemical detection [4], mass spectrometry [5], and nuclear magnetic resonance imaging [6] to name a few. Each technique has its own strengths and weaknesses with regard to the nature of the desired measurement. Some of these techniques, such as microscopy, give spatial information or the physical distribution of analytes, while other techniques, such as mass spectrometry, yield the chemical identities from cellular samples. We have chosen the use of droplet microfluidics for our single-cell studies because of the advantages droplets offer in sample handling.

In this chapter, the particular experimental requirements, such as handling exceedingly small volumes and detecting miniscule amounts of analyte, necessary for working with individual cells and droplets have been discussed. Furthermore, how microfluidics, particularly droplet-based microfluidics, is especially well suited to meet the needs presented by single-cell analysis is described, and the ways droplets are generated, methods for controlling the interior droplet environment, encapsulation of cells within droplets, and optical methods for control and sensitive detection have been covered.

## 9.2
## Droplet Properties

With a multitude of techniques and applications developed since the late 1980s, the field of microfluidics has become a mature area of research for studying small-scale phenomena in a fluid environment [7], and has even been used to study single cells in restricted volumes [8]. One particularly exciting subarea of microfluidics, termed,

*Chemical Cytometry.* Edited by Chang Lu

ISBN: 978-3-527-32495-8

*droplet microfluidics*, has significantly grown due to its demonstrated usefulness with handling very small samples, in a high-throughput manner [9–12]. The utility can be seen with applications such as protein crystallization, biological assays, and encapsulation of reagents.

A droplet consists of a small volume of aqueous fluid that is surrounded by an immiscible oil; when working with cells, the aqueous phase is a biological buffer or the appropriate growth media for the cell type. There exists more variety with respect to the oil, natural oils (soybean oil), organics (decanol and mineral oil), silicone oil, and fluorinated oils, being used. Additionally, surfactants are commonly used to aid in formation and stabilization of the droplets.

The primary challenges of single-cell studies are interfacing with the sample and detecting the desired analyte. Common cellular sample preparations, whether from collected tissues or grown in culture, work with macroscopic portions, and are on the order of tens of thousands to millions of cells. A reduction in handled sample size is not feasible with current preparation protocols. Also, there is a significant challenge presented by the limited amount of analyte to be detected [13]: an average mammalian cell with a diameter of 10 μm comprises a total volume of ~500 fl. Highly sensitive detection schemes are needed to accommodate this, especially when a protein or biomolecule of interest makes up only a miniscule fraction of the total cellular volume. To further exacerbate the problem, if the cell is lysed in the interest of detecting the contents, the analytes are subject to rapid dilution and further reduction of sample concentration.

Droplet microfluidics addresses these issues by encapsulating the cell, thereby defining the local environment. Furthermore, encapsulation provides a "handle" by which the droplet sample can be mixed with labeling or fixing reagents, moved, sorted, and collected. Dilution of the sample is also prevented, and can even be concentrated to aid in signal detection.

## 9.3 Droplet Generation

There are two primary methods for creating droplets: streaming and discrete droplet generation. Streaming droplet generation, or continuous droplet generation, is best suited for high-throughput applications, due to the fact that droplet streams are created on the order of hertz to kilohertz, allowing a great many samples to be potentially processed. The two main architectures for achieving such a rapid generation are known as the *T-channel* or *T-junction* and *flow focusing* [14].

The T-channel design comprises a primary channel through which flows the inert carrier fluid (silicone or another oil); perpendicular to the primary channel is the aqueous channel that supplies the sample to be encapsulated. As seen in Figure 9.1, the oil supplies a shearing force that continuously forms aqueous droplets; moreover, the interfacial tension between the two liquids, as well as their relative flow rates, determines the frequency and size of formed droplets [14, 15].

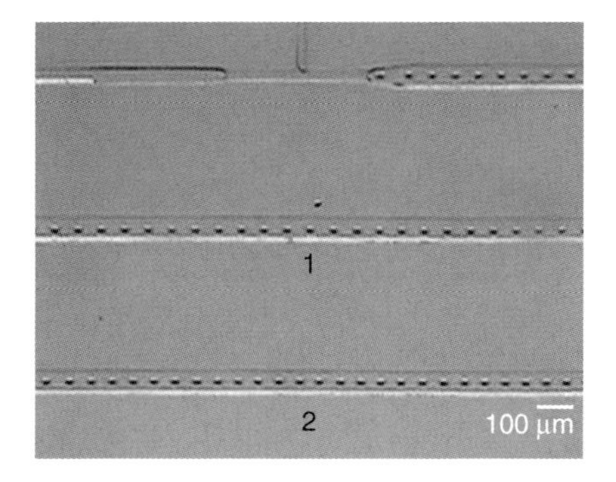


**Figure 9.1** Image showing a microfluidic T-junction and the generation of a steady-state stream of aqueous droplets.(Reproduction from [12], Copyright 2009 by American Chemical Society.)

Flow focusing–based droplet generation consists of an aqueous flow channel that is sandwiched between two parallel oil channels, where the three streams converge at an orifice, and the oil streams compress and destabilize the aqueous thread to form a droplet train. With fast oil and aqueous flow rates, the unstable jet creates droplets with diameters that are proportional to the aqueous jet diameter. Conversely, slower flow rates produce dripping, where the droplet diameter is smaller than the orifice, and droplet diameter decreases as a function of decreasing flow-rate ratio [14, 16].

Discrete or discontinuous droplet generation consists of creating individual droplets on demand, in a determined location. This bestows a benefit of enhanced control over how the droplet is utilized; the stationary droplet can easily be left in the initial position or moved for detection, as well as manipulated (i.e., fused with another droplet to add labeling reagents or reactive substrates). This technique is well suited to working with limited samples and long duration detection schemes, such as those found with single-cell samples.

One technique for discrete droplet generation is achieved with pressure pulses [17]. As seen in Figure 9.2, a stable interface between the aqueous and oil phase is maintained at the opening between an inlet channel and the central chamber. A pressure pulse is applied via a precision microinjector and the aqueous phase advances toward the main chamber, where it protrudes with a hemispherical shape that eventually detaches from the neck of the aqueous "finger," thereby forming a droplet. This process can be repeated from different inlets to produce droplets with a variety of encapsulated contents for multistep sample preparations.

In addition to using pressure, droplets can be generated using an applied electric field (18, 19). With this method, a short duration (milliseconds) high field (kilovolts) is applied across the water–oil interface, with the result of an aqueous jet being formed. The jet breaks into droplets due to Rayleigh instability, with the droplet size being a function of the amplitude and duration of the applied pulse, as well as the channel dimensions that define the interface.

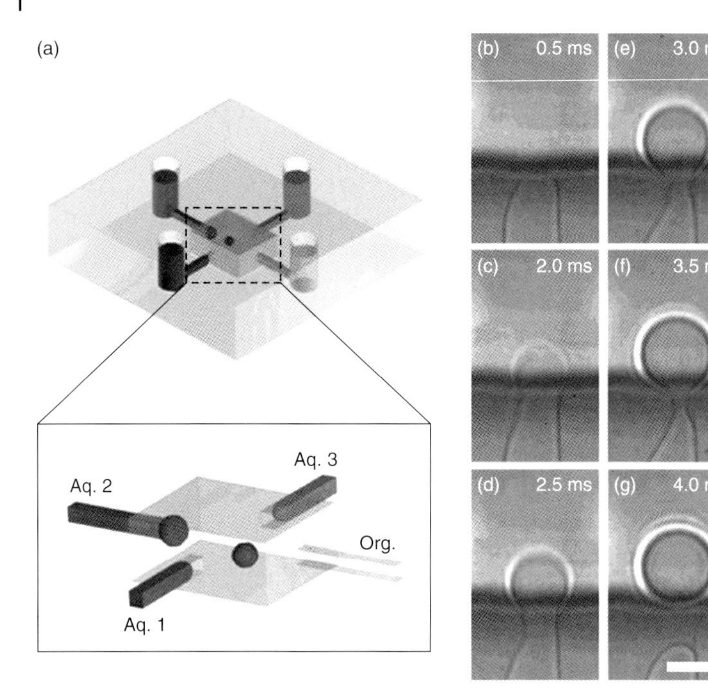


**Figure 9.2** (a) Schematic depicting single-droplet generation. The inset shows the inlets with aqueous solutions (Aq.) of different chemical compositions; Org, organic oil phase. (b–g) Images showing the generation of a water droplet. The aqueous and immiscible phase initially formed a stable interface at the opening of the inlet channel (b). By increasing the pressure of the aqueous phase (c and d), a droplet is formed at the inlet opening, and by quickly reducing the backing pressure of the aqueous phase (e and f), the aqueous neck shrinks, which led to the break off of the droplet (g). The scale bar is 4 μm. (Reproduction from [17], Copyright 2006 by American Chemical Society.)

## 9.4
## Cell Encapsulation

Although it is possible for clusters of cells or even multicellular organisms [20] to be encapsulated in droplets, there is a focus to have only one cell per droplet [21–25], to better characterize and understand the information gleaned from downstream processing: derivatization, sorting, and detecting. When an aqueous droplet is generated, whatever comprises the dispersed phase (buffer, nutrients, and suspended cells) becomes confined in the volume defined by the droplet boundary, thereby granting incredible control over a cell's local environment. With the appropriate conditions such as a gas permeable oil (fluorinated oil), a biocompatible surfactant, and a large enough surrounding volume of buffer, cells can replicate in droplets [26] or be collected later for culture [27].

Encapsulated cells are surrounded by an aqueous envelope of media or buffer, and depending on the nature of the desired experiment, the volumes can range from

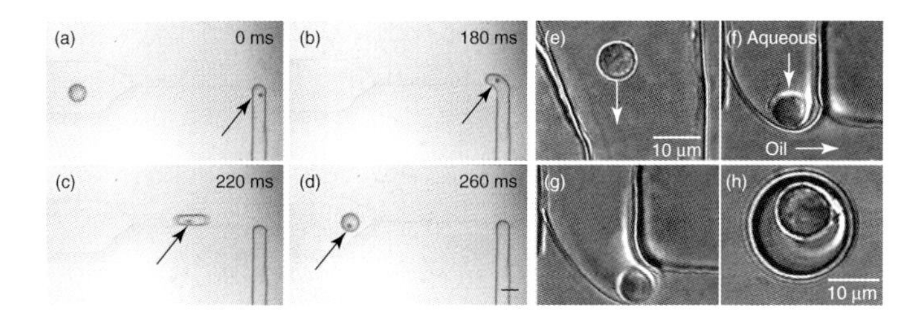


**Figure 9.3** Two sequences of images showing a B lymphocyte being encapsulated in an aqueous droplet. (a) The cell is transported to the interface by bulk fluid flow in the aqueous channel. (b and c) The droplet is sheared off creating a droplet encapsulated cell (d). Scale bar represent 30 µm for (d). (e–h) Closer view of encapsulating a single B lymphocyte. An optical trap was used to place the cell at the interface in (f). (Reproduction from [28, 29], Copyright 2005 and 2007 by American Chemical Society.)

nanoliters down to femtoliters; we match the droplet volume closely (picoliters) to the cell size to minimize sample dilution. Figure 9.3 shows the encapsulation process for both streaming and on demand: once a cell is at the interface, the droplet is sheared off with the cell inside, with the process being continuous and not all droplets containing cells. For guaranteed cell encapsulation, an optical trap can be used to position the cell at the interface, as seen in Figure 9.3e and f. The cell, cellular secretions, and any chemicals the aqueous solution contains are also encapsulated. These metabolic products and cellular material itself can be used in droplet for fluorescence readout or be introduced to another sensitive detection scheme, both of which are described later in the chapter.

## 9.5 Droplet Manipulation

When using droplets to confine cells, physical control over the droplet is important: to gain information, it is necessary to move the sample to a detection area, or to further interact with the sample by adding or removing contents from the droplet. There are a number of ways to apply force to droplets to satisfy the previously mentioned needs: hydrodynamic flow [30], dielectrophoresis [31], thermal gradients [32], and electrowetting [33] are some of the ways by which droplets can be moved. For this work, the technique of optical trapping to manipulate droplets has been chosen [17, 28, 34–39].

Optical trapping, also known as *optical tweezers*, takes the advantage of radiation pressure, or the force exerted on an object when exposed to photons (light). By focusing laser light through a high numerical aperture (NA) objective, a high-intensity-gradient region is created at the laser focal volume. This intensity gradient exerts a force on objects that have a higher refractive index than the surrounding medium [36]. Because it is generated by a laser, the optical trap seamlessly integrates with the transparent materials used for microfluidic devices and the microscope optics used for imaging and detection.

Although common, optical tweezers are not the best trap for aqueous droplets in oil. Aqueous droplets are not trapped by the standard single-beam gradient trap, due to the refractive index of water being lower than the surrounding oil (except for the case of fluorinated oils, which presents other challenges), thereby producing a repulsive instead of attractive force [17]. Our solution was to employ a Laguerre Gaussian trap (Figure 9.4), commonly referred to as a *vortex trap* (17, 37). This trap is created by sending the laser beam through a computer generated hologram (CGH), which changes the phase of the propagating light beam. This screw dislocation of the phase results in a trap with a torroidal region of intensity and a dark core. Aqueous droplets are trapped in the dark core, with the added benefit of the trap inhibiting unwanted droplet fusion by acting as a repellant force field [37].

Given that droplet fusion is necessary for work with encapsulated cells, a method to fuse droplets in a controlled manner using the vortex trap was developed [37]. First, two traps were generated by sending the trapping beam through a polarization beam-splitting cube. Next, a dove prism to break the symmetry of the beams was used, followed by lateral translation of the CGH with respect to the beam, to partially open the traps. The trapped droplets appear to follow the dark core movement, until the opening in the trap brings the droplets into contact and coalescence. This technique provides complete control on droplet positioning, as well as droplet fusion, which is important in the preparation of droplet samples for reaction and readout.

Besides using optical vortex traps to manipulate droplets, hydrodynamic flow to move and trap droplets was also used [40]. When working with more than one sample, it is useful if the other samples can be stored, for later use. Droplet docking

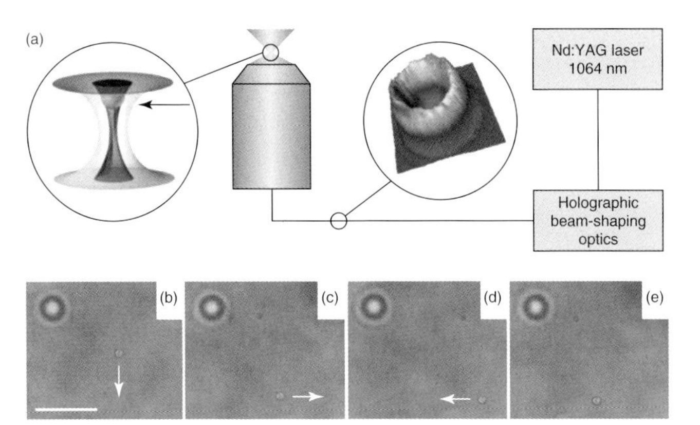


**Figure 9.4** (a) The vortex trap was formed by sending a Gaussian laser beam ($TEM_{00}$) through a microfabricated hologram to form the Laguerre Gaussian (LG) beam, or optical vortex, after which the desired LG mode was selected and spatially filtered, then used for vortex trapping. The arrow indicates the trapping position of a droplet in the vortex trap. (b–e) Images depicting the trapping and translation of an aqueous droplet (in focus). The arrow denotes the direction of translation of the vortex trap, illustrated using the surface bound droplet in the top left as a reference (out of focus). The scale bar in (b) represents 10 μm. (Reprinted with permission from [34].)

is a way to spatially index samples based on the order of filling, so that sequential individual observation, or simultaneous observation, is possible (Figure 9.9h). Additionally, docking is a suitable option if optical trapping is not available.

## 9.6 Droplet Concentration Control

While it is possible to set the initial concentrations of chemicals in the aqueous phase, we have discovered another way to control droplet–content concentration by using the vortex trap [34, 35]. Besides trapping aqueous droplets, the vortex trap can be used to modulate the droplet volume, which effects a change in concentration within the droplet. As can be seen in Figure 9.5, the droplet is trapped in the dark core. There is a slight overlap of the trap with the interface of the droplet, where the droplet is slightly heated (≤1 K) with the use of a laser beam. This gentle heating increases the solubility of water in oil, in a small localized region, which results in droplet shrinkage. Droplet shrinkage in oil for concentration has also

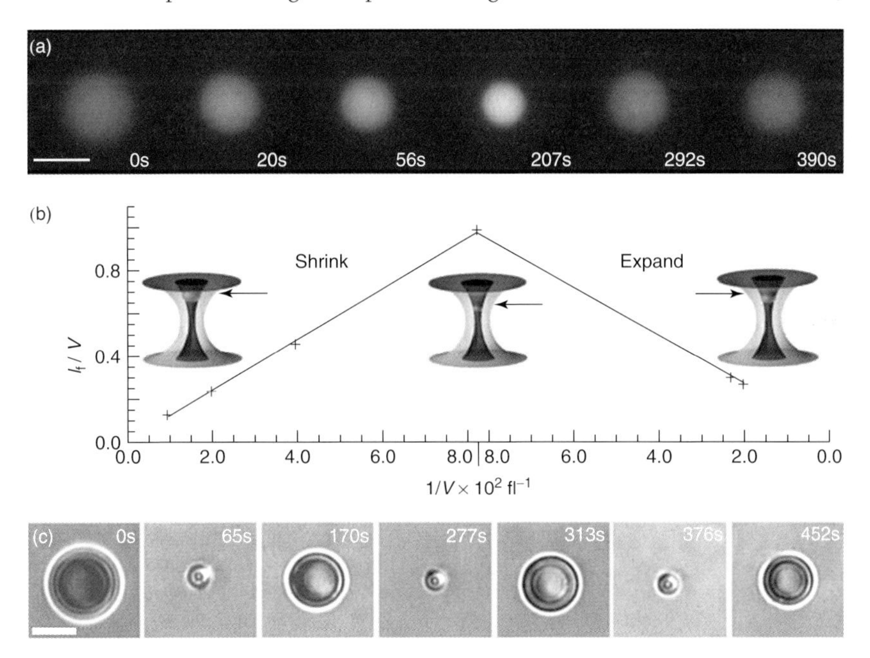


**Figure 9.5** (a) Fluorescence images of an aqueous droplet containing ~100 μM dye (Alexa 488) as the droplet went through one cycle of shrinkage and expansion. The scale bar represents 3 μm. (b) Plot of normalized fluorescent intensity per unit volume versus the respective reciprocal volume, depicting species conservation; the insets illustrate the change in the trapping position of the droplet (arrow) in the axial direction as the droplet changes in volume. (c) Images showing three consecutive cycles of droplet shrinkage and expansion, which demonstrates the reversibility of the effect. The scale bar represents 5 μm. (Reprinted with permission from [34].)

been demonstrated without the need of a trap [41]. Upon turning the trap off, the droplet cools and regains volume. This process can be repeated many times over, and can even be used to enlarge a droplet (the droplet to be enlarged is placed in the vicinity of a droplet that is shrunk with the trap).

The encapsulated biomolecules do not partition into oil due to their high ionic charge or molecular weight, which has allowed us to see orders of magnitude sample enrichment. An additional benefit is that the slight temperature change ($\leq$1 K) does not damage entrapped cells or small molecules. Overall, this technique provides a great method for controlling sample concentrations, a unique feat, which would be very challenging to achieve on a larger scale.

## 9.7 Temperature Control of Droplets

Besides using optical traps to manipulate droplet concentrations and locations, thermoelectric devices to adjust the temperature of droplets were also utilized [29]. In Figure 9.6, we used a T-channel device with a thermoelectric cooler to

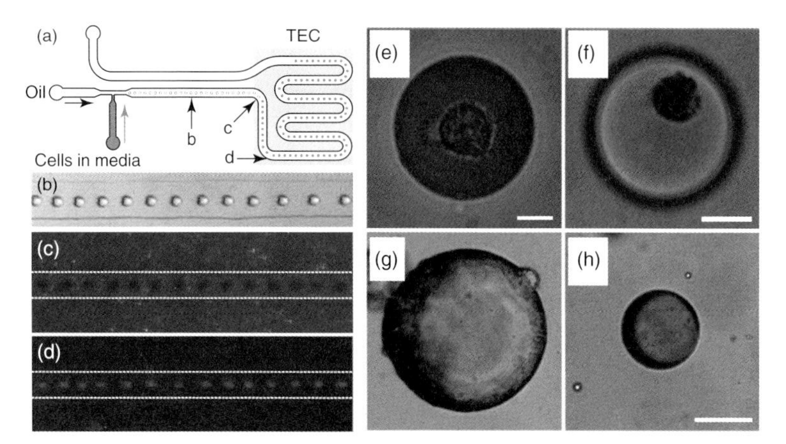


**Figure 9.6** (a-d) Freezing of droplet stream. (a) A schematic showing the T-channel geometry that was used for droplet generation; the serpentine channel downstream of the T-channel and under which the thermoelectric cooler (TEC) is placed was made long to ensure the transit time of droplets for a sufficient duration across the TEC so that the droplets would freeze and remain fully frozen. (b) Bright-field image showing a stream of unfrozen droplets. (c and d) Images of a stream of droplets over the TEC before (c) and after (d) freezing. Labels in (a) show the positions along the channel where the images shown in (b–d) were taken. The main channel in this experiment was 100 μm wide and 90 μm high. (e–h) Behavior of cells and droplets after freezing. (e and f) A cell in a droplet composed of 50% cell media and 50% PBS containing trypan blue (at 0.4% w/v in PBS) when alive (e) and dead (f). (g) A thawed droplet in AS 4 silicone oil and (h) in mineral oil. Note the matrix surrounding the droplet in (g). Both scale bars (e and f) represent 10 μm, and the one in (h) is 40 μm. (Reproduction from [29], Copyright 2007 by American Chemical Society.)

lower the temperature on a region of the chip; as droplets took longer to traverse the serpentine bends, they cooled and eventually froze. Because of the marked difference in the freezing temperature of aqueous solutions and oil (−100 °C for silicone oils and ~0 °C for aqueous samples), flow was not disturbed within the chip. Furthermore, a broad temperature range within which to operate is made available, when considering the upper temperature threshold, which would be limited by fragility of the sample (the boiling temperature of water being 100 °C and higher for oil).

To further investigate the usefulness of controlling droplet temperatures, droplets containing cells were frozen, and subsequently their viability as a function of time, oil, and with cryoprotectants was monitored (Figure 9.7). With the addition of DMSO to the cell media, cell viability after thawing was improved. With this

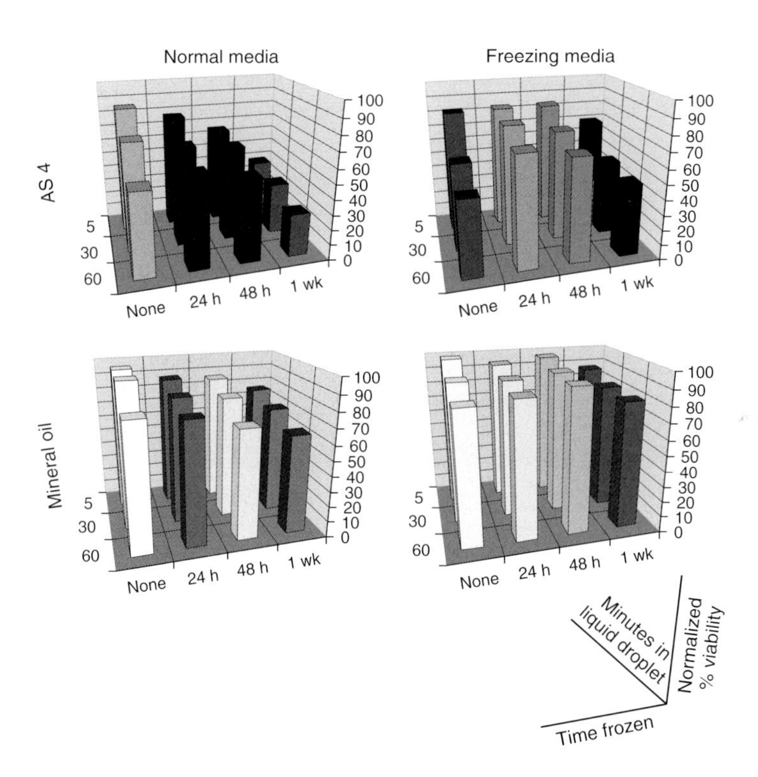


**Figure 9.7** Cell viability after being frozen in aqueous droplets. The left column shows data from cells in normal media (50/50 cell media and PBS containing trypan blue (at 0.4% w/v in PBS)), whereas the right column shows results for cells in freezing media (which contained an additional 5% DMSO supplemented to the cell media). The top row is data for cells frozen in droplets in AS 4 silicone oil and the bottom row shows data for cells frozen in droplets in mineral oil. Normalized cell viability was calculated by dividing the number of cells viable at the various time points in question by the number of viable cells in the original sample. (Reproduction from [29], Copyright 2007 by American Chemical Society.)

technology, temperature can be controlled regionally on fluidic devices for both cooling and heating, which would enable kinetic control of biomolecules.

## 9.8 Detection in Droplets

Fluorescence microscopy capable of detecting single molecules with immunolabeling has become one of the best techniques for the visual readout of cells. In Figure 9.8, we take advantage of fluorescence detection while conducting an enzymatic assay [28]. By releasing the enzyme $\beta$-galactosidase from an encapsulated cell surrounded by a fluorogenic substrate, we observed with time that there is an increase in signal, which demonstrates that the enzyme remained active, and that the fluorescent product generated did not partition into the oil phase. Generally, diffusion would reduce the fluorescent signal making such a measurement very challenging. Through encapsulation, a vast number of enzymatic assays are made possible.

Although droplet-confined detection is convenient, the number of fluorescent dyes that can be used simultaneously is limited; moreover, having access to more sophisticated detection schemes such as CE would provide an even richer source of information. It is with this motivation that we developed a means to

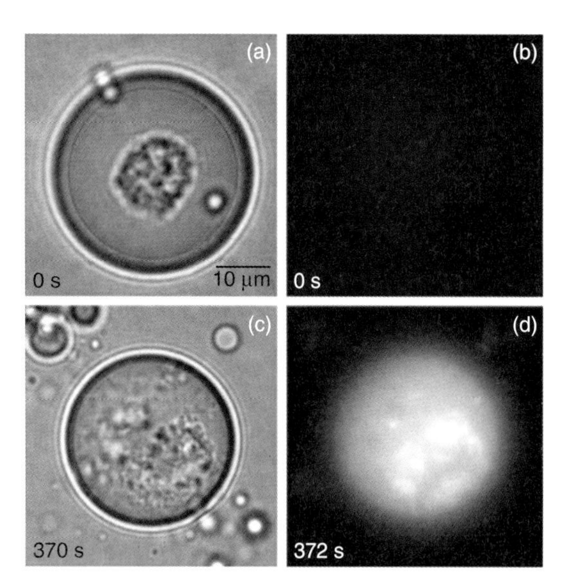


**Figure 9.8** Single-cell enzymatic assay within an aqueous droplet in soybean oil. (a) A mast cell was encapsulated in an aqueous droplet that contained the fluorogenic substrate FDG. (b) Prior to photolysis of the cell, there was little fluorescent product within the droplet because the intracellular enzyme $\beta$-galactosidase was physically separated from FDG by the cell membrane. (c and d) After laser-induced cell lysis (c), $\beta$-galactosidase catalyzed the formation of the product fluorescein, which caused the droplet to become highly fluorescent (d). The scale bar in (a) applies to all panels. (Reproduction from [28], Copyright 2005 by American Chemical Society.)

introduce droplets as sample aliquots in CE separation [42]. Figure 9.9 shows how we use an immiscible fluid partition to separate droplet generation from the CE separation channel. To initiate the separation, the droplet is introduced into the separation channel by fusing it with the immiscible barrier; the results are shown in Figure 9.9g. To further take advantage of the inherent advantages of droplets, a method for collecting the electrophoretically separated bands into droplets was devised (Figure 9.9h) [40]. This leads to the enrichment and encapsulation of the analyte of interest, with the added benefit of being docked, so that observation or retrieval is simplified.

With the limited amount of sample presented by a single cell, it is imperative that the detection method be as sensitive as possible so as to solve the "needle in a haystack" problem of finding low-copy-number molecules. For this, a method for measuring electrophoretic mobility of individual molecules in continuous flowing CE separations was developed [43]. The critical part of the technique is the use of two line-confocal detection beams (Figure 9.10), which illuminate a 2-μm-wide channel, where the transit time of the fluorescent molecule between the beams is cross correlated (Figure 9.10c). This yields 94% detection efficiency for single dye molecules, and makes the study and counting of rare molecules in cell samples a reality.

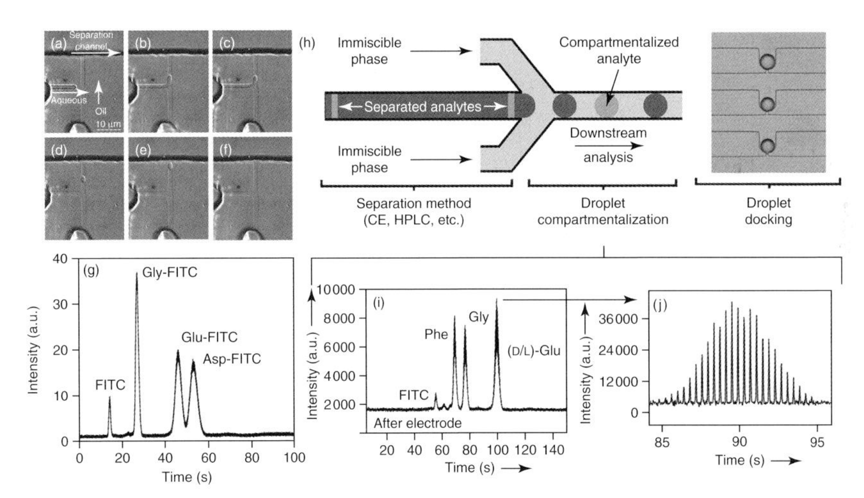


**Figure 9.9** (a-f) Images showing the generation and transport of a single aqueous droplet to the CE separation channel. (g) Separation of fluorescein isothiocyanate (FITC)-labeled amino acid (glycine (Gly), glutamic acid (Glu), and aspartic acid (Asp)) contents of a single 10 fl volume droplet; the applied voltage was $500\,V\,cm^{-1}$ and the separation distance was 2 cm. (h–j) Droplet compartmentalization of CE separation: (h) schematic illustrating the technique for encapsulating the outflow of a chemical separation into individual droplets, which can be docked and stored on chip for further analysis; (i) electropherogram showing the separation of a mixture of amino acids; (j) s blowup of the glutamate peak showing the encapsulation of separated band into a series of droplets. (Reproduction from [40, 42], Copyright 2009 and 2006 by American Chemical Society).

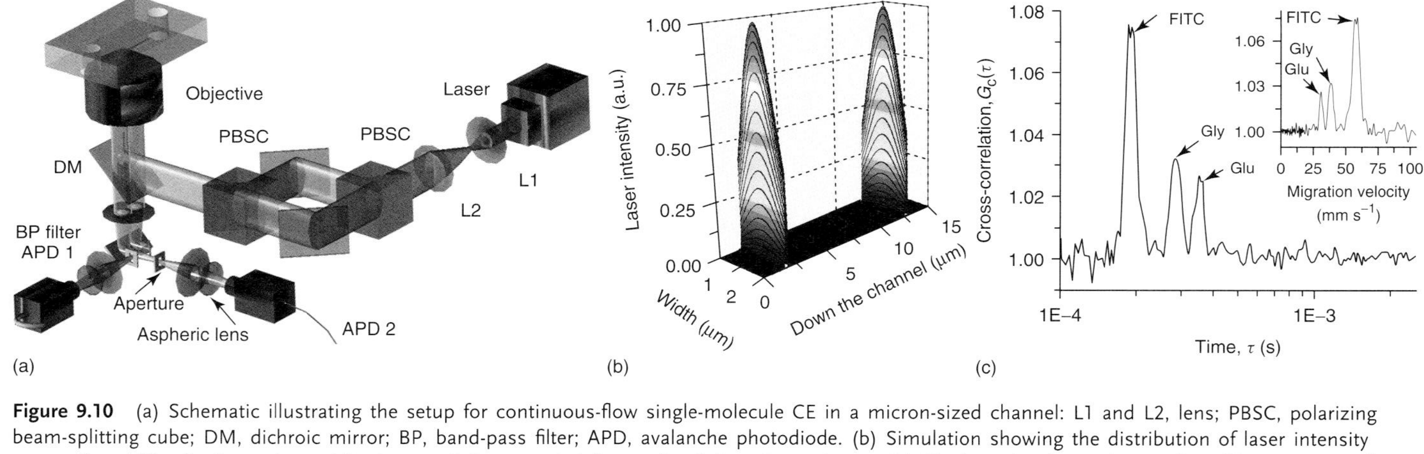


**Figure 9.10** (a) Schematic illustrating the setup for continuous-flow single-molecule CE in a micron-sized channel: L1 and L2, lens; PBSC, polarizing beam-splitting cube; DM, dichroic mirror; BP, band-pass filter; APD, avalanche photodiode. (b) Simulation showing the distribution of laser intensity across the width of a 2 μm channel for two spatially separated line-confocal detection volumes. (c) Single-molecule continuous-flow CE separation of a mixture of FITC, FITC-labeled glycine, and FITC-labeled glutamate. The inset shows the same data as a function of migration velocity. (Reprinted with permission from [43].)

## 9.9 Conclusions

The comprehensive chemical analysis of single cells is a demanding technical challenge. We believe that droplet-based microfluidics is particularly well suited for this task. This chapter highlights techniques that have been developed recently for the generation, manipulation, and analysis of single droplets. With continued development, refinement, and integration of these techniques into a single robust platform, we envision that droplets will play a central role in the next generation of high-sensitivity single-cell analytical techniques.

## References


1. Xie, W.J., Xu, A.S., and Yeung, E.S. (2009) Determination of NAD(+) and NADH in a single cell under hydrogen peroxide stress by capillary electrophoresis. *Anal. Chem.*, **81**, 1280–1284.
2. Herzenberg, L.A., Parks, D., Sahaf, B., Perez, O., Roederer, M., and Herzenberg, L.A. (2002) The history and future of the fluorescence activated cell sorter and flow cytometry: a view from Stanford. *Clin. Chem.*, **48**, 1819–1827.
3. Vukojevic, V., Heidkamp, M., Ming, Y., Johansson, B., Terenius, L., and Rigler, R. (2008) Quantitative single-molecule imaging by confocal laser scanning microscopy. *Proc. Natl. Acad. Sci. U.S.A.*, **105**, 18176–18181.
4. Adams, K.L., Puchades, M., and Ewing, A.G. (2008) In vitro electrochemistry of biological systems. *Annu. Rev. Anal. Chem.*, **1**, 329–355.
5. Rubakhin, S.S. and Sweedler, J.V. (2008) Quantitative measurements of cell-cell signaling peptides with single-cell MALDI MS. *Anal. Chem.*, **80**, 7128–7136.
6. Pinkernelle, J., Teichgraber, U., Neumann, F., Lehmkuhl, L., Ricke, J., Scholz, R., Jordan, A., and Bruhn, H. (2005) Imaging of single human carcinoma cells in vitro using a clinical whole-body magnetic resonance scanner at 3.0 T. *Magn. Reson. Med.*, **53**, 1187–1192.
7. West, J., Becker, M., Tombrink, S., and Manz, A. (2008) Micro total analysis systems: latest achievements. *Anal. Chem.*, **80**, 4403–4419.
8. Huang, B., Wu, H.K., Bhaya, D., Grossman, A., Granier, S., Kobilka, B.K., and Zare, R.N. (2007) Counting low-copy number proteins in a single cell. *Science*, **315**, 81–84.
9. Huebner, A., Sharma, S., Srisa-Art, M., Hollfelder, F., Edel, J.B., and Demello, A.J. (2008) Microdroplets: a sea of applications? *Lab Chip*, **8**, 1244–1254.
10. Song, H., Chen, D.L., and Ismagilov, R.F. (2006) Reactions in droplets in microflulidic channels. *Angew. Chem. Int. Ed.*, **45**, 7336–7356.
11. Teh, S.Y., Lin, R., Hung, L.H., and Lee, A.P. (2008) Droplet microfluidics. *Lab Chip*, **8**, 198–220.
12. Chiu, D.T. and Lorenz, R.M. (2009) Chemistry and biology in femtoliter and picoliter volume droplets. *Acc. Chem. Res.*, **42**, 649–658.
13. Chiu, D.T. (2003) Micro- and nano-scale chemical analysis of individual sub-cellular compartments. *TrAC Trends Anal. Chem.*, **22**, 528–536.
14. Christopher, G.F. and Anna, S.L. (2007) Microfluidic methods for generating continuous droplet streams. *J. Phys. D: Appl. Phys.*, **40**, R319–R336.
15. Thorsen, T., Roberts, R.W., Arnold, F.H., and Quake, S.R. (2001) Dynamic pattern formation in a vesicle-generating microfluidic device. *Phys. Rev. Lett.*, **86**, 4163–4166.

16 Anna, S.L., Bontoux, N., and Stone, H.A. (2003) Formation of dispersions using "flow focusing" in microchannels. *Appl. Phys. Lett.*, **82**, 364–366.
17 Lorenz, R.M., Edgar, J.S., Jeffries, G.D.M., and Chiu, D.T. (2006) Microfluidic and optical systems for the on-demand generation and manipulation of single femtoliter-volume aqueous droplets. *Anal. Chem.*, **78**, 6433–6439.
18 He, M., Kuo, J.S., and Chiu, D.T. (2006) Effects of ultrasmall orifices on the electrogeneration of femtoliter-volume aqueous droplets. *Langmuir*, **22**, 6408–6413.
19 He, M.Y., Kuo, J.S., and Chiu, D.T. (2005) Electro-generation of single femtoliter- and picoliter-volume aqueous droplets in microfluidic systems. *Appl. Phys. Lett.*, **87**, 031916.
20 Clausell-Tormos, J., Lieber, D., Baret, J.C., El-Harrak, A., Miller, O.J., Frenz, L., Blouwolff, J., Humphry, K.J., Koster, S., Duan, H., Holtze, C., Weitz, D.A., Griffiths, A.D., and Merten, C.A. (2008) Droplet-based microfluidic platforms for the encapsulation and screening of mammalian cells and multicellular organisms. *Chem. Biol.*, **15**, 427–437.
21 Chabert, M. and Viovy, J.L. (2008) Microfluidic high-throughput encapsulation and hydrodynamic self-sorting of single cells. *Proc. Natl. Acad. Sci. U.S.A.*, **105**, 3191–3196.
22 Edd, J.F., Di Carlo, D., Humphry, K.J., Koster, S., Irimia, D., Weitz, D.A., and Toner, M. (2008) Controlled encapsulation of single-cells into monodisperse picolitre drops. *Lab Chip*, **8**, 1262–1264.
23 Huebner, A., Olguin, L.F., Bratton, D., Whyte, G., Huck, W.T.S., de Mello, A.J., Edel, J.B., Abell, C., and Hollfelder, F. (2008) Development of quantitative cell-based enzyme assays in microdroplets. *Anal. Chem.*, **80**, 3890–3896.
24 Boedicker, J.Q., Li, L., Kline, T.R., and Ismagilov, R.F. (2008) Detecting bacteria and determining their susceptibility to antibiotics by stochastic confinement in nanoliter droplets using plug-based microfluidics. *Lab Chip*, **8**, 1265–1272.
25 Roman, G.T., Chen, Y.L., Viberg, P., Culbertson, A.H., and Culbertson, C.T. (2007) Single-cell manipulation and analysis using microfluidic devices. *Anal. Bioanal. Chem.*, **387**, 9–12.
26 Holtze, C., Rowat, A.C., Agresti, J.J., Hutchison, J.B., Angile, F.E., Schmitz, C.H.J., Koster, S., Duan, H., Humphry, K.J., Scanga, R.A., Johnson, J.S., Pisignano, D., and Weitz, D.A. (2008) Biocompatible surfactants for water-in-fluorocarbon emulsions. *Lab Chip*, **8**, 1632–1639.
27 Koster, S., Angile, F.E., Duan, H., Agresti, J.J., Wintner, A., Schmitz, C., Rowat, A.C., Merten, C.A., Pisignano, D., Griffiths, A.D., and Weitz, D.A. (2008) Drop-based microfluidic devices for encapsulation of single cells. *Lab Chip*, **8**, 1110–1115.
28 He, M.Y., Edgar, J.S., Jeffries, G.D.M., Lorenz, R.M., Shelby, J.P. and Chiu, D.T. (2005) Selective encapsulation of single cells and subcellular organelles into picoliter- and femtoliter-volume droplets. *Anal. Chem.*, **77**, 1539–1544.
29 Sgro, A.E., Allen, P.B., and Chiu, D.T. (2007) Thermoelectric manipulation of aqueous droplets in microfluidic devices. *Anal. Chem.*, **79**, 4845–4851.
30 Tan, Y.C., Fisher, J.S., Lee, A.I., Cristini, V., and Lee, A.P. (2004) Design of microfluidic channel geometries for the control of droplet volume, chemical concentration, and sorting. *Lab Chip*, **4**, 292–298.
31 Hunt, T.P., Issadore, D., and Westervelt, R.M. (2008) Integrated circuit/microfluidic chip to programmably trap and move cells and droplets with dielectrophoresis. *Lab Chip*, **8**, 81–87.
32 Cordero, M.L., Burnham, D.R., Baroud, C.N., and McGloin, D. (2008) Thermocapillary manipulation of droplets using holographic beam shaping: microfluidic pin ball. *Appl. Phys. Lett.*, **93**, 034107.
33 Kuo, J.S., Spicar-Mihalic, P., Rodriguez, I., and Chiu, D.T. (2003) Electrowetting-induced droplet movement in an immiscible medium. *Langmuir*, **19**, 250–255.
34 Jeffries, G.D.M., Kuo, J.S., and Chiu, D.T. (2007) Dynamic modulation of

chemical concentration in an aqueous droplet. *Angew. Chem. Int. Ed.*, **46**, 1326–1328.

35 Jeffries, G.D.M., Kuo, J.S., and Chiu, D.T. (2007) Controlled shrinkage and re-expansion of a single aqueous droplet inside an optical vortex trap. *J. Phys. Chem. B*, **111**, 2806–2812.

36 Kuyper, C.L. and Chiu, D.T. (2002) Optical trapping: a versatile technique for biomanipulation. *Appl. Spectrosc.*, **56**, 300a–312a.

37 Lorenz, R.M., Edgar, J.S., Jeffries, G.D.M., Zhao, Y.Q., McGloin, D., and Chiu, D.T. (2007) Vortex-trap-induced fusion of femtoliter-volume aqueous droplets. *Anal. Chem.*, **79**, 224–228.

38 Zhao, Y.Q., Edgar, J.S., Jeffries, G.D.M., McGloin, D., and Chiu, D.T. (2007) Spin-to-orbital angular momentum conversion in a strongly focused optical beam. *Phys. Rev. Lett.*, **99**, 073901.

39 Zhao, Y.Q., Milne, G., Edgar, J.S., Jeffries, G.D.M., McGloin, D., and Chiu, D.T. (2008) Quantitative force mapping of an optical vortex trap. *Appl. Phys. Lett.*, **92**, 161111.

40 Edgar, J.S., Milne, G., Zhao, Y., Pabbati, C.P., Lim, D.S., and Chiu, D.T. (2009) Compartmentalization of chemically separated components into droplets. *Angew. Chem. Int. Ed.* **48**, 2719–2722.

41 He, M.Y., Sun, C.H., and Chiu, D.T. (2004) Concentrating solutes and nanoparticles within individual aqueous microdroplets. *Anal. Chem.*, **76**, 1222–1227.

42 Edgar, J.S., Pabbati, C.P., Lorenz, R.M., He, M.Y., Fiorini, G.S., and Chiu, D.T. (2006) Capillary electrophoresis separation in the presence of an immiscible boundary for droplet analysis. *Anal. Chem.*, **78**, 6948–6954.

43 Schiro, P.G., Kuyper, C.L., and Chiu, D.T. (2007) Continuous-flow single-molecule CE with high detection efficiency. *Electrophoresis*, **28**, 2430–2438.

# 10
# Probing Exocytosis at Single Cells Using Electrochemistry

*Yan Dong, Michael L. Heien, Michael E. Kurczy, and Andrew G. Ewing*

## 10.1
## Introduction

Advances in methodology have allowed chemical measurements to be made with decreasing amounts of analyte and at smaller spatial dimensions. Indeed, investigations at single cells have given unique chemical and biological insight at this fundamental level. In the central nervous system, cells communicate by the transmission of chemical signals between cells (neurotransmission). A neuron can form a specialized structure, called a *synapse*, through which neurons communicate with each other and other specialized cells. This communication is achieved when a cell releases signaling molecules. Following mass transport through the extracellular space between the cells, the molecules can bind to receptors, where they are recognized and a signal is transduced. The initial release event is called *exocytosis*, and involves the fusion of a vesicle, filled with neurotransmitter molecules, with the cell plasma membrane. The signaling molecules in the vesicles are confined by a lipid bilayer; fusion with the plasma membrane creates a fusion pore, and the contents of the vesicle are released into the extracellular space. Mechanistic studies of this process are central to understanding neurotransmission. Therefore, methods have been developed to detect, measure, and characterize the release of molecules under both normal and experimental conditions.

One method used to measure exocytosis involves the direct oxidation of molecules; released electroactive molecules can be oxidized at the surface of an electrode. Typically, an electrode is placed next to a cell, held at a constant potential sufficient to oxidize the analyte (typically $<1.0$ V vs a Ag/AgCl reference electrode). The current generated by the oxidation is amplified, measured, and related to the number of molecules oxidized using Faraday's law. Thus, electrochemical methods are amiable to these measurements and many investigations with this approach have been carried out yielding important information about exocytosis and neurobiology [1–14]. In this chapter, the topic to exocytosis is narrowed, a general overview of electrochemical methods to detect and measure exocytosis at cells is given, and then some examples primarily from our own small part of this ever-expanding field are outlined.

*Chemical Cytometry*. Edited by Chang Lu

ISBN: 978-3-527-32495-8

## 10.2
## Measurement Requirements

Unraveling the dynamics of exocytosis is challenging because of the temporal and spatial domain in which these events occur. The volume of a synaptic vesicle can be as small as zeptoliters, and can contain as few as several thousand molecules (zeptomoles of measurable material). Exocytosis is also a rapid process. A fusion event typically lasts from 0.1 to 100 ms, thus requiring high temporal fidelity in order to accurately measure the event. Finally, it is possible for cells to release multiple chemicals [15] making identification of the compound a challenge as well.

The small scale of these measurements requires a method with a high signal-to-noise ratio. When designing a biological experiment it can be helpful to take cues from nature to replicate the task. The postsynaptic cell in a synapse is able to "detect" the small number of signaling molecules released by the presynaptic cell because the restricted volume of the synapse minimizes dilution, the surface of the cell is sensitive to the neurotransmitter, and the signal depends on concentration at the surface not the amount released. The electrode used in an electrochemical measurement of these events must therefore be extremely close to the surface of the cell to match the distance of the synaptic cleft (submicron) to measure signals above the baseline noise. The capacitance of the electrode is a source of noise, and is dependent on the electrode area, whereas the desired signal due to faradaic current is dependent on the amount of neurotransmitter released. This makes it important to restrict the area of the electrode surface to the confines of the surface of the cell being analyzed. Any superfluous surface area will contribute to the noise current without adding faradaic current. Thus, the small scale of these experiments requires small electrodes.

## 10.3
## Electrode Fabrication

Carbon-fiber microelectrodes are the most common type of electrode used to investigate release from single cells. The construction of a typical carbon-fiber microelectrode for single-cell measurements is carried out as follows. Briefly, a 5–10-μm-diameter carbon fiber is aspirated into a glass capillary and the capillary–fiber assembly is pulled with a pipette puller. The result is two tapered glass capillaries that share a single carbon fiber. The fiber is cut to yield two cylindrical electrodes, each has a carbon fiber that extends beyond the tapered glass. The fiber is cut back to the glass and is sealed by simply dipping the tip of the electrode into a high-quality epoxy and allowing it to cure. The electrode surface is then beveled at 45° to expose the carbon surface. The angle of the tip also allows the electrode to be manipulated from the periphery of the cell while measuring from the top of the cell. The pulled and beveled electrode is a small elliptical disk (diameter typically 5–10 μm). The disk electrode is ideal because the total surface area is exposed to the active cell surface, thus minimizing the capacitive current, and the flat surface beveled at 45° makes it possible to place the electrode surface extremely close to the cell surface. Both these factors lead to the improved signal-to-noise ratio that is required to make such measurements.

## 10.4
## Measurements at Single Cells

There are two modes of electrochemical techniques used to measure release from single cells with carbon-fiber microelectrodes: amperometry and cyclic voltammetry. Amperometric measurements use a constant potential; the oxidation current is continuously monitored as the signaling molecules are released. In cyclic voltammetry, the potential is periodically scanned and the current is recorded as a function of voltage. The curve produced, or cyclic voltammogram (CV), represents the characteristic oxidation and reduction potentials of the molecule at the electrode surface. This is a powerful method, which can be used to identify a released molecule by comparing the CV to a standard CV. It can also be employed to discriminate molecules in a mixture [16]. In one report, chromaffin cells were stimulated, and successive CVs were collected at a carbon-fiber microelectrode (Figure 10.1). Data were collected at a rate of 50 CVs per second, allowing resolution of individual release events. The CVs can be plotted as "color plots," where the abscissa is time, the ordinate is applied potential, and the current is represented in false color. Principal component regression can be used to chemically identify the released molecules as either epinephrine or norepinephrine.

Both amperometry and cyclic voltammetry can be used to collect data at a rate sufficient to temporally resolve individual exocytosis events. However, amperometry has exceptional temporal resolution affording the ability to obtain precise information about the kinetics of the release event. The rise time of the peak is related to the time required for the fusion pore to open, the peak width at half the maximum is used to measure the duration of the event, the amplitude gives the maximum flux of signaling molecules released, and the decay time is related to the diffusion of the molecules. These metrics have been usefully used to compare the effects of exogenous agents such as drug treatments [17–19], neurotoxins, and nonnative lipids [3, 10].

In an experiment, electrochemical measurements are carried out by placing a carbon-fiber microelectrode on the cell surface while the cell is stimulated, causing exocytosis. Each of the resultant spikes corresponds to single release events. This was first demonstrated by Wightman *et al.* [8, 12]. In this experiment, bovine chromaffin cells were induced to undergo exocytosis by exposure to nicotine, carbamoylcholline, or potassium ion delivered with a pressure pulse from a micropipette. The released chemicals were identified as catecholamines with cyclic voltammetry. Catecholamines are an important class of neurotransmitter that includes dopamine and adrenaline (epinephrine). An increase in the frequency of the spikes was clearly observed following stimulation of vesicle fusion. Fusion events depend on the availability of $Ca^{2+}$ [20] and if the $Ca^{2+}$ was removed from the

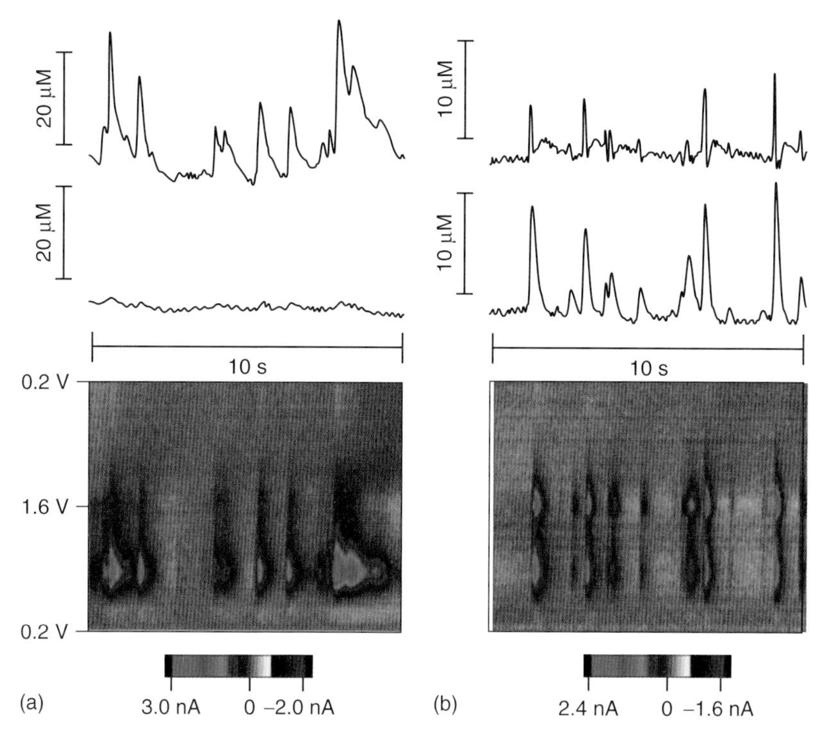


**Figure 10.1** Vesicular release events measured at individual cells. (a) Lower panel: color representation of release from a single cell. The upper trace is the concentration of norepinephrine assigned by principal component regression and the middle trace is the epinephrine assignment. (b) Lower panel: release measured at another cell. The upper trace is the norepinephrine prediction and the middle trace is the epinephrine assignment. (Reproduced with permission [16].)

buffer no spikes were observed. Amperometric spikes measured in the presence of $Ca^{2+}$ were found to have distributions of areas around a mean. This matches well with the description of exocytosis as discreet chemical release events from a population of similar sized vesicles. More evidence for quantized release was offered as it was shown that the intensity of the stimulation had no effect on the area of the spikes; it did, however, have a direct relationship to the frequency of the events. Finally, the area of the peak was used to calculate the amount of material detected; the values obtained agreed with the average number of catecholamine molecules expected to be housed inside of a single chromaffin cell vesicle.

Cyclic voltammetry can also be used to identify two catecholamines co-released from individual adrenal medullary chromaffin cells. These cells were known to release both epinephrine and norepinephrine [21] priming the Wightman group to use cyclic voltammetry to determine the chemical composition of individual events [15]. They found that the cells fall into three subcategories: cells that primarily release epinephrine, cells that primarily release norepinephrine, and a small group of cells that release a combination of the two. This work both showcases the chemical specificity of cyclic voltammetry and the importance of single-cell investigations. Subgroups would have been unobservable if the total population was investigated simultaneously.

## 10.5 Fusion Pore Dynamics

The fusion pore is the molecular structure that connects two cellular membrane compartments during their fusion. All vesicle trafficking occurs through the formation and dispersion of fusion pores. Through fusion pores intracellular compartments transport materials, and by controlling the exchange of phospholipids and proteins through the lips of fusion pores, the cells have the capability to selectively maintain the composition of an intracellular compartment. Exocytosis is a specialized form of vesicle trafficking and an essential process in nerve and endocrinal cells whereby $Ca^{2+}$ enters the cell and triggers the release of neurotransmitters, neuropeptides, or hormones from the secretory vesicles. Release happens through a fusion pore, the exocytotic fusion pore, which forms a continuous aqueous connection between the vesicle lumen and the extracellular space for the extrusion of vesicle contents. As the fusion pore forms, the vesicle membrane merges with the plasma membrane. Since the fusion is created between the extracellular and intracellular compartments, it is well suited to study with electrophysiological, electrochemical, and fluorescent methods. Owing to the use of these methods, our understanding of fusion pore dynamics has developed dramatically during the past decades and, indeed, our knowledge about fusion pore broadens our present views of exocytosis.

The earliest effort to profile fusion pores employed electron microscopy of quick-freezing frog neuromuscular junctions [22–24]. These pictures indicated narrow fluid connections between the vesicle and the cell exterior with diameters as small as 20 nm [24]. In addition to the data from nerve terminals, Chandler and Heuser [25] captured fusion pores in degranulating mast cells by quick-freezing techniques. Their data showed membrane-lined pores of 20–100 nm in diameters, which provided aqueous channels connecting the granule interior with extracellular space. More recently, the advent of atomic force microscope provides new insights of the structure of fusion pores [26]. The early work of Heuser and Reese stressed the existence of fusion pores. Electrical recording from living cells in the act of secretion ultimately shed light on the nature and properties of the initial fusion pore.

### 10.5.1 Studying Fusion Pore in Living Cells

The dynamics of the fusion pore have been mainly investigated at the level of single cells by two techniques: patch-clamp measurements of the electrical capacitance of cell membrane [27–29] and the amperometric detection of catecholamine neurotransmitters with carbon fibers [1, 6, 12, 30]. While patch-clamp detects changes of cell membrane area and conductance due to vesicular fusion, the electrochemical method measures the currents produced during oxidation of released secretory products from each exocytotic event. Amperometric transients from exocytosis are often preceded by a small pedestal, called the *foot*, of the spike [1, 6]. Combined patch-clamp and amperometric measurements have confirmed that the foot represents neurotransmitter release through the fusion pore in beige mouse mast cells [1]. Furthermore, in chromaffin cells the catecholamine released during the amperometric foot and during the main portion of the spike is the same [13], thus confirming the existence of a dynamic fusion pore.

In our lab, amperometry and transmission electron microscopy have been used to determine the dynamics of fusion pore (Figure 10.2). Release through a stable

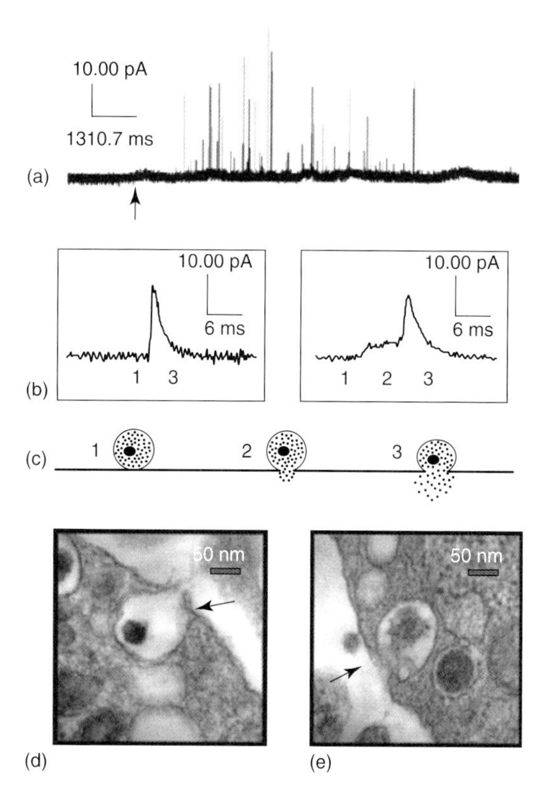


**Figure 10.2** (a) Representative amperometric data from a single PC12 cell. The arrow under the trace represents the time of stimulus (100 mM $K^+$) application. (b) Examples of individual amperometric current transients. The trace on the left has no discernable foot signal; that on the right is preceded by a foot. (c) A schematic diagram illustrating the flux of neurotransmitter through the fusion pore. The stages of fusion are numerically coordinated with the associated regions of the amperometric traces in (b). (d and e) Representative TEM images of PC12 cell dense-core vesicles fusing with the plasma membrane on stimulation with 100 mM $K^+$. Dark arrows indicate vesicles that appear to be undergoing exocytosis. Scale bars are 50 nm. (Reproduced with permission [31].)

fusion pore is distinguished in amperometric records of exocytosis as a prespike foot. The integrated area under the foot portion of the spike is representative of the number of molecules released through the fusion pore, prior to full fusion. The duration of release during the foot is indicative of the lifetime, or stability, of the fusion pore structure. Finally, the frequency with which amperometric feet are observed is a direct measure of the frequency with which vesicles release neurotransmitter through a stable exocytotic fusion pore as an intermediate state, as opposed to explosive vesicular fusion. Data from PC12 (rat adrenal pheochromocytoma) cells indicate that vesicular volume before secretion is strongly correlated with the characteristics of amperometric foot events [31]. Reserpine and L-3,4-dihydroxyphenylalanine (L-DOPA) have been used to decrease and increase, respectively, the volume of single PC12 cell vesicles [7]. Exposure of PC12 cells to 100 μM L-DOPA for 90 min significantly decreased the observed frequency of foot events to 72% of control [31]. In contrast, when PC12 cells were treated for 90 min with 100 nM reserpine, there was an increase in the frequency of foot events to 155% of the measurements done at control cells. Under control conditions, an average of ~33.7 ± 0.5% of the exocytotic events from PC12 cells exhibited a prespike feature. Thus, a clear trend exists in relating the frequency of foot events to vesicular volume. Further analysis indicated that (i) both foot duration and foot area are directly related to the physical size of the vesicle (Figure 10.3) and (ii) the percentage of the total contents released in the foot portion of the event is dependent on vesicle size.

Interestingly, smaller vesicles display a foot in the amperometric record more frequently than the control. We hypothesize that this result is largely an effect of the membrane tension differential that exists across the fusion pore. Immediately following vesicular fusion with the plasma membrane and pore formation, the vesicular dense core swells apparently resulting in release of core-bound transmitter putting pressure on the vesicular membrane. Because the dense core of a reserpine-treated vesicle largely fills the vesicle, core expansion should increase the vesicular tension to a greater extent than in the control case, where the core does not occupy as much of the overall vesicular volume. The resultant tension differential across the pore induces transient pore stabilization until membrane flow relieves this tension and full distention of the vesicle can occur. Thus, the fusion pores of smaller vesicles would appear to be more significantly stabilized, and amperometric feet occur in the data record more frequently.

### 10.5.2
### Studying Fusion Pore in Artificial Cells

Besides living cells, lipid models have contributed a great deal to our understanding about membrane fusion. A common model has involved vesicles containing channel proteins that are driven by osmotic pressure to fuse with a planar lipid bilayer [32, 33]. An adaptation of this model has been used to demonstrate transient opening of fusion pores in protein-free membranes, suggesting that indeed proteins might not be needed for this process [34]. Liposomes have been described as artificial cells and have also been used to examine membrane fusion [35–38]. Recently, we

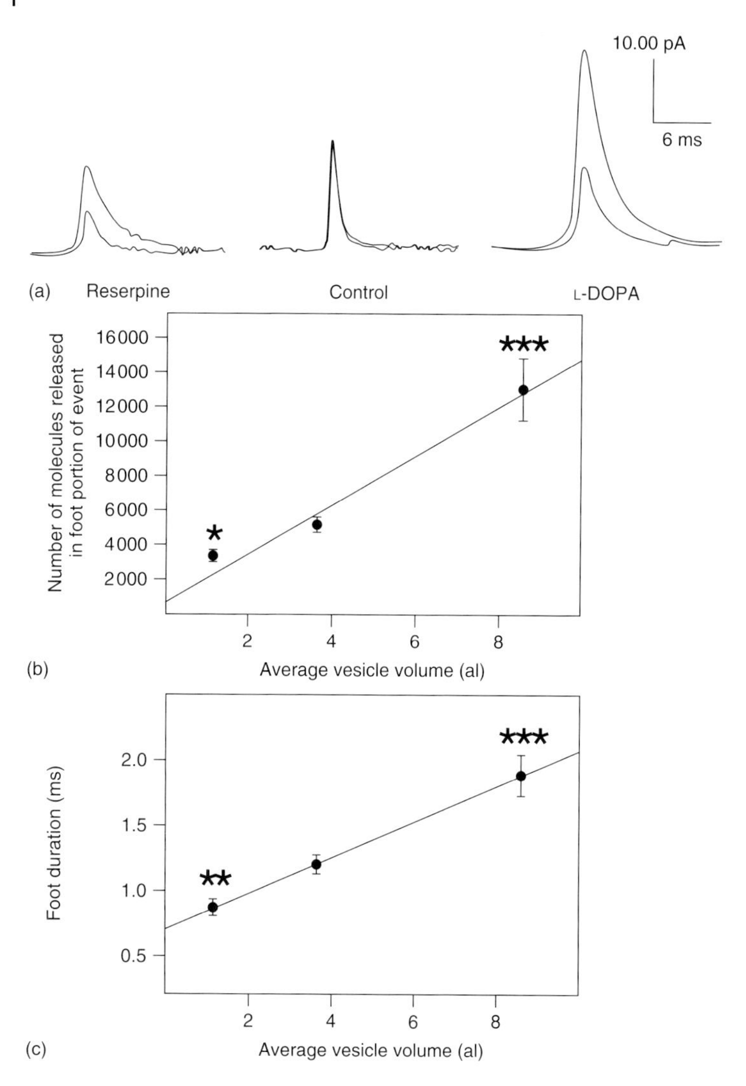


**Figure 10.3** Both the number of molecules released through the exocytotic fusion pore and the time course of said release are dependent on vesicular size. (a) Averaged amperometric current transients for one reserpine-treated, saline-treated, or L-DOPA-treated PC12 cell before (black) and after (gray) the 90-min incubation period. The scale bar is the same for all three averaged transients. (b) Mean foot area values shown as a function of vesicle volume. Under control conditions, a cellular average of 5695 ± 751 molecules was released through the fusion pore (before full fusion) per event. (c) Mean foot duration values (the time lapse between the onset of the foot and the inflection point between the foot and the full fusion event) shown as a function of vesicle volume. Under control conditions, the average cellular time course for release through the fusion pore was 1.3 ± 0.1 ms. Error bars represent the mean ± SEM of the foot characteristic values for the different experimental conditions. $^{*}p < 0.05$, $^{**}p < 0.01$, and $^{***}p < 0.001$ versus control, respectively (*t*-test). (Reproduced with permission [31].)

use electroinjection technology to develop a protein-free liposome system as an artificial cell that undergoes exocytosis [39–41]. Fluorescence microscopy and amperometry were used to detect leakage of transmitter through a nanoscopic fusion pore and quantal release during the final stage of exocytosis.

As shown in Figure 10.4, a vesicle is formed inside a surface-immobilized liposome via electroinjection. The vesicle is connected to the artificial cell membrane by a lipid nanotube. This nanotube initially resembles an elongated fusion pore. Microinjection of fluid into the pulled lipid nanotube leads to a local increase in membrane tension. To reduce this tension difference, lipid material flows from regions of lower tension (outer membrane) along the nanotube toward higher tension, forming the membrane of the small vesicle. Data from the artificial cell system suggest that the time course for leakage (foot length) through the fusion pore is governed by injection flow rate, as determined by pipette pressure, and is proportional to the size of a vesicle immediately before release (Figure 10.5a–c). The total amount of material leaking through the lipid nanotube (foot area) is proportional to vesicle size at the stage directly before release but is independent of injection flow rate (Figure 10.5d), indicating that nanotube length is the critical parameter determining total leakage. However, at any given vesicle size the rate of leakage through the lipid nanotube (foot area/foot length) varies with flow rate (pipette pressure) and vesicle size (Figure 10.5e). These observations can be explained by considering the sources of liquid flow inside the nanotube (Figure 10.5f). Lipid is transported from the artificial cell into the vesicle during expansion, resulting in shear flow of the liquid column inside the nanotube and toward the vesicle. In contrast, solution pressure from the pipette results in Poiseuille flow from the interior of the vesicle that opposes the direction of the shear flow. These results support the conclusion that the rate of fluid transport through the nanotube in the model cell depends on the length of the lipid nanotube and the factors affecting the counterbalance of shear and Poiseuille flow in the nanotube.

The data from the artificial cells are strikingly similar to what is observed during exocytosis in cells. Thus, the artificial cell model can be used to examine fundamental aspects of exocytosis with a great deal of control of many experimental degrees of freedom. These include membrane and solution composition, differential pH values across vesicle membranes, temperature, and vesicle size.

### 10.5.3 Flickering Fusion Pore

Early capacitance measurements indicated that the fusion pores in mast cells from *beige* mice have a lifetime of up to several seconds and may rapidly fluctuate about a small mean diameter ("flicker"), or even close transiently, before expanding irreversibly [42, 43]. Simultaneous capacitance and amperometric measurements have confirmed that transmitter starts to trickle out during capacitance flickering – that is, while the pore is still narrow and before the vesicle has completely collapsed into the plasma membrane [1]. Later on, data from bovine adrenal chromaffin cells further confirm the existence of flickering fusion pores [44]. Interestingly, flickering

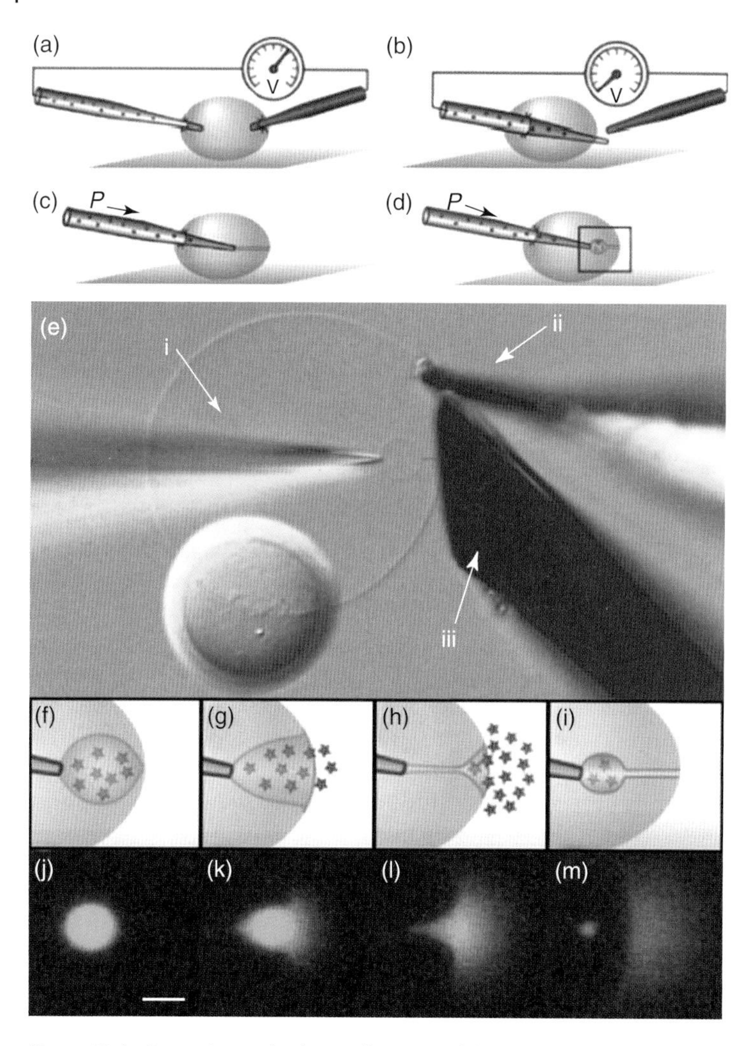


**Figure 10.4** Formation and release of vesicles in an artificial cell. (a–d) Schematics of a microinjection pipette electroinserted into the interior of a unilamellar liposome and then through the opposing wall, pulled back into the interior, followed by spontaneous formation of a lipid nanotube and formation of a vesicle from flow out of the tip of the micropipette. (e) Nomarski image of a unilamellar liposome, with a multilamellar liposome attached as a reservoir of lipid: (i) microinjection pipette, (ii) electrode for electroinsertion, and (iii) 30-μm-diameter amperometric electrode beveled to a 45° angle. A small red line depicts the location of the lipid nanotube, which is difficult to observe in the computer image with a 20× objective, illustrating a vesicle with connecting nanotube inside a liposome. (f–i) Fluid injection at a constant flow rate results in growth of the newly formed vesicle with a simultaneous shortening of the nanotube until the final stage of exocytosis takes place spontaneously and a new vesicle is formed with the attached nanotube. (j–m) Fluorescence microscopy images of fluorescein-filled vesicles showing formation and final stage of exocytosis matching the events in (f–i). Scale bar represents 10 μm. (Reproduced with permission [39].)

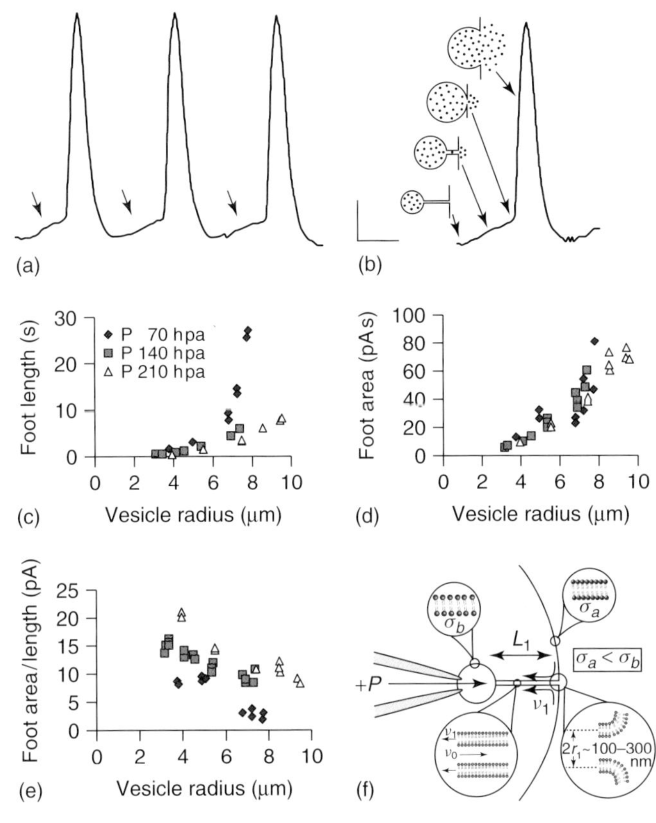


**Figure 10.5** Amperometric monitoring of release via an artificial fusion pore. (a) Amperometric detection of release from a 5-μm-radius vesicle showing prespike feet (arrows), indicating catechol transport through the lipid nanotube or fusion pore. Scale bar is 80 pA × 500 ms. (b) Time correlation of vesicle growth, transport of transmitter through the lipid nanotube, and the final stage of exocytosis with amperometric detection. (c–e) Plots of foot length (c), foot area (d), and the ratio of foot area over foot length (e) observed with amperometry for vesicles fusing with an artificial cell at three different pressures used to inflate the vesicles. (f) Schematic model of the factors affecting flow in the vesicle and nanotube of the artificial cell. (Reproduced with permission [39].)

fusion pores are also seen in small synaptic vesicles (SSVs) [45]. In rat cultured ventral midbrain neurons, SSV fusion pores flicker either once or multiple times in rapid succession, with each flicker estimated to release ~25–30% of vesicular dopamine (Figure 10.6). Flickering of the fusion pore results in the release of a larger fraction of an SSV's neurotransmitter content. Indeed, flickering of fusion pores in SSVs and large dense-core vesicles (LDCVs) shows different profiles (Table 10.1). The presynaptic terminals of midbrain dopamine neurons contain a relatively small number of SSVs. Fusion pore flickering may be particularly important for such synapses to prevent the loss of SSVs during full fusion and the relatively slow process of endocytosis and recycling.

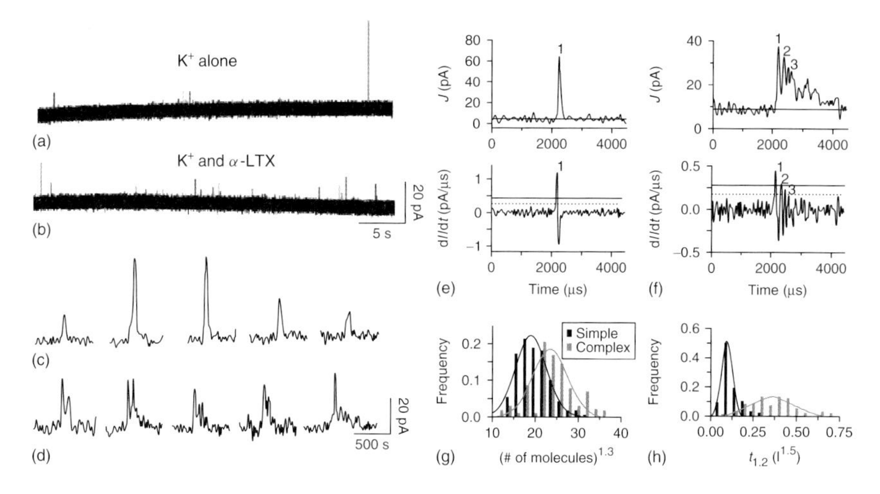


**Figure 10.6** Dopamine release from axonal varicosities of rat ventral midbrain dopamine neurons. (a and b) Representative segment of current trace showing dopamine release from neurons stimulated with $K^+$ alone (a) or with $K^+/\alpha$-latrotoxin (b). The stimulus was given earlier in a portion of the trace that has been omitted because of the paucity of events. (c and d) Representative examples of simple events (c) and complex events (d). Simple events each have a single rising and falling slope, whereas complex events have multiple flickers, each with distinct rising and falling phases. (e and f) The upper panels show examples of amperometric current traces; the lower panels show the first derivative ($dI/dt$) of the currents. In the amperometric traces, the mean background current is indicated by a solid line (upper panel). To be considered an "event," the $dI/dt$ must cross a 4.5 × rms threshold (solid line, lower panel). (e) Events with derivatives that cross the 3.0 × rms threshold (dotted line) only once in a rising trajectory are "simple." (f) Events that cross the 3.0 × rms threshold multiple times are "complex." The corresponding flickers (1–3) are indicated in the current trace. (g and h) Histograms of simple versus complex event characteristics obtained from amperometric recordings after $K^+/\ \alpha$-LTX stimulation ($n = 532$ simple events and $n = 130$ complex events from eight sites). (Reproduced with permission [45].)

**Table 10.1** Characteristics of flickering of fusion pores in SSVs and LDCVs.

| | Duration (μs) | Frequency (Hz) | Fraction of vesicle's neurotransmitter (%) |
|---|---|---|---|
| SSV | 100–150 | 4000 | 25–30 |
| LDCV | 100 000–500 000 | 170 | <1 |

## 10.6 Conclusions

Electrochemistry has been intensively used to investigate exocytotic vesicle fusion over the past two decades since the discovery of the technique by Wightman and coworkers [8, 12]. However, the underlying molecular mechanisms of this process still remain poorly understood. The majority of evidence indicates that membrane merger proceeds via a lipidic rather than a proteinaceous fusion pore [46–52], as the lipidic components are much more abundant than proteins in the membrane and the energy requirements for bilayer merger are intimately linked with lipid rearrangements. Proteins, on the other hand, are thought to be critical in targeting, priming, and vesicle docking and in the triggering and modulation of the fusion process [53–58]. Despite the realization that both lipids and proteins are involved in the process of membrane fusion, only lately work in the field has focused more equally on both components. Studying the contributions of both lipids and proteins to the fusion process will provide greater insight into the mechanism underlying membrane fusion and favor our understanding of synaptic transmission and the synaptic vesicle cycle.

## Acknowledgments

The authors gratefully acknowledge the work of the many colleagues and coworkers whose work is referenced in this article. The National Institutes of Health (USA) and the Swedish Science Foundation have funded, in part, work discussed here. AGE is a Marie Curie Chair, funded by the 6th Framework of the EU.

## References

1 Alvarez de Toledo, G., Fernandez-Chacon, R., and Fernandez, J.M. (1993) Release of secretory products during transient vesicle fusion. *Nature*, **363**, 554–558.

2 Amatore, C., Arbault, S., Bonifas, I., Lemaitre, F., and Verchier, Y. (2007) Vesicular exocytosis under hypotonic conditions shows two distinct populations of dense core vesicles in bovine chromaffin cells. *ChemPhysChem*, **8**, 578–585.

3 Amatore, C., Arbault, S., Bouret, Y., Guille, M., Lemaitre, F., and Verchier, Y. (2006) Regulation of exocytosis in chromaffin cells by trans-insertion of lysophosphatidylcholine and arachidonic acid into the outer leaflet of the cell membrane. *ChemBioChem*, **7**, 1998–2003.

4 Anderson, B.B., Chen, G., Gutman, D.A., and Ewing, A.G. (1999) Demonstration of two distributions of vesicle radius in the dopamine neuron of Planorbis corneus from electrochemical data. *J. Neurosci. Methods*, **88**, 153–161.

5 Chen, G., Gavin, P.F., Luo, G., and Ewing, A.G. (1995) Observation and quantitation of exocytosis from the cell body of a fully developed neuron in Planorbis corneus. *J. Neurosci.*, **15**, 7747–7755.

6 Chow, R.H., von Ruden, L., and Neher, E. (1992) Delay in vesicle fusion revealed by electrochemical monitoring of single secretory events in adrenal chromaffin cells. *Nature*, **356**, 60–63.

7 Colliver, T.L., Pyott, S.J., Achalabun, M., and Ewing, A.G. (2000) VMAT-Mediated changes in quantal size and vesicular volume. *J. Neurosci.*, **20**, 5276–5282.

8 Leszczyszyn, D.J., Jankowski, J.A., Viveros, O.H., Diliberto, E.J.Jr., Near, J.A., and Wightman, R.M. (1990) Nicotinic receptor-mediated catecholamine secretion from individual chromaffin cells. Chemical evidence for exocytosis. *J. Biol. Chem.*, **265**, 14736–14737.

9 Sombers, L.A., Maxson, M.M., and Ewing, A.G. (2005) Loaded dopamine is preferentially stored in the halo portion of PC12 cell dense core vesicles. *J. Neurochem.*, **93**, 1122–1131.

10 Uchiyama, Y., Maxson, M.M., Sawada, T., Nakano, A., and Ewing, A.G. (2007) Phospholipid mediated plasticity in exocytosis observed in PC12 cells. *Brain Res.*, **1151**, 46–54.

11 Westerink, R.H., de Groot, A., and Vijverberg, H.P. (2000) Heterogeneity of catecholamine-containing vesicles in PC12 cells. *Biochem. Biophys. Res. Commun.*, **270**, 625–630.

12 Wightman, R.M. *et al.* (1991) Temporally resolved catecholamine spikes correspond to single vesicle release from individual chromaffin cells. *Proc. Natl. Acad. Sci. U.S.A.*, **88**, 10754–10758.

13 Wightman, R.M., Schroeder, T.J., Finnegan, J.M., Ciolkowski, E.L., and Pihel, K. (1995) Time course of release of catecholamines from individual vesicles during exocytosis at adrenal medullary cells. *Biophys. J.*, **68**, 383–390.

14 Finnegan, J.M., Pihel, K., Cahill, P.S., Huang, L., Zerby, S.E., Ewing, A.G., Kennedy, R.T., and Wightman, R.M. (1996) Vesicular quantal size measured by amperometry at chromaffin, mast, pheochromocytoma, and pancreatic beta-cells. *J. Neurochem.*, **66**, 1914–1923.

15 Ciolkowski, E.L., Cooper, B.R., Jankowski, J.A., Jorgenson, J.W., and Wightman, R.M. (1992) Direct observation of epinephrine and norepinephrine cosecretion from individual adrenal-medullary chromaffin cells. *J. Am. Chem. Soc.*, **114**, 2815–2821.

16 Heien, M.L., Johnson, M.A., and Wightman, R.M. (2004) Resolving neurotransmitters detected by fast-scan cyclic voltammetry. *Anal. Chem.*, **76**, 5697–5704.

17 Camacho, M., Machado, J.D., Montesinos, M.S., Criado, M., and Borges, R. (2006) Intragranular pH rapidly modulates exocytosis in adrenal chromaffin cells. *J. Neurochem.*, **96**, 324–334.

18 Haynes, C.L., Buhler, L.A., and Wightman, R.M. (2006) Vesicular Ca(2+)-induced secretion promoted by intracellular pH-gradient disruption. *Biophys. Chem.*, **123**, 20–24.

19 Pothos, E.N. *et al.* (2002) Stimulation-dependent regulation of the pH, volume and quantal size of bovine and rodent secretory vesicles. *J. Physiol.*, **542**, 453–476.

20 Holz, R.W., Senter, R.A., and Frye, R.A. (1982) Relationship between Ca2+ uptake and catecholamine secretion in primary dissociated cultures of adrenal medulla. *J. Neurochem.*, **39**, 635–646.

21 Livett, B.G. (1984) Adrenal medullary chromaffin cells in vitro. *Physiol. Rev.*, **64**, 1103–1161.

22 Heuser, J.E., Reese, T.S., Dennis, M.J., Jan, Y., Jan, L., and Evans, L. (1979) Synaptic vesicle exocytosis captured by quick freezing and correlated with quantal transmitter release. *J. Cell Biol.*, **81**, 275–300.

23 Fesce, R., Grohovaz, F., Hurlbut, W.P., and Ceccarelli, B. (1980) Freeze-fracture studies of frog neuromuscular junctions during intense release of neurotransmitter. III. A morphometric analysis of the number and diameter of intramembrane particles. *J. Cell Biol.*, **85**, 337–345.

24 Heuser, J.E. and Reese, T.S. (1981) Structural changes after transmitter release at the frog neuromuscular junction. *J. Cell Biol.*, **88**, 564–580.

25 Chandler, D.E. and Heuser, J.E. (1980) Arrest of membrane fusion events in mast cells by quick-freezing. *J. Cell Biol.*, **86**, 666–674.

26 Schneider, S.W., Sritharan, K.C., Geibel, J.P., Oberleithner, H., and Jena, B.P. (1997) Surface dynamics in living acinar cells imaged by atomic force microscopy: identification of plasma membrane structures involved in exocytosis. *Proc. Natl. Acad. Sci. U.S.A.*, **94**, 316–321.

27 Neher, E. and Marty, A. (1982) Discrete changes of cell membrane capacitance observed under conditions of enhanced secretion in bovine adrenal chromaffin cells. *Proc. Natl. Acad. Sci. U.S.A.*, **79**, 6712–6716.

28 Haller, M., Heinemann, C., Chow, R.H., Heidelberger, R., and Neher, E. (1998) Comparison of secretory responses as measured by membrane capacitance

and by amperometry. *Biophys. J.*, **74**, 2100–2113.

29 Dernick, G., Alvarez de Toledo, G., and Lindau, M. (2003) Exocytosis of single chromaffin granules in cell-free inside-out membrane patches. *Nat. Cell Biol.*, **5**, 358–362.

30 Chen, T.K., Luo, G., and Ewing, A.G. (1994) Amperometric monitoring of stimulated catecholamine release from rat pheochromocytoma (PC12) cells at the zeptomole level. *Anal. Chem.*, **66**, 3031–3035.

31 Sombers, L.A., Hanchar, H.J., Colliver, T.L., Wittenberg, N., Cans, A., Arbault, S., Amatore, C., and Ewing, A.G. (2004) The effects of vesicular volume on secretion through the fusion pore in exocytotic release from PC12 cells. *J. Neurosci.*, **24**, 303–309.

32 Woodbury, D.J. (1999) Building a bilayer model of the neuromuscular synapse. *Cell Biochem. Biophys.*, **30**, 303–329.

33 Zimmerberg, J., Cohen, F.S., and Finkelstein, A. (1980) Fusion of phospholipid vesicles with planar phospholipid bilayer membranes. I. Discharge of vesicular contents across the planar membrane. *J. Gen. Physiol.*, **75**, 241–250.

34 Chanturiya, A., Chernomordik, L.V., and Zimmerberg, J. (1997) Flickering fusion pores comparable with initial exocytotic pores occur in protein-free phospholipid bilayers. *Proc. Natl. Acad. Sci. U.S.A.*, **94**, 14423–14428.

35 Evans, E., Bowman, H., Leung, A., Needham, D., and Tirrell, D. (1996) Biomembrane templates for nanoscale conduits and networks. *Science*, **273**, 933–935.

36 Hoekstra, D. (1990) Fluorescence assays to monitor membrane fusion: potential application in biliary lipid secretion and vesicle interactions. *Hepatology*, **12**, 61S–66S.

37 Kahya, N., Pecheur, E.I., de Boeij, W.P., Wiersma, D.A., and Hoekstra, D. (2001) Reconstitution of membrane proteins into giant unilamellar vesicles via peptide-induced fusion. *Biophys. J.*, **81**, 1464–1474.

38 Takei, K., Haucke, V., Slepnev, V., Farsad, K., Salazar, M., Chen, H., and De Camilli, P. (1998) Generation of coated intermediates of clathrin-mediated endocytosis on protein-free liposomes. *Cell*, **94**, 131–141.

39 Cans, A.S., Wittenberg, N., Karlsson, R., Sombers, L., Karlsson, M., Orwar, O., and Ewing, A. (2003) Artificial cells: unique insights into exocytosis using liposomes and lipid nanotubes. *Proc. Natl. Acad. Sci. U.S.A.*, **100**, 400–404.

40 Karlsson, A., Karlsson, R., Karlsson, M., Cans, A.S., Stromberg, A., Ryttsen, F., and Orwar, O. (2001) Networks of nanotubes and containers. *Nature*, **409**, 150–152.

41 Karlsson, M., Nolkrantz, K., Davidson, M.J., Stromberg, A., Ryttsen, F., Akerman, B., and Orwar, O. (2000) Electroinjection of colloid particles and biopolymers into single unilamellar liposomes and cells for bioanalytical applications. *Anal. Chem.*, **72**, 5857–5862.

42 Almers, W. and Tse, F.W. (1990) Transmitter release from synapses: does a preassembled fusion pore initiate exocytosis? *Neuron*, **4**, 813–818.

43 Monck, J.R. and Fernandez, J.M. (1994) The exocytotic fusion pore and neurotransmitter release. *Neuron*, **12**, 707–716.

44 Zhou, Z., Misler, S., and Chow, R.H. (1996) Rapid fluctuations in transmitter release from single vesicles in bovine adrenal chromaffin cells. *Biophys. J.*, **70**, 1543–1552.

45 Staal, R.G., Mosharov, E.V., and Sulzer, D. (2004) Dopamine neurons release transmitter via a flickering fusion pore. *Nat. Neurosci.*, **7**, 341–346.

46 Churchward, M.A., Rogasevskaia, T., Brandman, D.M., Khosravani, H., Nava, P., Atkinson, J.K., and Coorssen, J.R. (2008) Specific lipids supply critical negative spontaneous curvature–an essential component of native Ca2+-triggered membrane fusion. *Biophys. J.*, **94**, 3976–3986.

47 Churchward, M.A., Rogasevskaia, T., Hofgen, J., Bau, J., and Coorssen, J.R. (2005) Cholesterol facilitates the native mechanism of Ca2+-triggered membrane fusion. *J. Cell Sci.*, **118**, 4833–4848.

48 Coorssen, J.R., Blank, P.S., Albertorio, F., Bezrukov, L., Kolosova, I., Chen, X., Backlund, P.S. Jr., and Zimmerberg, J. (2003) Regulated secretion: SNARE density, vesicle fusion and calcium dependence. *J. Cell Sci.*, **116**, 2087–2097.

49 Graziani, A., Rosker, C., Kohlwein, S.D., Zhu, M.X., Romanin, C., Sattler, W., Groschner, K., and Poteser, M. (2006) Cellular cholesterol controls TRPC3 function: evidence from a novel dominant-negative knockdown strategy. *Biochem. J.*, **396**, 147–155.

50 Takamori, S. *et al.* (2006) Molecular anatomy of a trafficking organelle. *Cell*, **127**, 831–846.

51 Taverna, E., Saba, E., Rowe, J., Francolini, M., Clementi, F., and Rosa, P. (2004) Role of lipid microdomains in P/Q-type calcium channel (Cav2.1) clustering and function in presynaptic membranes. *J. Biol. Chem.*, **279**, 5127–5134.

52 Whalley, T., Timmers, K., Coorssen, J., Bezrukov, L., Kingsley, D.H., and Zimmerberg, J. (2004) Membrane fusion of secretory vesicles of the sea urchin egg in the absence of NSF. *J. Cell Sci.*, **117**, 2345–2356.

53 Chen, X., Arac, D., Wang, T.M., Gilpin, C.J., Zimmerberg, J., and Rizo, J. (2006) SNARE-mediated lipid mixing depends on the physical state of the vesicles. *Biophys. J.*, **90**, 2062–2074.

54 Dennison, S.M., Bowen, M.E., Brunger, A.T., and Lentz, B.R. (2006) Neuronal SNAREs do not trigger fusion between synthetic membranes but do promote PEG-mediated membrane fusion. *Biophys. J.*, **90**, 1661–1675.

55 Gauthier, B.R. and Wollheim, C.B. (2008) Synaptotagmins bind calcium to release insulin. *Am. J. Physiol. Endocrinol. Metab.*, **295**, E1279–E1286.

56 Sudhof, T.C. and Rothman, J.E. (2009) Membrane fusion: grappling with SNARE and SM proteins. *Science*, **323**, 474–477.

57 Szule, J.A. and Coorssen, J.R. (2003) Revisiting the role of SNAREs in exocytosis and membrane fusion. *Biochim. Biophys. Acta*, **1641**, 121–135.

58 Ungermann, C., Sato, K., and Wickner, W. (1998) Defining the functions of trans-SNARE pairs. *Nature*, **396**, 543–548.

# 11
# Electrochemical Determination of Enzyme Activity in Single Cells

Wenrui Jin

## 11.1
## Introduction

Enzymes are important biological components in cells. They control the balance of cytochemicals and actively participate in cell proliferation. Electrochemical detection (ECD) is a powerful tool for single-cell analysis [1–3]. However, enzymes present in single cells are not natively electroactive. Moreover, the amount of some enzymes in single cells is low, in the order of zeptomole (zmol, $10^{-21}$ mol). It is difficult to determine such a low amount using conventional ECD methods. The strategy for the determination of enzyme activity in single cells is to detect electroactive products of substrates of enzyme-catalyzed reactions. Usually, in single-cell analysis of enzymes based on ECD, the procedure is as follows: introduce a single whole cell into a microreactor such as a capillary or a microwell; lyse the cell to release the enzymes; separate the isoenzymes; allow the enzyme to convert its substrates to its products through enzyme-catalyzed reaction, and then electrochemically detect the electroactive products. The detected electrochemical signal is used to quantify enzyme activity in the cell by comparing it with that obtained for standard solutions of the enzyme. In this chapter, we summarize our work on the ECD of enzyme activity in single cells.

## 11.2
## Electrochemical Detection Coupled with Capillary Electrophoresis

Capillary electrophoresis (CE) has many inherent features in its operation, such as extremely small sample size, high separation speed and efficiency, and biocompatible environments [1, 4–6], which make it suitable for analysis of single cells. CE with sensitive laser-induced fluorescence (LIF) detection [7–12] and laser-based particle-counting microimmunoassay [13] have been successfully applied to determine enzyme activity in single cells. CE coupled with ECD is an important mode [1, 4, 14] for the detection of electroactive components such as

*Chemical Cytometry*. Edited by Chang Lu

ISBN: 978-3-527-32495-8

neurotransmitters [15–18], glutathione [19–22], diclofenac (a drug) [23], amino acids [21, 24–26], ascorbic acid [27], and histamine [28] in single cells. The species detected so far are those at relatively high levels (femtomole–attomole, $10^{-15}$–$10^{-18}$ mol) inside the cells. In order to quantify enzyme activities as low as zeptomole in single cells, the conventional CE–ECD methods should be improved.

In CE-based single-cell analysis, the electroosmotic flow rate for the running buffer is large enough to draw a cell at the capillary inlet into the capillary. The cytolysis process must preserve the integrity of the relevant compounds under their lysis conditions while releasing them from the cell. Usually, chemical lysis is accomplished, sometimes by means of a high voltage. Erythrocytes can be lysed easily in the CE running buffer [20, 29]. Some chemical reagents such as NaOH, SDS, and some organic solvents that can lyse cells may denature the enzyme. Therefore, these reagents are not used for the determination of enzyme activity in single-cell analysis. When the cell injected into the capillary cannot be lysed by the CE running buffer in the presence of high voltage, ultrasonication can be used [30]. In order to perform the on-capillary enzyme-catalyzed reaction, the enzyme substrates in the CE running buffer can be injected together with the cell or after introducing the cell into the capillary. The enzyme can convert the enzyme substrate into its product at a relatively high reaction rate. As a catalyst the enzyme is not consumed during the reaction, which provides amplification of the signal with prolonged reaction time. As a result of the enzyme amplification, a significant amount of the product can be produced for the final ECD. In single-cell analysis for enzyme activity, the velocity of the on-capillary enzyme-catalyzed reaction is a key factor. The wall of the capillary limits the diffusion of the reagents and the velocity of the enzyme-catalyzed reaction. To enhance the reaction velocity, different steps such as increasing the reaction temperature, prolonging the reaction time, and adding activator of the enzyme-catalyzed reaction should be adopted. However, when the reaction time is prolonged, the compounds, including the substrates, products, and enzymes, are also diluted. Obviously, the dilution factor increases with prolonged reaction time. Thus, long reaction time leads to widened electrophoretic peak and reduced peak height, and thus decreases the sensitivity of the method. When the product concentration yielded in the capillary is high enough, the product is delivered to the outlet of the capillary, where a working electrode is aligned, and the product is detected by the electrode. Usually, ECD is carried out with a three-electrode system that consists of a working electrode, a reference electrode such as a saturated calomel electrode (SCE) or an Ag/AgCl electrode, and an auxiliary electrode such as a Pt electrode. The Pt electrode also serves as the ground for the high potential drop across the capillary. Since the signal is very low in single-cell analysis, the electrochemical cell with the three electrodes is housed in a Faraday cage in order to minimize interference from noise from external sources. Carbon-fiber disk bundle electrodes (CFDBEs) are the most commonly used working electrodes due to their chemical inertia and easy alignment to the capillary (10–25 μm i.d.). During the detection of the product and the recording of its electropherogram, a constant potential, which depends on the product, is applied to the working electrode. A calibration curve is used to quantify enzyme activity in individual cells. For different activity or amount of enzymes in the single cells, the method used is different. In the following section, we provide examples of three different single-cell analyses corresponding to glucose-6-phosphate dehydrogenase (G6PDH) with low activity at $10^{-11}$–$10^{-10}$ unit ($10^{-21}$–$10^{-20}$ mol) level, alkaline phosphatase (ALP) isoenzymes with middle activity at $10^{-7}$–$10^{-6}$ unit ($10^{-19}$–$10^{-18}$ mol) level and lactate dehydrogenase (LDH) isoenzyme with high activity (LDH) at $10^{-6}$–$10^{-5}$ unit ($10^{-18}$–$10^{-17}$ mol) level.

### 11.2.1 Determination of Activity of Glucose-6-phosphate Dehydrogenase (G6PDH) in Single Human Erythrocytes

The amount of G6PDH in a single human erythrocyte with a diameter of 7 μm and a volume of 87 fl, ($10^{-15}$ l) can be as low as zeptomole. Therefore, CE–ECD for single-cell analysis should focus on increasing the sensitivity of the method [29]. G6PDH can catalyze enzyme substrates nicotinamide adenine dinucleotide ($NAD^+$) and glucose 6-phosphate (Glc6P) to reduced nicotinamide adenine dinucleotide (NADH), and glucose acid 6-phosphate (GlcA6P) via Reaction 11.1. The activity of G6PDH can be measured by determining the oxidation current generated by the NADH using a CFDBE held at 1.05 V and a CE running buffer consisting of $3.0 \times 10^{-2}$ mol $l^{-1}$ Tris–HCl (pH 7.8), $1 \times 10^{-3}$ mol $l^{-1}$ Glc6P and $1 \times 10^{-3}$ mol $l^{-1}$ $NAD^+$ as the substrates, and $1 \times 10^{-3}$ mol $l^{-1}$ $Mg^{2+}$ as activator of the enzyme-catalyzed reaction. Four steps can be used to increase the sensitivity for G6PDH activity determination. First, a capillary with an internal diameter as small as 10 μm is used to decrease the background noise. Second, the inside of the detection end of the capillary is etched to a horn shape. Doing so leads to the width at half height of the electrophoretic peak of G6PDH to become narrower and the limit of detection (LOD) to be reduced by approximately five times, as compared with detection using a normal detection end of the capillary. Third, the temperature for the enzyme-catalyzed reaction is increased to 37 °C to enhance the reaction rate. Fourth, $Mg^{2+}$ as the activator is added to the running buffer to activate the enzyme-catalyzed reaction. Under these conditions, LODs of the activity concentration and activity of G6PDH for the reaction time of 30 min are $6.2 \times 10^{-4}$ unit/ml (corresponding to $1.2 \times 10^{-11}$ mol $l^{-1}$) and 68 punit (corresponding to $1.3 \times 10^{-21}$ mol). The extremely low LOD is comparable to that of CE–LIF detection [7] and CE–laser-based particle-counting microimmunoassay [13] for analysis of enzymes in single cells. Typical electropherograms of three single erythrocytes for incubation of 100 s at 37 °C are shown in Figure 11.1. The activities of G6PDH determined in single cells are 9.0–76 $\times 10^{-11}$ unit (corresponding to 1.7–14 $\times 10^{-21}$ mol).

$$\text{Glc6P} + \text{NAD}^+ \xrightleftharpoons{\text{G6PDH}} \text{GlcA6P} + \text{NADH} + \text{H}^+ \qquad (11.1)$$

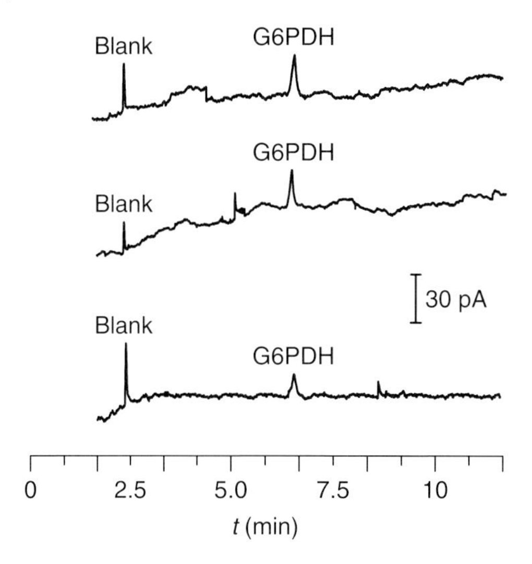


**Figure 11.1** Electropherograms of G6PDH in three single erythrocytes. (Adapted from [29], with permission from ACS Publications.)

### 11.2.2
### Separation and Determination of Activity of Alkaline Phosphatase (ALP) Isoenzymes in Single BALB/c Fibrolast Cells of Mouse Bone Marrow

ALP isoenzymes present in individual BALB/c fibrolast cells of mouse bone marrow are at zeptomole level. When CE–ECD is used to determine the activities of ALP isoenzymes, separation of the isoenzymes should be considered [31]. ALP catalyzes the conversion of electroinactive disodium phenyl phosphate (DPP) to electroactive phenol in alkaline conditions according to Reaction 11.2. The concentration of phenol produced is proportional to the activity of ALP. The ALP activity can be measured by determining the oxidation current generated by the phenol at the CFDBE held at 1.05 V. In this method, a phosphate buffer cannot be used because phosphate is a substrate of ALP. Therefore, $5.0 \times 10^{-2}$ mol l$^{-1}$ $Na_2B_4O_7$–$3.0 \times 10^{-2}$ mol l$^{-1}$ NaOH (pH 9.8) containing $1.0 \times 10^{-3}$ mol l$^{-1}$ DPP as the CE running buffer is used for CE–ECD of ALP isoenzymes. The linear ranges of the activity concentration and the activity of ALP are 3.00–900 unit/l and $4.50 \times 10^{-9}$–$1.35 \times 10^{-6}$ unit, respectively. The LOD of the activity concentration and the activity are 1.0 unit/l (corresponding to $2.4 \times 10^{-12}$ mol l$^{-1}$) and 1.5 nunit (corresponding to $3.6 \times 10^{-21}$ mol), respectively. Since the activity of ALP is higher than that of G6PDH, determination of ALP activity is easier compared to that of G6PDH. Thus, the steps taken for CE–ECD of G6PDH to increase detection sensitivity, such as adding activator of the enzyme-catalyzed reaction, decreasing the inside diameter of the capillary as well as using a warm water bath at 37 °C and a horn-shaped capillary end are not needed here. However, the separation of the isoenzymes should still be considered because all isoenzymes can catalyze

DPP to phenol. Therefore, the isoenzymes released from a single cell have to be preseparated by applying a high voltage and then incubated with DPP in the running buffer prior to ECD. Figure 11.2 shows the electropherograms of three single cells. Three well-separated peaks, corresponding to the blank and the ALP isoenzymes I and II, appear on the electropherograms. Despite irreproducible blank peaks, resulting from different injection volumes of the buffer during introduction of single cells, the two ALP isoenzymes are well resolved. The activities of ALP isoenzymes determined in single cells are in the range from 2.1 to $4.5 \times 10^{-7}$ unit and 1.0 to $3.9 \times 10^{-7}$ unit for ALP isoenzymes I and II, respectively.

$$\mathrm{DPP} + \mathrm{H_2O} \overset{\mathrm{ALP}}{\rightleftharpoons} \mathrm{Na_2HPO_4} + \mathrm{Phenol} \qquad (11.2)$$

### 11.2.3
### Separation and Determination of Activity of Lactate Dehydrogenase (LDH) Isoenzymes in Single Rat Glioma Cells

LDH in rat glioma cells has three isoenzymes [32]. Moreover, their activities in the cells are very high. The CE–ECD assay for the determination of activity of LDH isoenzymes in individual cells should consider these two characteristics [30]. The enzyme-catalyzed reaction shown in Reaction 11.3 can be used to determine the activity of LDH isoenzymes. The product of the catalysis reaction NADH can be amperometrically measured by a CFDBE held at 1.00 V versus an SCE as

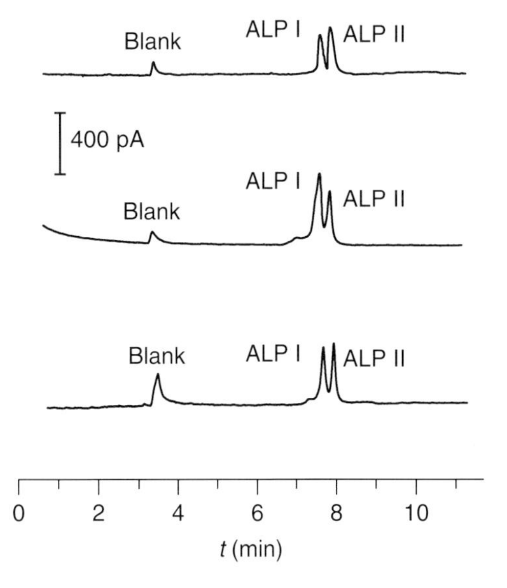


**Figure 11.2** Electropherograms of ALP isoenzymes in three individual BALB/c mouse bone marrow fibroblast cells. (Adapted from [31], with permission from Wiley-VCH, Weinheim.)

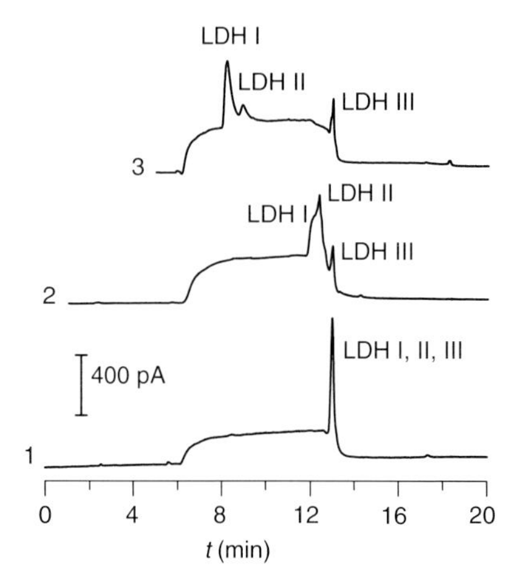


**Figure 11.3** Electropherograms of LDH isoenzymes in individual glioma cells at different preseparation times of (1) 0, (2) 1, and (3) 5 min before incubation. (Adapted from [30], with permission from Elsevier Science.)

mentioned above. LDH activity can be detected by measuring NADH. A CE running buffer consisting of $5.0 \times 10^{-2}\,\text{mol}\,\text{l}^{-1}$ Tris–HCl, $5.0 \times 10^{-2}\,\text{mol}\,\text{l}^{-1}$ lactate, and $5.0 \times 10^{-3}\,\text{mol}\,\text{l}^{-1}$ $NAD^+$ (pH 9.3) is used for CE–ECD in single-cell analysis. In this method, the cell injected into the capillary is lysed by ultrasonication. Before ECD, the isoenzymes are separated by applying a separation voltage across the capillary. Since the activity of LDH in single cells is very high, even short reaction time of 2 min can yield enough amount of NADH for ECD. Figure 11.3 shows the electropherograms of single glioma cells at different preseparation times. It can be noted that when the preseparation time is less than 5 min, the three zones corresponding to the three LDH isoenzymes cannot be separated (curves 1 and 2). When the preseparation time is increased to 5 min, three peaks are observed. Additionally, a plateau on the electropherogram after ~7 min is observed. The enzyme-catalyzed reaction caused by the LDH zone migrating in the capillary is responsible for the plateau [33]. Owing to the peaks of the LDH isoenzymes I and II cannot be well separated. Their peak areas are measured based on the product of the peak current and the width at the peak half height according to the method described in [30]. The activities of LDH isoenzymes determined in single cells are 2.4–11, 0.60–7.9, and $0.81–3.8 \times 10^{-6}$ unit.

$$\text{Lactate} + \text{NAD}^+ \xrightleftharpoons{\text{LDH}} \text{Pyruvate} + \text{NADH} \tag{11.3}$$

## 11.3 Voltammetry

Voltammetry and amperometry have been used to measure secretion or exocytosis of neurotransmitters from different kinds of cells or vesicles[2, 34–41], oxygen consumption [42], and photosynthetic activity [43] at single cells and study on the oxidative burst from fibroblasts [44–46]. In all these studies, measurements are performed outside a single cell. Thus, the intracellular biological molecules do not affect the measurements. Voltammetry and amperometry using microelectrodes (MEs) can also be applied for concentration measurements of contents in single cells [47–52]. However, since the detecting ME is implanted into the cell, it is surrounded by a number of high-molecular weight species present in the cell cytoplasm, which might be easily adsorbed on the electrode surface and, thus, foul the electrode and cause a deterioration of the voltammetric or amperometric response of the electrode, making these *in vivo* experiments very difficult [51]. To obtain accurate results, a linear-average calibration method [49] or a pulse voltammetry for minimizing electrode fouling [51] is suggested. Microwells with MEs can also be used to monitor compounds from single cells [53–56], or transport into a single cell [57]. In these electrochemical measurements, the detecting MEs are in solutions, where an intact single cell is placed and the measurements are carried out outside the cell. Thus, there is no electrode fouling from the adsorption of intracellular biological macromolecules. Electrode fouling from biological macromolecules is a thorny problem for quantitative determination of complements in single cells. This problem has to be overcome when enzyme activity in single cells is to be determined by voltammetry. Using voltammetry for the determination of activity of peroxidase (PO) inside single neutrophils and single acute promyelocytic leukemia (APL) cells has been reported [58]. In this method, a nanoliter-scale microwell coupled with a positionable dual ME consisted of an Au disk working electrode with a 20 μm diameter and an Ag/AgCl reference electrode with a 120 μm diameter and 2 mm length is used by combining enzyme-catalyzed reaction. In the presence of $H_2O_2$, PO converts enzyme substrates hydroquinone ($H_2Q$) into its product benzoquinone (BQ) according to Reaction 11.4. The steady-state current of BQ on the voltammograms is used for quantification of PO activity in the single cells.

$$H_2Q + H_2O_2 \xrightleftharpoons{PO} BQ + 2H_2O \qquad (11.4)$$

The experimental setup shown in Figure11.4 is used in the single-cell voltammetry. The microwell array is simply constructed using chemical etching without photolithographic techniques. The constructed microwells are of 2–112 nl with 200–710 μm diameter at the top, 150–620 μm diameter at the bottom and 100–320 μm in depth. In order to lyse the cells easily, they are chemically perforated with digitonin, which bound to the cholesterol on the cell membrane and micropores are formed. A simple and rapid method of pushing a microscope slide on the microwell array is adopted to introduce single cells into the microwells. The single cells in the microwell array are subjected to freeze-thawing for cytolysis in a constant-humidity chamber to obtain the single-cell extract containing intracellular substances involved in PO. After the single-cell extracts in the microwells are allowed to evaporate, 20 (for neutrophils) or 100 nl (for APL cells) of PBS containing $2.0 \times 10^{-3}$ $H_2Q$ and $2.0 \times 10^{-3}$ $H_2O_2$ is added to redissolve the intracellular substances. Then, a dual electrode is inserted into the microwell with a depth

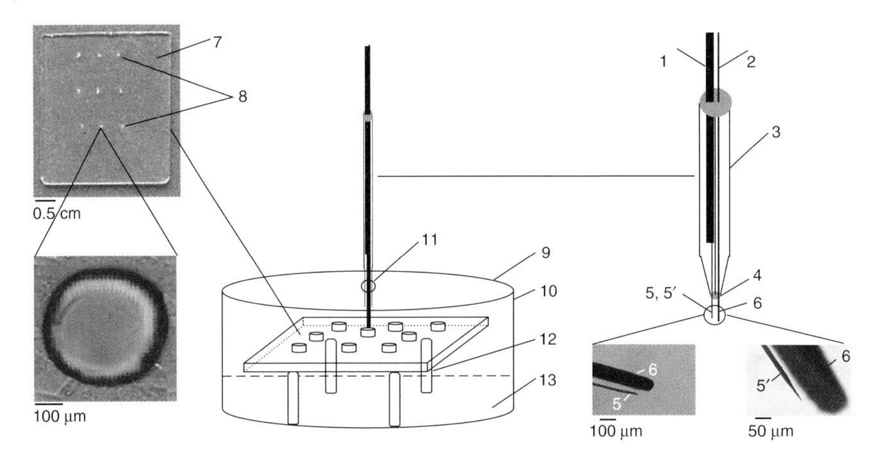


**Figure 11.4** Schematic diagram of the setup with microwell array and dual electrode in a constant-humidity chamber for single-cell analysis: 1, copper lead; 2, Ag wire; 3, dual electrode; 4, epoxy resin; 5, micrometer-sized Au disk working electrode; 5′, nanometer-sized Au disk working electrode; 6, Ag/AgCl reference electrode; 7, wax; 8, microwell array; 9, cover; 10, Petri dish as a constant-humidity chamber; 11, hole; 12, stand; 13, water. (Adapted from [58], with permission from ACS Publications.)

of ~100 μm in the solution by means of a scanning electrochemical microscope (SECM). Subsequently, a voltammogram is recorded by scanning potential from 0.05 to −0.4 V as the blank. Finally, the solution is incubated for 10 min and the voltammogram is recorded again. After subtracting the blank, the steady-state current is used for quantification. Figure 11.5 shows the linear scan voltammograms of a single lysed neutrophil and a single lysed APL cell. Since both PO and horseradish

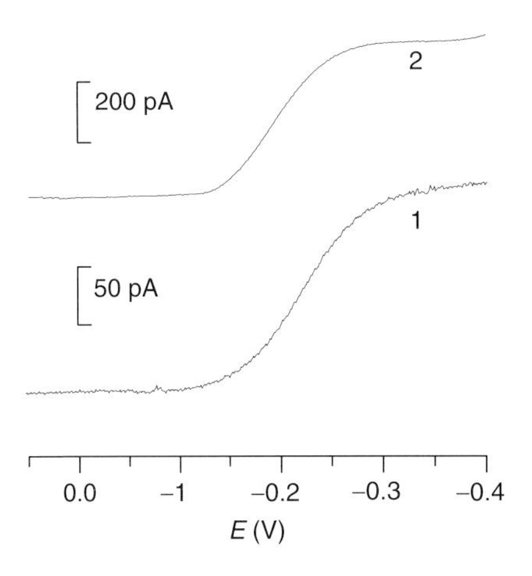


**Figure 11.5** Linear scan voltammograms of (1) a single lysed neutrophil and (2) a single lysed APL cell after incubation for 10 min and subtracting blank. (Adapted from [58], with permission from ACS Publications.)

peroxidase (HRP) are PO and the activity of all POs is defined and determined with the same method, HRP can serve as a standard to quantify PO activity. Thus, PO activity in single cells can be quantified using the standard calibration curve of HRP. The PO activities in individual neutrophils and individual APL cells are determined to be in the range of $0.65{-}2.1 \times 10^{-8}$ unit and $0.93{-}3.1 \times 10^{-7}$ unit.

In the single-cell analysis method using voltammetry, five factors should be considered. First, the working electrode size that affects the current signal should be selected. When a 20-μm-diameter disk working electrode is used, a sufficient ratio of signal to noise can be obtained. In this case, the area of the reference electrode of the dual ME must be much larger (at least $10^3$ times) than that of the working electrode. Additionally, inserting the dual electrode into the microwell is also difficult, because the dual electrode can break easily. The use of a SECM can overcome this difficulty. Second, solution evaporation must be considered when measurements are performed with microwells. To prevent quick evaporation of the solution in the microwells, experiments on single-cell analysis should be carried out in a constant-humidity chamber. Third, the concentration of the product of the enzyme-catalyzed reaction in the detected solution should be high enough so that the electrochemical signal can be detected by the micrometer-sized working electrode. A high concentration can be achieved through the enzyme-catalyzed reaction by increasing the reaction time and decreasing the detected solution volume. Fourth, the concentrations of the biological macromolecules such as proteins from a single cell in the detected solution should be low enough so that these macromolecules do not foul the working electrode. This requires the detected solution volume to be large enough to prevent the macromolecules from fouling the working electrode. Therefore, controlling the solution volume in the microwell, which depends on the enzyme activity in single cells, is very important. The higher the enzyme activity in a single cell, the larger the solution volume can be. This is why the solution volume is different for neutrophils (20 nl) and APL cells (100 nl). Finally, oxygen reduction in the solution is also detected simultaneously. In this case, the recorded voltammogram includes the oxygen reduction current. Therefore, the steady-state current after subtracting blank corresponding to the oxygen reduction current is used for quantification.

## 11.4 Scanning Electrochemical Microscopy (SECM)

SECM with MEs as its tip is a useful tool for determining biochemical components in single cells [3, 59]. It has been successfully used to examine and investigate cellular viability [59], photosynthetic electron transport [60], *in vivo* topography [60–64], photosynthetic activity [43, 61, 65], respiratory activity [65–70], redox and acid–base reactivity [71–73], the mechanism of charge transfer reactions [74], image fields of different types of cells [75], and interactions between silver nanoparticles and cells [76]. Neurotransmitter secretion [77] and nitric oxide released [78, 79] from single cells, the local permeation of the nuclear membrane by mediators [80], the antimicrobial effects of silver ion [81], and drug metabolism [3] have been followed and monitored by SECM as well. Recently, an electrochemical method using SECM was developed for quantitative determination of PO activity in single neutrophils by scanning an Au ME over a nitrocellulose film–covered microreactor with micropores [82]. In this method, a single-cell extract is prepared in a ~17 nl microwell with PBS containing $2.0 \times 10^{-3}$ mol l$^{-1}$ $H_2Q$ and $2.0 \times 10^{-3}$ mol l$^{-1}$ $H_2O_2$. In order to allow released PO from the cell to convert $H_2Q$ into BQ, the solution in the microwell is incubated for a certain time. Then, the microwell is covered with a porous nitrocellulose film to create the microreactor. After PBS containing $H_2Q$ and $H_2O_2$ is added over the microreactor, the Au ME held at −0.3 V is moved along the central line across the microreactor (Figure 11.6). In this case, only small molecules such as BQ can diffuse out from the microreactor interior through the micropores on the nitrocellulose film to the solution on top of the microreactor, while the larger PO molecules remain inside the microreactor due to the small micropore size. The BQ amount above the microreactor reflects the PO activity of the single cell. During the scan of the Au ME, BQ is electrochemically reduced at the ME and a scan curve with a peak is recorded. The peak current on the scan curve is proportional to the PO activity of the single cell. A typical scan curve of a single neutrophil is shown in Figure 11.6 (inset). The PO activity in the single cell can be obtained on the basis of a calibration curve.

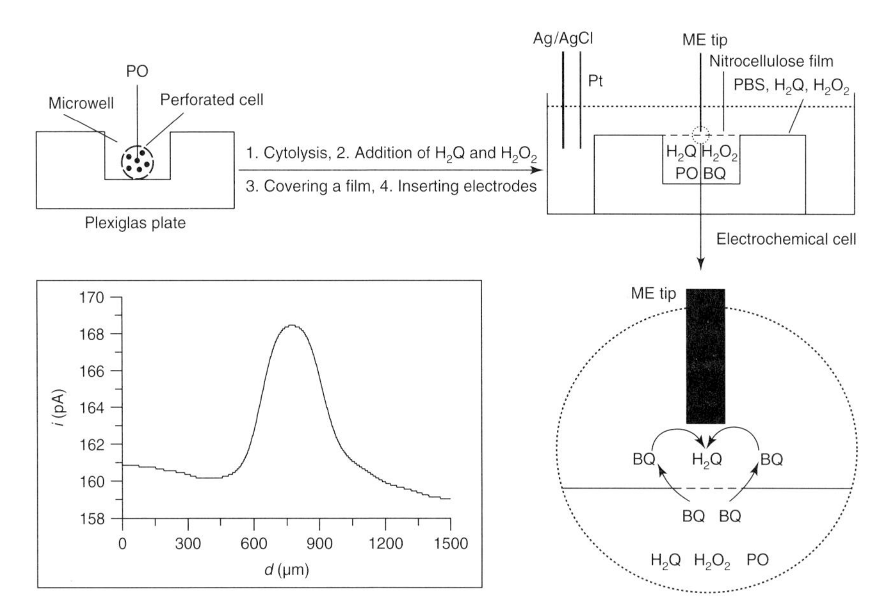


**Figure 11.6** Schematic diagram showing electrochemical measurement of PO activity in single cells by scanning a microelectrode coupled with a microreactor. Inset: a typical scan curve of a single neutrophil. (Adapted from [82], with permission from ACS Publications.)

In this method, the micropore size on the film is an important factor for the determination of PO activity. Micropores on the nitrocellulose film should hold back the PO molecules while letting small $H_2Q$, $H_2O_2$, and BQ molecules through. Another key factor is the diameter of the microreactor. When the diameter of the microreactor is 350 μm, a concave image is observed because the nitrocellulose film on the microreactor is caved in. When the diameter of the microwell is larger than 350 μm, sometimes the microwell could not be sealed completely with the nitrocellulose film and solution leakage could occur. When the diameter of the microreactor is smaller than 200 μm, symmetrical peak-shaped images are obtained. Additionally, the microwell volume should be as small as possible, to increase the enzyme concentration in the single-cell extract. In this method, the detected signal decreases with ME scanning time, because a portion of BQ diffuses out from the microreactor and leaves the microreactor surface. Owing to film resistance, the signal detected using this method is lower than that directly detected in the microwell without the film using linear scan voltammetry explained earlier. However, this method has two obvious advantages. First, there is no electrode fouling. The ME as the working electrode and the sample solution are completely separated by the nitrocellulose film with micropores. The micropores are so small that only small molecules that do not foul the ME can diffuse out from the microreactor and make contact with the ME. Therefore, the ME is very clean during the measurements. Second, oxygen in the solution does not interfere with the determination of enzyme activity. This is because not only oxygen reduction on Au electrodes is very weak, but the oxygen reduction also appears as a constant baseline on the scan curve, which does not affect the accurate measurement of the peak current.

SECM can also quantify PO activity in single intact neutrophils [83]. The strategy of the single-cell analysis method using SECM is shown in Figure 11.7. The SECM measurements are performed for the perforated cells immobilized on a silanized coverslip in PBS containing $1.0 \times 10^{-3}$ mol l$^{-1}$ $H_2Q$ and $1.0 \times 10^{-3}$ mol l$^{-1}$ $H_2O_2$. In this case, large PO molecules are retained inside the cell interior [84, 85]. Small molecules $H_2Q$ and $H_2O_2$ diffuse through the micropores into the cell interior. There, $H_2Q$ is converted into BQ by intracellular PO. While BQ diffuses from the cell interior onto the cell surface through the micropores with a steady flux, the BQ near the cell surface is detected by an Au ME held at −0.3 V. When the ME as the SECM tip is scanned laterally over the cell or along the central line over the cell, a 3D image or a 2D scan curve with a peak is obtained. Figure 11.8a shows the 3D images of two single neutrophils using SECM. In the SECM measurements, due to a negative potential applied to the ME, oxygen in the solution can be reduced simultaneously at the Au ME, which contributes to the baseline on the images of PO activity. In SECM for single intact cells, the contribution of cell topography to the scan curve should be noted. In this case, the cell is also an insulator protruding from the substrate. When the ME tip is scanned above the cell, the cell as an insulator can block the diffusion of oxygen as mediator from the solution's interior to the cell surface. Oxygen reduction current over the cell will decrease as compared to that over the substrate without the cell based on the SECM negative feedback

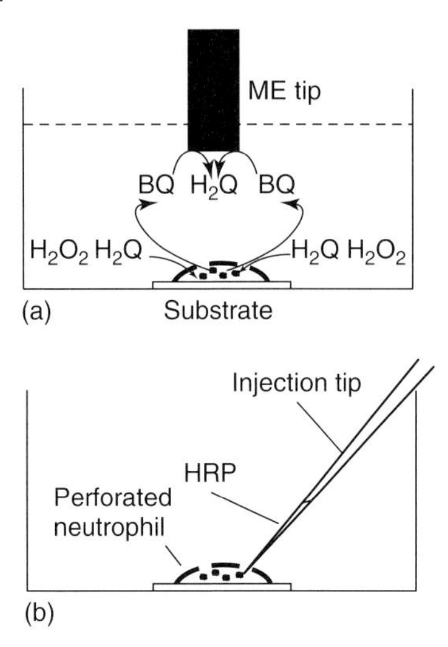


**Figure 11.7** Schematic representation of quantification of PO activity in a single intact neutrophil by SECM coupled with ultramicroinjection. (a) Scanning a ME tip over a perforated neutrophil and (b) injecting HRP standard solution into the perforated neutrophil. (Adapted from [83], with permission from RSC Publishing.)

mode [86]. Thus a scan curve with a small negative peak corresponding to the cell topography is recorded. When the intracellular standard addition method is used to quantify PO activity in single intact cells, the effect of oxygen reduction current can be eliminated. In order to save time and eliminate the effect of electrode surface change, the 2D scan curve along the cell central line over a perforated neutrophil is used to quantify PO activity in single cells.

For SECM of single intact cells, the electrochemical response may depend on membrane features such as the membrane area, which is determined by the cell size, the number, size, and thickness of the micropores and the kinetic behavior of intermembrane transport of enzyme substrates through micropores. To avoid the effect of these factors on the quantification of PO activity, the intracellular standard addition method is used. In the current method, the volume of the standard solution injected into a cell is a key factor. If the volume of the injected solution is too large, not only permanent membrane damages but also membrane rupture on the cells may be induced. The cell volume will also change before and after injection of the standard solution. Thus, variation of the cell volume must be considered. However, accurate measurement of variation of cell volume for an ~10-μm-diameter neutrophil is very difficult. To solve this problem, a negligible volume of the standard solution (<10 fl) as compared with the neutrophil volume (500–900 fl) is injected. Submicrometer-sized micropipette tips are used to inject such a small volume of the enzyme standard solution into the cell. At the same time, when a tip is inserted into and withdrawn from the cell submicrometer-sized micropipette tips can also prevent cell trauma. In addition, the amount of the

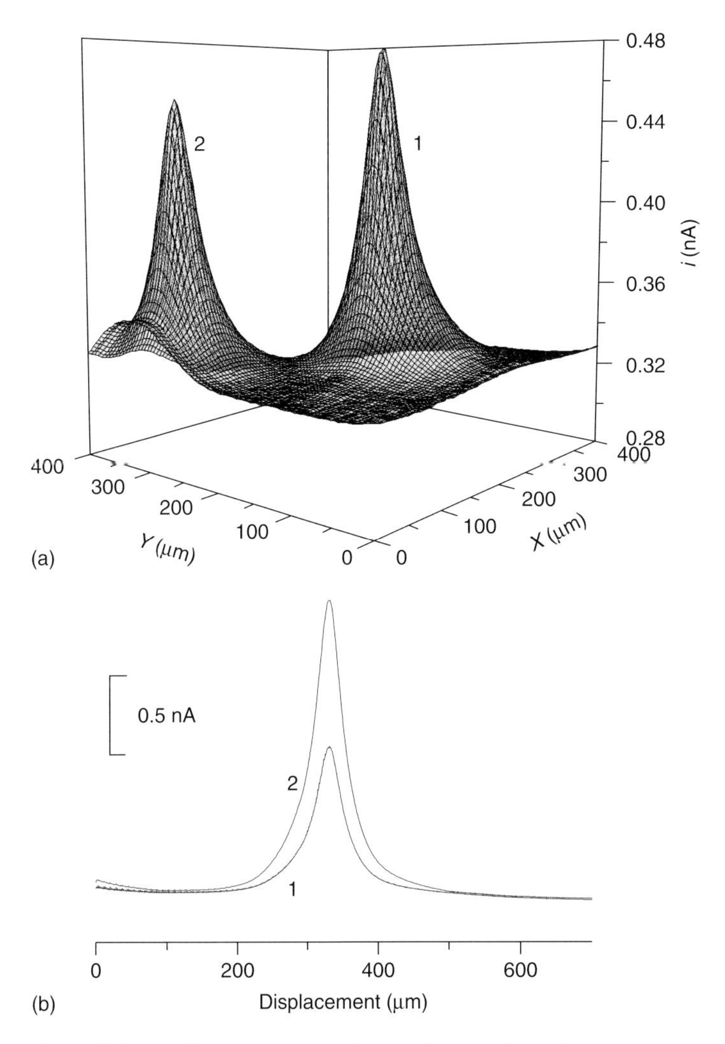


**Figure 11.8** (a) Three-dimensional image of two perforated neutrophils labeled 1 and 2 and (b) the scan curves (1) before and (2) after injection of HRP standard solution into a perforated neutrophil. (Adapted from [83], with permission from RSC Publishing.)

standard enzyme injected into the cell, which depends on the injection volume and the concentration of the standard enzyme solution, should match with that of the natural cellular enzyme in the cell. Figure 11.8b shows typical scan curves along the central line over a perforated cell before and after ultramicroinjection of the enzyme standard solution.

This method has several major advantages. First of all, the determined PO is retained in the cell interior during the determination process, meaning that there is no sample dilution. Secondly, the signal is amplified by the intracellular enzyme-catalyzed reaction. These two advantages lead to a very high sensitivity for the determination of enzyme activity in single intact cells. The peak current for a single intact neutrophil using this method is much higher than that for a lysed cell diluted in a microwell using nitrocellulose film–covered microreactor with micropores as described above. Thirdly, there is no electrode fouling from adsorption of intracellular biological molecules, because the ME tip and the biological molecules are completely separated by the cell membrane. Fourthly, the electroactive compounds that can be directly oxidized at the ME did not interfere with the determination of PO activity in single cells due to the negative detected potential. Finally, oxygen in the detected solution and the negative feedback response due to the difference between ME-to-cell distance and ME-to-substrate do not affect the measurement of PO activity in single cells. This is because the peak current difference before and after ultramicroinjection of the enzyme standard solution for the same cell is used to quantify the PO activity in single intact cells. In this case, the current corresponding to oxygen reduction and the feedback response is subtracted.

## 11.5
## High-throughput ECD

In a majority of the reported papers on single-cell analysis of enzymes, a single cell must be lysed in the microsampler such as capillary or microwell to release enzymes before detection. Usually, the microsampler must be treated to remove cellular debris and intracellular substances adsorbed on the wall of the microsampler before the next single-cell determination. This makes the analysis rate, that is, the cell throughput, rather low and limits the practical application of the technique. To perform a high-throughput method for single-cell analysis, a variety of methods such as use of a multipurpose single-cell injector [87], continuous cell introduction method for erythrocytes [88], and a microfluidic chip-based high-throughput single-cell analysis [89] have been developed. Recently, an electrochemical high-throughput method for analysis of PO activity in single neutrophils without cytolysis has been reported [85]. In this method, neutrophils in a reaction tube (Fig. 11.9a) are first perforated with digitonin (Figure 11.9b). The perforated cells with PBS containing $5.0 \times 10^{-4}\,\mathrm{mol\,l^{-1}}$ $H_2Q$ and $8.0 \times 10^{-4}\,\mathrm{mol\,l^{-1}}$ $H_2O_2$ at 4 °C in a syringe are continuously introduced by nitrogen pressure into a capillary in a warm bath of 37 °C (Figure 11.9c), to increase the rate of the intracellular enzyme-catalyzed reaction in the capillary. $H_2Q$ and $H_2O_2$ diffuse into the cell interior, where PO converted $H_2Q$ into BQ (Figure 11.9c and d). The electroactive BQ diffuses out from the cell interior to the cell surface through the micropores and forms a BQ zone around the cell (Figure 11.9d and e). This process proceeds inside moving cells, and BQ zones are formed around every perforated moving cell. The BQ zones with the moving cells are continuously delivered to the capillary outlet under hydraulic flow and detected by a CFDBE. Figure 11.10 shows a segment of 32 min on the elution curve of 39 intact neutrophils, indicating an average detection rate of >1 cell/min. For the high-throughput single-cell analysis, cell concentration should

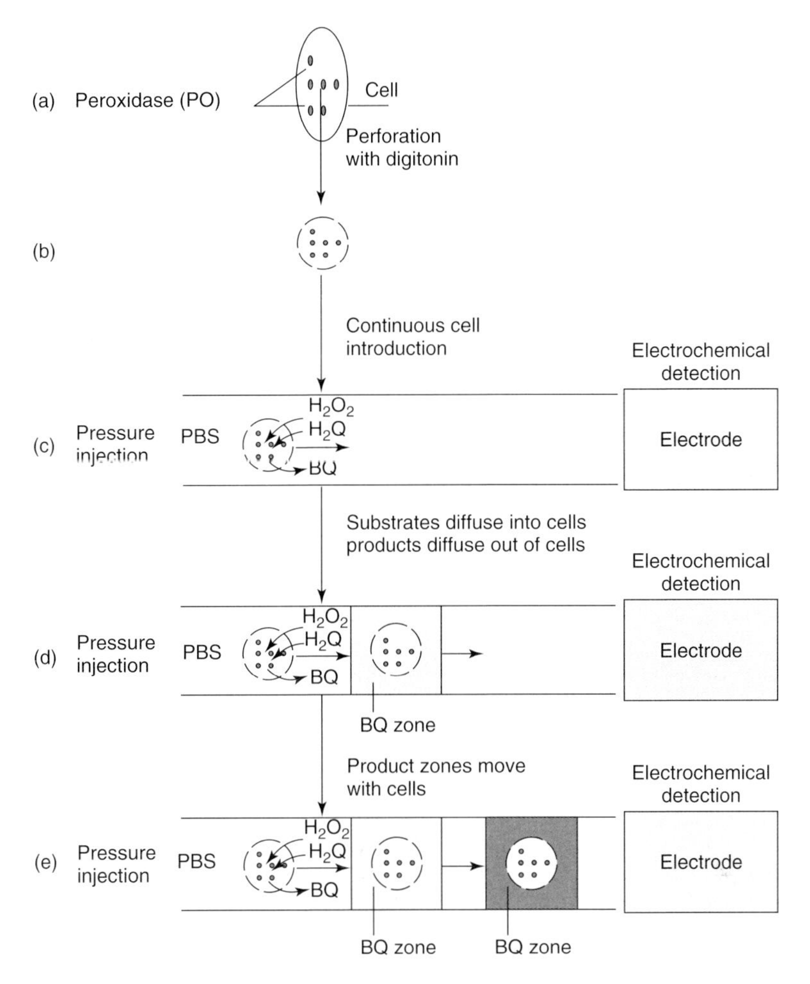


**Figure 11.9** Schematic diagram showing the process of high-throughput ECD of PO activity in single cells. (Adapted from [85], with permission from ACS Publications.)

be controlled. The lower the cell concentration, the larger the interval between the peaks, and thus the lower the analysis throughput. On the other hand, higher cell concentrations will lead to overlapping peaks. Additionally, the temperature of the perforated cell suspension in the syringe should be low. When the temperature of the whole detection system is 37 °C, the blank current from BQ produced in the syringe increases the baseline with increasing run time. This is because of the direct oxidation of $H_2Q$ to BQ by oxygen, peroxide itself, and trace metals, as well as PO inside the perforated cells in the syringe. The BQ amount in the syringe can be decreased by decreasing the enzyme-catalyzed reaction rate. When both cell suspension and PBS containing $H_2Q$ and $H_2O_2$ at 4 °C are used and the syringe with the cell suspension is placed in an ice bath, the blank current can

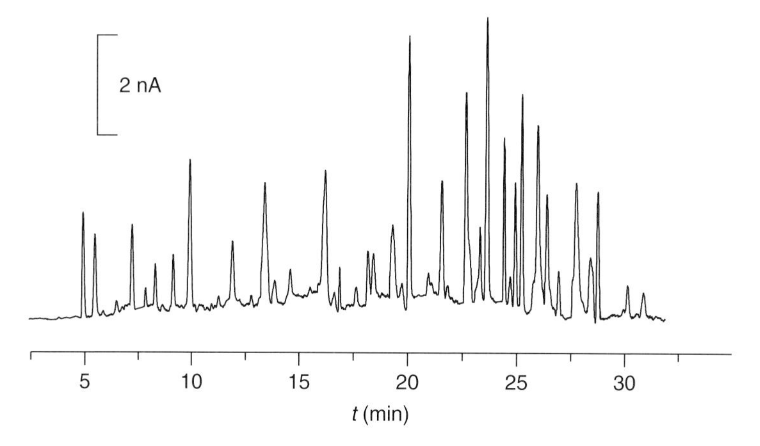


**Figure 11.10** Continuously recorded elution curve of human perforated neutrophils with a cell concentration of $4.0 \times 10^3$ cells/ml. (Adapted from [85], with permission from ACS Publications.)

be neglected. In this method, the biological macromolecules that can be adsorbed on the capillary wall or on the working electrode do not escape from cells and do not interfere with the determination of enzyme activity. Not only the precision of the results can be improved, but also the high-throughput single-cell analysis for enzyme activity can be realized without capillary treatment between runs.

## 11.6 Perspective

Without question, ECD is a powerful technique for quantifying enzyme activity in single cells. The ECD techniques described in this chapter can be applied to determine other enzymes in single cells, when suitable enzyme substrates are chosen. The ECD strategy of enzyme activity can also be applied to other detection techniques to obtain high sensitivity and improve reproducibility of method. Although ECD has been successfully used to quantify electroinactive enzymes in single cells, new methods should be developed. Increasing throughput and automation of analysis process like flow cytometry should first be considered; otherwise, the technique cannot be effectively applied in life science. Additionally, electrode fouling should be avoided. The analysis of intact cells without cytolysis is the best way to solve these problems.

## Acknowledgment

This project was supported by the National Natural Science Foundation of China (Grant Nos. 20475033, 20675047, 90713016, and 20705016).

## References


1 Ewing, A.G., Mesaros, J.M., and Gavin, P.F. (1994) Electrochemical detection in microcolumn separations. *Anal. Chem.*, **66**, 527A–537A.

2 Wightman, R.M., Jankowski, J.A., Kennedy, R.T., Kawagoe, K.T., Schroeder, T.J., Leszczyszyn, D.J., Near, J.A., Diliberto, E.J. Jr., and Viveros, O.H. (1991) Temporally resolved catecholamine spikes correspond to single vesicle release from individual chromaffin cells. *Proc. Natl. Acad. Sci. U.S.A.*, **88**, 10754–10758.

3 Bard, A.J., Li, X., and Zhan, W. (2006) Chemically imaging living cells by scanning electrochemical microscopy. *Biosens. Bioelectron.*, **22**, 461–472.

4 Jankowski, J.A., Tracht, S., and Sweedler, J.V. (1995) Assaying single cells with capillary electrophoresis. *Trends Anal. Chem.*, **14**, 170–176.

5 Yeung, E.S. (1999) Study of single cells by using capillary electrophoresis and native fluorescence detection. *J. Chromatogr. A*, **830**, 243–262.

6 Woods, L.A., Roddy, T.P., and Ewing, A.G. (2004) Capillary electrophoresis of single mammalian cells. *Electrophoresis*, **25**, 1181–1187.

7 Xue, Q. and Yeung, E.S. (1994) Variability of intracellular lactate dehydrogenase isoenzymes in single human erythrocytes. *Anal. Chem.*, **66**, 1175–1178.

8 Tan, W. and Yeung, E.S. (1995) Simultaneous determination of enzyme activity and enzyme quantity in single human erythrocytes. *Anal. Biochem.*, **226**, 74–79.

9 Xue, Q. and Yeung, E.S. (1996) Determination of lactate dehydrogenase isoenzymes in single lymphocytes from normal and leukemia cell lines. *J. Chromatogr. B*, **677**, 233–240.

10 Meredith, G.D., Sims, C.E., Soughayer, J.S., and Allbritton, N.L. (2000) Measurement of kinase activation in single mammalian cells. *Nat. Biotechnol.*, **18**, 309–312.

11 Zarrine-Afsar, A. and Krylov, S.N. (2003) Use of capillary electrophoresis and endogenous fluorescent substrate to monitor intracellular activation of protein kinase A. *Anal. Chem.*, **75**, 3720–3724.

12 Li, H., Sims, C.E., Kaluzova, M., Stanbridge, E.J., and Allbritton, N.L. (2004) A quantitative single-cell assay for protein kinase B reveals important insights into the biochemical behavior of an intracellular substrate peptide. *Biochemistry*, **43**, 1599–1608.

13 Rosenzweig, Z. and Yeung, E.S. (1994) Laser-based particle-counting microimmunoassay for the analysis of single human erythrocytes. *Anal. Chem.*, **66**, 1771–1776.

14 Wallingford, R.A. and Ewing, A.G. (1987) Capillary zone electrophoresis with electrochemical detection. *Anal. Chem.*, **59**, 1762–1766.

15 Wallingford, R.A. and Ewing, A.G. (1988) Capillary zone electrophoresis with electrochemical detection in 12.7 microns diameter columns. *Anal. Chem.*, **60**, 1972–1975.

16 Olefirowicz, T.M. and Ewing, A.G. (1990) Capillary electrophoresis in 2 and 5 microns diameter capillaries: application to cytoplasmic analysis. *Anal. Chem.*, **62**, 1872–1876.

17 Bergquist, J., Tarkowski, A., Ekman, R., and Ewing, A. (1994) Discovery of endogenous catecholamines in lymphocytes and evidence for catecholamine regulation of lymphocyte function via an autocrine loop. *Proc. Natl. Acad. Sci. U.S.A.*, **91**, 12912–12916.

18 Hu, S., Pang, D., Wang, Z., Cheng, J., Li, Z., Fan, Y., and Hu, H. (1996) Single nerve cell analysis by capillary Electrophoresis with Amperometrlc detection. *Chem. J. Chin. Univ.*, **17**, 1207–1209.

19 Jin, W., Dong, Q., Ye, X., and Yu, D. (2000) Assay of glutathione in individual mouse peritoneal macrophages by capillary zone electrophoresis with electrochemical detection. *Anal. Biochem.*, **285**, 255–259.

20 Jin, W., Li, W., and Xu, Q. (2000) Quantitative determination of glutathione in single human erythrocytes by capillary zone electrophoresis with

electrochemical detection. *Electrophoresis*, **21**, 774–779.

21 Jin, W., Li, X., and Gao, N. (2003) Simultaneous determination of tryptophan and glutathione in individual rat hepatocytes by capillary zone electrophoresis with electrochemical detection at a carbon fiber bundle–Au/Hg dual electrode. *Anal. Chem.*, **75**, 3859–3864.

22 Wang, W., Xin, H., Shao, H., and Jin, W. (2003) Determination of glutathione in single human hepatocarcinoma cells by capillary electrophoresis with electrochemical detection. *J. Chromatogr. B*, **789**, 425–429.

23 Dong, Q. and Jin, W. (2001) Monitoring diclofenac sodium in single human erythrocytes introduced by electroporation using capillary zone electrophoresis with electrochemical detection. *Electrophoresis*, **22**, 2786–2792.

24 Weng, Q. and Jin, W. (2001) Determination of free intracellular amino acids in single mouse peritoneal macrophages after naphthalene-2,3-dicarboxaldehyde derivatization by capillary zone electrophoresis with electrochemical detection. *Electrophoresis*, **22**, 2797–2803.

25 Dong, Q., Wang, X., Zhu, L., and Jin, W. (2002) Method of intracellular naphthalene-2,3-dicarboxaldehyde derivatization for analysis of amino acids in a single erythrocyte by capillary zone electrophoresis with electrochemical detection. *J. Chromatogr. A*, **959**, 269–279.

26 Weng, Q. and Jin, W. (2003) Assay of amino acids in individual human lymphocytes by capillary zone electrophoresis with electrochemical detection. *Anal. Chim. Acta*, **478**, 199–207.

27 Jin, W. and Jiang, L. (2002) Measurement of ascorbic acid in single human neutrophils by capillary zone electrophoresis with electrochemical detection. *Electrophoresis*, **23**, 2471–2476.

28 Weng, Q., Xia, F., and Jin, W. (2002) Measurement of histamine in individual rat peritoneal mast cells by capillary zone electrophoresis with electrochemical detection. *J. Chromatogr. B*, **779**, 347–352.

29 Sun, X. and Jin, W. (2003) Catalysis-electrochemical determination of zeptomole enzyme and its application for single-cell analysis. *Anal. Chem.*, **75**, 6050–6055.

30 Wang, W., Han, S., and Jin, W. (2006) Determination of lactate dehydrogenase isoenzymes in single rat glioma cells by capillary electrophoresis with electrochemical detection. *J. Chromatogr. B*, **831**, 57–62.

31 Sun, X., Jin, W., Li, D., and Bai, Z. (2004) Measurement of alkaline phosphatase isoenzymes in individual mouse bone marrow fibroblast cells based on capillary electrophoresis with on-capillary enzyme-catalyzed reaction and electrochemical detection. *Electrophoresis*, **25**, 1860–1866.

32 Seeds, N.W. (1975) Expression of differentiated activities in reaggregated brain cell cultures. *J. Biol. Chem.*, **250**, 5455–5458.

33 Yang, W., Yu, A., and Chen, H. (2001) Studies on an on-line reaction of micro lactate dehydrogenase in capillary electrophoresis using electrochemical detection. *Chem. J. Chin. Univ.*, **22**, 547–551.

34 Kennedy, R.T., Huang, L., Atkinson, M.A., and Dush, P. (1993) Amperometric monitoring of chemical secretions from individual pancreatic beta-cell. *Anal. Chem.*, **65**, 1882–1887.

35 Chen, T., Luo, G., and Ewing, A.G. (1994) Amperometric monitoring of stimulated catecholamine release from rat pheochromocytoma (PC12) cells at the zeptomole level. *Anal. Chem.*, **66**, 3031–3035.

36 Pihel, K., Hsieh, S., Jorgenson, J.W., and Wightman, R.M. (1995) Electrochemical detection of histamine and 5-hydroxytryptamine at isolated mast cells. *Anal. Chem.*, **67**, 4514–4521.

37 Kennedy, R.T., Huang, L. and Aspinwall, C.A. (1996) Extracellular pH Is Required for rapid release of insulin from Zn-insulin precipitates in $\beta$-cell secretory vesicles during exocytosis. *J. Am. Chem. Soc.*, **118**, 1795–1796.

38 Xin, Q. and Wightman, R.M. (1998) Simultaneous detection of catecholamine exocytosis and Ca2+ release from

single bovine chromaffin cells using a dual microsensor. *Anal. Chem.*, **70**, 1677–1681.

**39** Hochstetler, S.E., Puopolo, M., Gustincich, S., Raviola, E., and Wightman, R.M. (2000) Real-time amperometric measurements of zeptomole quantities of dopamine released from neurons. *Anal. Chem.*, **72**, 489–496.

**40** Huang, W., Cheng, W., Zhang, Z., Pang, D., Wang, Z., Cheng, J., and Cui, D. (2004) Transport, location, and quantal release monitoring of single cells on a microfluidic device. *Anal. Chem.*, **76**, 483–488.

**41** Wu, W., Huang, W., Wang, W., Wang, Z., Cheng, J., Xu, T., Zhang, R., Chen, Y., and Liu, J. (2005) Monitoring dopamine release from single living vesicles with nanoelectrodes. *J. Am. Chem. Soc.*, **127**, 8914–8915.

**42** Jung, S., Gorski, W., Aspinwall, C.A., Kauri, L.M., and Kennedy, R.T. (1999) Oxygen microsensor and its application to single cells and mouse pancreatic islets. *Anal. Chem.*, **71**, 3642–3649.

**43** Yasukawa, T., Uchida, I., and Matsue, T. (1999) Microamperometric measurements of photosynthetic activity in a single algal protoplast. *Biophys. J.*, **76**, 1129–1135.

**44** Arbault, S., Pantano, P., Jankowski, J.A., Vuillaume, M., and Amatore, C. (1995) Monitoring an oxidative stress mechanism at a single human fibroblast. *Anal. Chem.*, **67**, 3382–3390.

**45** Amatore, C., Arbault, S., Bruce, D., de Oliveira, P., Erard, M., and Vuillaume, M. (2000) Analysis of individual biochemical events based on artificial synapses using ultramicroelectrodes: cellular oxidative burst. *Faraday Discuss.*, (116), 319–333.

**46** Amatore, C., Arbault, S., Bruce, D., de Oliveira, P., Erard, M., and Vuillaume, M. (2001) Characterization of the electrochemical oxidation of peroxynitrite: relevance to oxidative stress bursts measured at the single cell level. *Chem. Eur. J.*, **7**, 4171–4179.

**47** Meulemans, A., Poulain, B., Baux, G., Tauc, L., and Henzel, D. (1986) Micro carbon electrode for intracellular voltammetry. *Anal. Chem.*, **58**, 2088–2091.

**48** Abe, T., Lau, Y., and Ewing, A.G. (1991) Intracellular analysis with an immobilized-enzyme glucose electrode having a 2-μm diameter and subsecond response times. *J. Am. Chem. Soc.*, **113**, 7421–7423.

**49** Lau, Y., Chien, J., Wong, D.K.Y., and Ewing, A.G. (1991) Characterization of the voltammetric response at intracellular carbon ring electrodes. *Electroanalysis*, **3**, 87–95.

**50** Xue, J., Ying, X., Chen, J., Xian, Y., and Jin, L. (2000) Amperometric ultramicrosensors for peroxynitrite detection and its application toward single myocardial cells. *Anal. Chem.*, **72**, 5313 5321.

**51** Chen, T., Lau, Y., Wong, D.K.Y., and Ewing, A.G. (1992) Pulse voltammetry in single cells using platinum microelectrodes. *Anal. Chem.*, **64**, 1264–1268.

**52** Lau, Y., Abe, T., and Ewing, A.G. (1992) Voltammetric measurement of oxygen in single neurons using platinized carbon ring electrodes. *Anal. Chem.*, **64**, 1702–1705.

**53** Bratten, C.D., Cobbold, P.H., and Cooper, J.M. (1998) Single-cell measurements of purine release using a micromachined electroanalytical sensor. *Anal. Chem.*, **70**, 1164–1170.

**54** Cai, X., Klauke, N., Glidle, A., Cobbold, P., Smith, G.L., and Cooper, J.M. (2002) Ultra-low-volume, real-time measurements of lactate from the single heart cell using microsystems technology. *Anal. Chem.*, **74**, 908–914.

**55** Yasukawa, T., Glidle, A., Cooper, J.M., and Matsue, T. (2002) Electroanalysis of metabolic flux from single cells in simple picoliter-volume microsystems. *Anal. Chem.*, **74**, 5001–5008.

**56** Chen, P., Xu, B., Tokranova, N., Feng, X., Castracane, J., and Gillis, K.D. (2003) Amperometric detection of quantal catecholamine secretion from individual cells on micromachined silicon chips. *Anal. Chem.*, **75**, 518–524.

**57** Troyer, K.P. and Wightman, R.M. (2002) Dopamine transport into a single cell in a picoliter vial. *Anal. Chem.*, **74**, 5370–5375.

**58** Gao, N., Zhao, M., Zhang, X., and Jin, W. (2006) Measurement of enzyme activity in single cells by voltammetry using a microcell with a positionable dual electrode. *Anal. Chem.*, **78**, 231–238.

59 Yasukawa, T., Kaya, T., and Matsue, T. (2000) Characterization and imaging of single cells with scanning electrochemical microscopy. *Electroanalysis*, **12**, 653–659.

60 Tsionsky, M., Cardon, Z.G., Bard, A.J., and Jackson, R.B. (1997) Photosynthetic electron transport in single guard cells as measured by scanning electrochemical microscopy. *Plant Physiol.*, **113**, 895–901.

61 Yasukawa, T., Kaya, T., and Matsue, T. (1999) Dual imaging of topography and photosynthetic activity of a single protoplast by scanning electrochemical microscopy. *Anal. Chem.*, **71**, 4637–4641.

62 Liebetrau, J.M., Miller, H.M., Baur, J.E., Takacs, S.A., Anupunpisit, V., Garris, P.A., and Wipf, D.O. (2003) Scanning electrochemical microscopy of model neurons: imaging and real-time detection of morphological changes. *Anal. Chem.*, **75**, 563–571.

63 Kurulugama, R.T., Wipf, D.O., Takacs, S.A., Pongmayteegul, S., Garris, P.A., and Baur, J.E. (2005) Scanning electrochemical microscopy of model neurons: constant distance imaging. *Anal. Chem.*, **77**, 1111–1117.

64 Hirano, Y., Nishimiya, Y., Kowata, K., Mizutani, F., Tsuda, S., and Komatsu, Y. (2008) Construction of time-lapse scanning electrochemical microscopy with temperature control and its application to evaluate the preservation effects of antifreeze proteins on living cells. *Anal. Chem.*, **80**, 7349–9354.

65 Yasukawa, T., Kaya, T., and Matsue, T. (1999) Imaging of photosynthetic and respiratory activities of a single algal protoplast by scanning electrochemical microscopy. *Chem. Lett.*, **28**, 975–976.

66 Yasukawa, T., Kondo1, Y., Uchida, I., and Matsue, T. (1998) Imaging of cellular activity of single cultured cells by scanning electrochemical microscopy. *Chem. Lett.*, **27**, 767–768.

67 Shiku, H., Shiraishi, T., Ohya, H., Matsue, T., Abe, H., Hoshi, H., and Kobayashi, M. (2001) Oxygen consumption of single bovine embryos probed by scanning electrochemical microscopy. *Anal. Chem.*, **73**, 3751–3758.

68 Kaya, T., Torisawa, Y.S., Oyamatsu, D., Nishizawa, M., and Matsue, T. (2003) Monitoring the cellular activity of a cultured single cell by scanning electrochemical microscopy (SECM). A comparison with fluorescence viability monitoring. *Biosens. Bioelectron.*, **18**, 1379–1383.

69 Takii, Y., Takoh, K., Nishizawa, M., and Matsue, T. (2003) Characterization of local respiratory activity of PC12 neuronal cell by scanning electrochemical microscopy. *Electrochim. Acta*, **48**, 3381–3385.

70 Zhu, L., Gao, N., Zhang, X., and Jin, W. (2008) Accurately measuring respiratory activity of single living cells by scanning electrochemical microscopy. *Talanta*, **77**, 804–808.

71 Liu, B., Rotenberg, S.A., and Mirkin, M.V. (2000) Scanning electrochemical microscopy of living cells: different redox activities of nonmetastatic and metastatic human breast cells. *Proc. Natl. Acad. Sci. U.S.A.*, **97**, 9855–9860.

72 Liu, B., Cheng, W., Rotenberg, S.A., and Mirkin, M.V. (2001) Scanning electrochemical microscopy of living cells: Part 2. Imaging redox and acid/basic reactivities. *J. Electroanal. Chem.*, **500**, 590–597.

73 Cai, C., Liu, B., Mirkin, M.V., Frank, H.A., and Rusling, J.F. (2002) Scanning electrochemical microscopy of living cells. 3. Rhodobacter sphaeroides. *Anal. Chem.*, **74**, 114–119.

74 Liu, B., Rotenberg, S.A., and Mirkin, M.V. (2002) Scanning electrochemical microscopy of living cells. 4. Mechanistic study of charge transfer reactions in human breast cells. *Anal. Chem.*, **74**, 6340–6348.

75 Feng, W., Rotenberg, S.A., and Mirkin, M.V. (2003) Scanning electrochemical microscopy of living cells. 5. Imaging of fields of normal and metastatic human breast cells. *Anal. Chem.*, **75**, 4148–4154.

76 Chen, Z., Me, S., Shen, L., Du, Y., He, S., Li, Q., Liang, Z., Meng, X., Li, B.,

Xu, X., Ma, H., Huang, Y., and Shao, Y. (2008) Investigation of the interactions between silver nanoparticles and Hela cells by scanning electrochemical microscopy. *Analyst*, **133**, 1221–1228.

77 Hengstenberg, A., Blöchl, A., Dietzel, I.D., and Schuhmann, W. (2001) Spatially resolved detection of neurotransmitter secretion from individual cells by means of scanning electrochemical microscopy. *Angew. Chem. Int. Ed.*, **40**, 905–908.

78 Isik, S., Etienne, M., Oni, J., Blochl, A., Reiter, S., and Schuhmann, W. (2004) Dual microelectrodes for distance control and detection of nitric oxide from endothelial cells by means of scanning electrochemical microscope. *Anal. Chem.*, **76**, 6389–6394.

79 Borgmann, S., Radtke, I., Erichsen, T., Blochl, A., Heumann, R., and Schuhmann, W. (2006) Electrochemical high-content screening of nitric oxide release from endothelial cells. *ChemBioChem*, **7**, 662–668.

80 Guo, J. and Amemiya, S. (2005) Permeability of the nuclear envelope at isolated Xenopus oocyte nuclei studied by scanning electrochemical microscopy. *Anal. Chem.*, **77**, 2147–2156.

81 Holt, K.B. and Bard, A.J. (2005) Interaction of silver(I) ions with the respiratory chain of Escherichia coli: an electrochemical and scanning electrochemical microscopy study of the antimicrobial mechanism of micromolar Ag+. *Biochemistry*, **44**, 13214–13223.

82 Zhang, X., Sun, F., Peng, X., and Jin, W. (2007) Quantitative determination of enzyme activity in single cells by scanning microelectrode coupled with a nitrocellulose film-covered microreactor by means of a scanning electrochemical microscope. *Anal. Chem.*, **79**, 1256–1261.

83 Gao, N., Wang, X., Li, L., Zhang, X., and Jin, W. (2007) Scanning electrochemical microscopy coupled with intracellular standard addition method for quantification of enzyme activity in single intact cells. *Analyst*, **132**, 1139–1146.

84 Martins, A.M., Mendes, P., Cordeiro, C., and Freire, A.P. (2001) In situ kinetic analysis of glyoxalase I and glyoxalase II in Saccharomyces cerevisiae. *Eur. J. Biochem.*, **268**, 3930–3936.

85 Gao, N., Wang, W., Zhang, X., Jin, W., Yin, X., and Fang, Z. (2006) High-throughput single-cell analysis for enzyme activity without cytolysis. *Anal. Chem.*, **78**, 3213–3220.

86 Mirkin, M.V. and Horrocks, B.R. (2000) Electroanalytical measurements using the scanning electrochemical microscope. *Anal. Chim. Acta*, **406**, 119–146.

87 Krylov, S.N., Starke, D.A., Arriaga, E.A., Zhang, Z., Chan, N.W., Palcic, M.M., and Dovichi, N.J. (2000) Instrumentation for chemical cytometry. *Anal. Chem.*, **72**, 872–877.

88 Chen, S. and Lillard, S.J. (2001) Continuous cell introduction for the analysis of individual cells by capillary electrophoresis. *Anal. Chem.*, **73**, 111–118.

89 McClain, M.A., Culbertson, C.T., Jacobson, S.C., Allbritton, N.L., Sims, C.E., and Ramsey, J.M. (2003) Microfluidic devices for the high-throughput chemical analysis of cells. *Anal. Chem.*, **75**, 5646–5655.

# 12
# Single-cell Mass Spectrometry

*Ann M. Knolhoff, Stanislaw S. Rubakhin, and Jonathan V. Sweedler*

## 12.1
## Introduction

Mass spectrometry (MS) has been widely used to investigate a variety of biological systems, ranging from subcellular structures to entire organisms. This powerful analytical technique has enabled the characterization and identification of a plethora of analytes including elements, metabolites, peptides, and proteins. Furthermore, spatial, temporal, and chemical information can be obtained from bioanalytical MS measurements. One promising application of this information-rich approach has been in single-cell investigations – cellular measurements that have advanced our understanding of tissues and organisms at a fundamental level. These experiments have yielded detailed insights regarding many biological functions. Although there are multiple techniques capable of detecting analytes of interest within a cell, what sets MS apart is its ability to simultaneously detect and identify many analytes without preselection or tagging. Posttranslational modifications and prohormone processing can also be determined. Moreover, cell-to-cell differences in chemical heterogeneity can be characterized via MS.

Different cell types often vary greatly in their analyte amounts and concentrations, and even morphologically similar cells show differences in analyte profiles. Accordingly, single-cell studies demand methods that meet three key requirements: high-sensitivity, chemically rich information, and a wide dynamic range. Because MS is a highly sensitive technique, with detection limits for many MS-based approaches in the attomole range, it is well suited for probing single-cell biochemistry. In fact, efforts to improve detectors in MS instrumentation have resulted in detection efficiencies approaching 100% [1]. Finally, the wide dynamic range of MS facilitates the characterization of analytes over large variations in their concentration.

Single-cell MS has been used to examine many cell types from a variety of organisms, ranging from unicellular organisms such as bacteria and yeast, to complex animals such as mammals. The invertebrates *Aplysia californica* and *Lymnaea stagnalis* have been popular model systems for the development of

*Chemical Cytometry*. Edited by Chang Lu

ISBN: 978-3-527-32495-8

single-cell MS because of their well-defined neuronal systems in which cells are relatively easy to identify, isolate, and test physiologically [2–7]. These single-cell MS studies have resulted in the discovery of many cell-to-cell signaling molecules, including information on their location, release, and function. Smaller cells from a variety of insect models – the moth [8], fruit fly [9, 10], and cockroach [11] – have also been successfully investigated with single-cell MS. The success of these invertebrate studies has led to the application of this technology to more structurally complex mammalian tissues, with most cells being no larger than 10–20 μm in size. It follows that increasing the sensitivity of MS detection, as well as improving sample quality and preparation, must be emphasized in continued efforts to advance and optimize this methodology.

Given the small sample volumes and relatively low levels of analyte inherent to single-cell experiments, it is not surprising that their success depends heavily on sample quality. Exceptional attention must be given to cell isolation, handling, and analyte extraction. Choosing the right sample preparation technique, which is highly dependent upon the analytical method being used and the analyte of interest, is also important. Here, we discuss the three major ionization approaches used in single-cell MS studies: matrix-assisted laser/desorption ionization (MALDI) MS, secondary ion mass spectrometry (SIMS), and electrospray ionization (ESI), as well as the appropriate sample preparation process for each.

## 12.2 Mass Spectrometry

In MS, analytes are ionized in some fashion and then introduced into the mass analyzer, where the ions are focused and separated based on analyte properties, such as kinetic energy or momentum. From this, the mass-to-charge ratio ($m/z$) of the ion is calculated. One can determine the charge and experimental mass of the detected species by considering the isotopic distribution and position of the ions in the resulting mass spectrum. Modern mass spectrometers can characterize analytes with high mass accuracy, often in the low parts per million range. Maintaining high mass accuracy aids in analyte identification, in part by matching the measured and theoretical masses for specific analytes; however, one often cannot unequivocally identify an analyte based on its detected mass alone. Additional techniques can serve to confirm MS identifications; for example, using standards, implementing *in situ* and immunohistochemical approaches, and/or incorporating additional MS data. In an approach known as tandem mass spectrometry (MS/MS), the molecular ion of the analyte of interest can be fragmented and the identity determined by the resulting fragmentation pattern. If the concentration of the analyte is not high enough for tandem-assisted studies, which is often the case for single-cell analyses, samples can be pooled to increase the amount of analyte available for investigation.

The most common ionization methods used for complex biological samples are highlighted here. Note that a single ionization method is not always sufficient;

rather, they can provide complementary information. A more in-depth coverage of other aspects of mass spectrometers, such as mass analyzers and detectors, can be found elsewhere [12, 13].

### 12.2.1 Matrix-assisted Laser Desorption/Ionization

A great advantage of using MALDI MS is that it allows the direct analysis of individual cells and can ionize a wide range of analyte classes, including lipids and proteins. As mentioned previously, this technique has a high sensitivity with a dynamic range amenable to single-cell analysis. In MALDI, analytes are incorporated into a matrix, such as $\alpha$-cyano-4-hydroxycinnamic acid (CHCA) or 2,5-dihydroxybenzoic acid (DHB). These specific matrixes are commonly used to analyze peptides and proteins, but there are many other MALDI matrixes available, often ideal for specific analyte classes. During sample preparation, the matrix is dissolved in solvent and applied to the sample – the solvent evaporates and the analytes are incorporated into the matrix as it crystallizes. To ionize analytes, a laser, with a wavelength strongly absorbed by the matrix, irradiates the sample spot (Figure 12.1a). This causes rapid vaporization and ionization of the matrix molecules; the analyte is also vaporized and ionized during this process. MALDI is considered a "soft" ionization technique because intact molecules are ionized with little to no fragmented species observed. For a more detailed explanation of the ionization process, there are several reviews that thoroughly describe the specific mechanisms involved [14–17].

A number of factors must be considered when planning a MALDI MS experiment. For example, the MALDI matrix should be selected according to the class of analyte being examined. The matrix should also be soluble in the solvent chosen and used at concentrations known to lead to efficient MS detection. The solvent plays an important role in the sample preparation process; slower solvent evaporation results in more time for analyte extraction and incorporation into the MALDI matrix. Some substances may hinder MALDI matrix crystal formation, including organic buffers and inorganic salts. However, in comparison with other ionization methods, MALDI is fairly tolerant to these factors. Using a higher matrix concentration can aid in appropriate crystal formation in the presence of physiological salts [18]. The matrix concentration and volume applied to a sample should also be optimized for that sample. In addition, mixed matrixes can aid in obtaining better results with MS profiling [19]. Recrystallizing the MALDI matrix while on the sample can create better crystals and aid in the elimination of salt effects [20]; however, with single-cell sampling, if the matrix spot is too large, this results in dilution of the analyte, making its characterization problematic. Efforts to promote increased ionization with the addition of ion pairing agents to the matrix, or coating the sample with gold [21], have been reported. As mentioned, due to the small size of most cells and thus, the typically low amounts of available analytes, the quality of cell isolation, handling, and analyte extraction are often essential for experimental success. The sample preparation strategies for single-cell MALDI must be selected with regard to the specific organism and cell type under investigation.

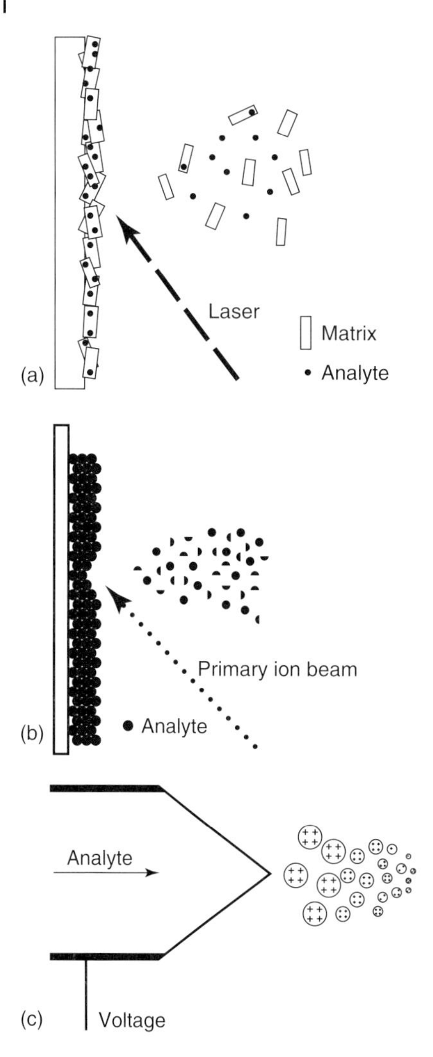


**Figure 12.1** Mass spectrometric ionization processes. (a) MALDI utilizes a pulsed laser to irradiate a mixture of analyte and matrix. (b) SIMS implements an ion gun to sputter secondary ions from the sample surface. (c) ESI requires the sample to be in solution where a voltage difference creates a Taylor cone of small droplets containing analytes.

#### 12.2.1.1 **Sample Preparation for Single-cell MALDI**

Several strategies have been implemented to obtain chemical information from single-cell samples with MALDI MS. Typically, individual cells are isolated from tissues or organs using enzymatic or nonenzymatic approaches and transferred onto the MALDI target where matrix is applied onto the cell. Early studies using single-cell MALDI MS were performed with neurons from the mollusks *L. stagnalis* and *A. californica* [2–7]. *Aplysia*, a marine sea slug, presents an additional challenge because its extracellular environment includes >450 mM NaCl. To address this issue, a procedure of washing *Aplysia* cells with an aqueous solution of DHB was introduced to remove excess salts, while maintaining the integrity of the cell membrane [22]. The concentration of inorganic salts can also be significantly decreased by microextraction procedures; for example, when analytes released from stimulated *Aplysia* bag cell neurons were captured by solid-phase extraction beads placed at the site of release [23, 24]. The beads are then washed to reduce the amount of salts in samples without significant loss of analyte, resulting in an increase in signal intensity. In situations where extracellular inorganic salts are present in lower amounts, sample conditioning steps can be omitted. For example, smaller insect cells have been isolated without enzyme treatment, exposed directly to the matrix, and analyzed with MALDI MS, with an outstanding profile of signaling peptides demonstrated [8, 10, 11]. Additional optimization of sample preparation and signal acquisition protocols is required when smaller cells with lower concentrations of analytes need to be investigated using MALDI MS.

Direct analyses of cells with MALDI MS benefit from the confinement of cellular analytes to a small surface area, especially with smaller cell types. This is necessary to reduce analyte spreading during sample preparation. One successful approach has used a small volume capillary for cell lysis with spatially defined analyte deposition [25]. In this case, red blood cells were lysed inside a capillary and spotted onto a MALDI target; the detected analytes were from hemoglobin, which is obviously highly abundant in red blood cells. Laser capture microdissection can also be employed for precise isolations of single mammalian cells from tissue sections for protein detection [26]. Another isolation protocol for single mammalian cell detection involves the addition of a glycerol-containing solution to stabilize the cell membranes and prevent cell lysis during isolation, without a detectable change in the biochemical profile of the cell [27–29]. After cells are incubated in a glycerol solution for approximately 15 min, a glass micropipette is used to isolate single cells (Figure 12.2). Next, the individual cells are transferred to the surface of a clean, indium-tin-oxide-coated microscope slide or glass coverslip. One must be careful at this point to make sure that the isolated cell does not include any excess glycerol because it may interfere with MALDI matrix crystal formation. Using another glass micropipette filled with MALDI matrix, nanoliter volumes of matrix can be applied to single cells by gently touching the tip to the surface, where spot sizes less than 50 μm are achievable. It is especially important to conduct multiple controls when such small samples are analyzed. For example, the extracellular solution should be analyzed as a control to make sure that the detected signals are truly coming from the studied cells and not the extracellular media, which may be contaminated by compounds from damaged cells.

#### 12.2.1.2 Recent Applications of Single-cell MALDI

To date, most single-cell MALDI reports relate to the investigation of the nervous system, in part because of the interest in understanding the cell-to-cell differences in key compounds in neurons. There have been multiple reports of single-cell MS of the molluskan nervous system, demonstrating peptide identification in

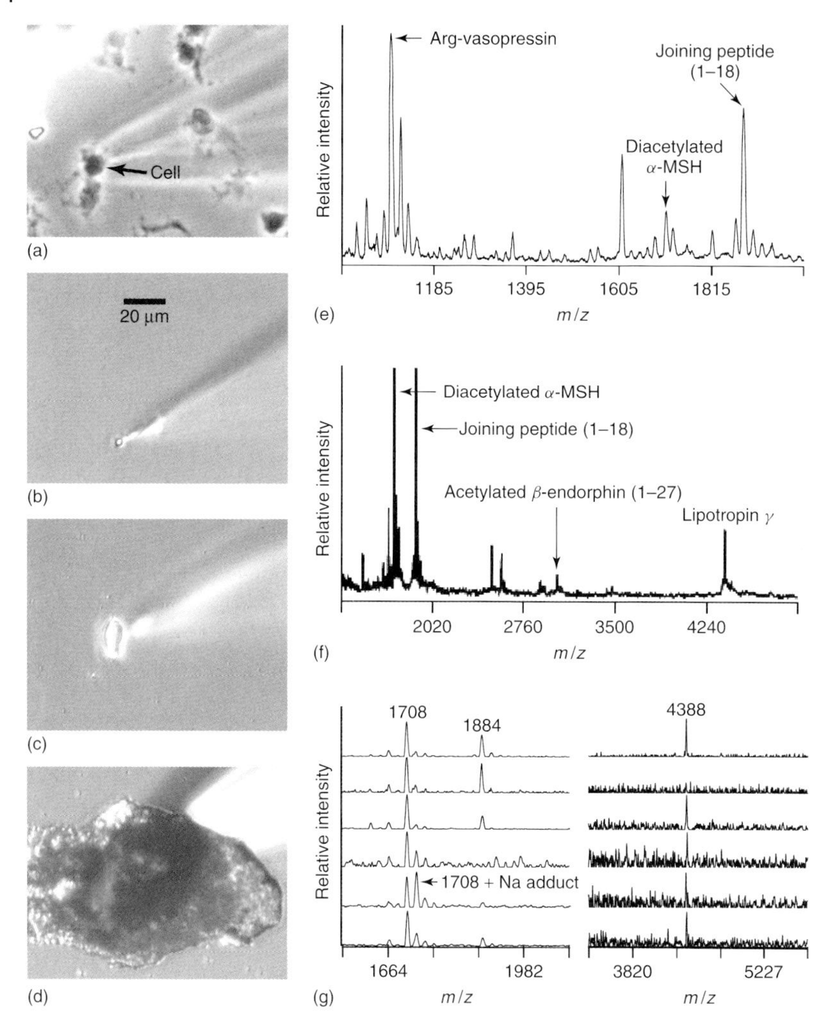


**Figure 12.2** Single-cell sample preparation and subsequent mass spectrometric profiling. (a) Attaching the cell to the glass micropipette. (b) Transfer and (c) deposition of the cell onto a clean glass surface. (d) Application of MALDI matrix to the spot containing the cell. (e) Representative mass spectrum of a pituitary blot and (f) a single pituitary cell. (g) Portions of the mass spectra obtained from six different, isolated pituitary cells. (Reproduced with permission from the American Chemical Society [28]).

these samples via MS/MS [5, 30]. Recently, similar work has been applied to insect neurons. For example, specific neurons from *Drosophila melanogaster*, with 10–20 µm diameter somas, were identified by cell-specific expression of fluorescent protein, isolated, profiled with MALDI MS, and several peptides identified from these single cells [9]. Moreover, previously unknown peptides can be discovered by detection and identification by MS/MS in various organisms. This approach also helps to elucidate peptide processing and posttranslational modifications that may be present. In terms of peptide discovery in other organisms, the first MS-based identification of a prohormone from the arthropods *Ixodes ricinus* and *Boophilus microplus* was accomplished recently using single-cell analysis; these ticks have neurons smaller than 30 μm [31].

In the Sweedler group, single-cell MALDI research has ranged from detecting peptides from *A. californica* in single isolated organelles [32] to spatial profiling of a single neuron [33] to quantifying peptides in individual cells [34]. These experiments have resulted in the discovery of a number of new neuropeptide prohormones [35–38], as well as information on unique peptide posttranslational modifications [4, 39, 40]. Moving toward smaller cell types, single mammalian cells from the rat pituitary have also been analyzed, where approximately 10 peptides were detected per cell, with more than 15 observed overall [28]. As seen in Figure 12.2, there is a relatively high signal intensity from a single cell. Although the identities of many peaks in the mass spectra have been determined, there are other detected analytes that await identification. This methodology can be applied to other cell types in other tissues and organs.

A limitation of direct single-cell analyses via MS has been quantitation, although this is beginning to be addressed. Quantitative measurements directly from tissues with MALDI are inherently difficult, in part because the incorporation of the analyte into the matrix needs to be reproducible among samples and the matrix spot should be homogeneous to ensure even analyte distribution and reproducible ionization. Despite the inherent challenges of analyte quantification in cells with MALDI, a relative quantitation protocol has been developed using single *Aplysia* neurons [34]. As shown in Figure 12.3, iTRAQ reagents were implemented to relatively quantitate components present within the cell. The differences in analyte concentrations among different cells are revealed in the MS/MS data. This protocol could certainly be implemented in other cell types to aid in functional studies.

Another exciting technological advance that has been applied to single-cell studies is mass spectrometric imaging (MSI). MSI allows one to visualize the spatial distribution of many analytes within a particular sample and is a growing field in bioanalytical MS [41–43]. In MALDI MSI, the sample stage is moved in small increments so that the laser ionizes each spot, thereby acquiring the mass spectra at an array of locations across the sample surface. Images of specific analyte distributions in the sample are created by extracting the signal at a particular mass as a function of the spatial positions of the sample stage. The spatial distribution of many analytes, in some cases hundreds, can be determined in a single experiment; however, MALDI imaging at the single-cell level is problematic because of limitations in the spatial resolution of MALDI MSI. Spatial resolution is determined by the laser spot size, the movement of the sample stage, or spreading of analytes. For optimal sensitivity, it is important to extract the analytes from the cell and fully incorporate them into the matrix. To achieve optimal spatial resolution, one needs to reduce lateral analyte spreading. Accordingly, spatial resolution and sensitivity are linked. Addressing this issue, Monroe *et al.* [44] developed a protocol whereby tissue sections are placed onto the surface of a layer

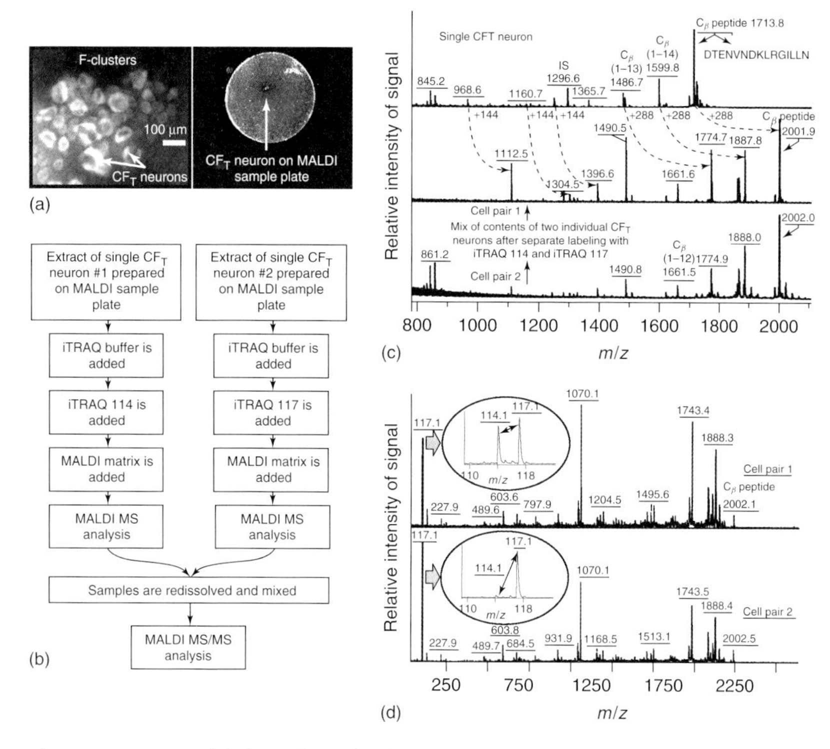


**Figure 12.3** iTRAQ labeling allows the determination of relative quantities of peptides present in single neurons. (a) F-cluster with individual cell bodies visible (left). Individual CFT neuron deposited on a MALDI sample plate (right). (b) Protocol for relative peptide quantitation in single neurons using iTRAQ reagents. (c) Top mass spectrum represents the peptide profile of an individual CFT neuron before labeling. Two bottom mass spectra are mixtures of two different pairs of CFT cells. (d) Tandem mass spectra of a peptide acquired from the mixed contents of two pairs of individual CFT neurons after separate labeling with iTRAQ 114 and iTRAQ 117. The insets show the mass range containing isotopic labels cleaved from labeled molecular ions during the fragmentation process. (Reproduced with permission from the American Chemical Society [34].)

of 40-µm glass beads embedded in Parafilm M. When the substrate is stretched, each bead has on the order of a single cell attached to it. After matrix is applied to the sample, recrystallization of the matrix enhances the quality of the matrix crystals. Furthermore, due to the hydrophobic nature of Parafilm M, analyte redistribution is minimized. Instead of trying to profile each individual bead, the entire sample can be profiled and the data stitched together to reconstruct an image of analyte localization from the original tissue sample [45].

Another approach using MALDI MSI on the cellular scale has been introduced by the Heeren group [46, 47]. Rather than decreasing the size of the laser spot to increase spatial resolution, a mass-microscope uses a large laser spot, but the spatial orientation of the ions is maintained during separation in the mass analyzer. This technique is capable of a spatial resolution of 4 μm with 500-nm pixel sizes. Although single-cell detection was not demonstrated, the scale for single-cell ionization and detection has certainly been achieved.

MSI has also been employed in measuring time-resolved release of peptides from neurons using a microfluidic-based collection approach [48]. In this work, a single *Aplysia* neuron is confined inside a channel where three connected channels are used to collect releasates from a neuron before stimulation, during chemical stimulation, and poststimulation. Each channel is also functionalized with a C18 layer to collect the released analytes. This approach for temporal isolation of released peptides could be further improved by including additional channels for better temporal resolution.

As mentioned previously, MALDI MS is more commonly used for detection and identification of larger molecules, but metabolites can also be studied with this approach. As an example, metabolites in yeast were detected with single-cell sensitivity by the Zenobi group [49]. As noted in the supplementary material of their report, several sample protocols and optimizations were employed to increase sensitivity. The optimized protocol included measuring the number of cells within a certain volume with flow cytometry and performing a cell extract on that sample. Matrix was sprayed onto a target and the extract of various volumes was spotted onto it with a piezo printer. It was determined that 390 pl was sufficient for obtaining metabolite signal from the extract, while the volume of a single yeast cell contains less than 100 fl. Although not specifically single-cell MALDI, these results demonstrate that small cells and their small molecule contents can be assayed using this approach.

Without a doubt, the figures of merit for single-cell MALDI MS are currently impressive, and they continue to be improved. As enhanced sample preparation protocols and more sensitive instruments are developed, the efficacy of such measurements will progress. This will allow a greater range of analytes to be measured from more cell types at higher spatial resolutions. High-throughput analyses of individual cells using MALDI MS will expand application of the technique to include fundamental research, clinical studies, and industrial investigations.

### 12.2.2 Secondary Ion Mass Spectrometry

Like MALDI, SIMS can directly profile samples and has been applied to single-cell studies. As one major difference, SIMS typically does not require the addition of a specific matrix to the sample. The ionization process for SIMS utilizes an energetic, focused ion beam to sputter secondary ions from the sample surface (Figure 12.1b). This technique is considered a "hard" ionization technique, which produces internal fragmentation of analyte molecules. SIMS is implemented for smaller molecular weight compounds, with the majority of detected analytes having molecular masses less than 500 Da. The upper mass threshold is limited by the sputtering and desorption processes of SIMS because the ejection of intact molecules is a lower probability event for larger molecular weight compounds.

Thus, the high mass limit depends on the amount of material present, as well as the sampling and sputtering processes. There are two modes of operation with SIMS: static and dynamic. Static SIMS can generally be considered as a nondestructive method because less than 1% of the surface is probed by the primary ion beam [50]. In contrast, dynamic SIMS uses a large number of primary ions where the surface layer is generally damaged and removed during analysis [51]. Experiments using this method are also known as depth-profiling studies.

A variety of ion beams can be implemented for sputtering with the analytical figures of merit dependant on the details of the ion beam. Atomic and diatomic sources available include $Au^+$, $Ar^+$, $Cs^+$, $Ga^+$, $In^+$, $N_2^+$, and $O_2^-$, but the use of cluster ion sources, such as $C_{60}^{x+}$, $SF_5^+$, $Ar_n^{++}$, $Cs_nI_y^+$, $Bi_n^{x+}$, and $Au_n^{x+}$, are becoming more popular. Cluster ion beams have gained increasing attention because they generate higher secondary ion yields with greater efficiencies than atomic sources, improving the limit of detection for larger molecules in addition to decreasing sample damage. One of the more impressive advantages of using SIMS is that analyte distribution can be imaged within a single cell, with spatial resolutions less than 50 nm possible [52]. SIMS has also been implemented in three-dimensional (3D) molecular analysis, where the ion beam initially rasters across the surface, generating secondary ions. A sputter beam removes this analyzed layer, thus exposing a new layer for further analysis. A detailed description of cluster sources, and their application to depth profiling, can be found elsewhere [53].

Another difference between MALDI and SIMS is that MALDI detects cellular content, and SIMS, as a surface technique, characterizes analytes on the sample surface. Because of the fragmentation that occurs with the SIMS process, the detected analytes are typically identified by characteristic mass fragments and their comparison to standards. MS/MS analyte identification is not common, but has been developed [54, 55].

#### 12.2.2.1 **Sample Preparation for Single-cell SIMS**

Most studies that employ SIMS for single-cell analyses focus on its imaging capability; thus, the overall goal of sample preparation in SIMS is preserving the original localization of analytes to obtain a representative view of processes occurring in functioning biological systems. Freeze-fracture procedures for sample preparation are used because the process is rapid and preserves cellular morphology [56]. However, this protocol tends to yield few cells for analysis and requires a specially designed cryochamber for sample handling.

Although inorganic salts can be imaged with SIMS, these salts can interfere with the detection of other analytes. If salt is present, which is common for cellular samples, it can dominate the spectrum. Ammonium acetate can be employed to remove salt and other interferants from the cell while still maintaining the integrity of the cell membrane in a reproducible fashion [57]. Furthermore, multiple fixation techniques can be used, such as rinsing the sample in water [58], 70% ethanol [59], trehalose and glycerol [60], or sucrose and water [21].

Other sample preparation steps have also proven useful for improving signal with SIMS. Charging can commonly occur, which results in poorly resolved peaks.

To alleviate this, a cell can be deposited or cultured on silicon, rather than on glass, or the entire sample can be coated with a thin layer of gold [21]. This sample coating provides the additional benefit of increasing secondary ion yields with less fragmentation. It has also resulted in the detectability of larger analytes; however, the resulting spectrum may include metal adducts and clusters, which can hamper data analysis. Adding a MALDI matrix for SIMS application has also been demonstrated to increase the detectable mass range, but has not yet been implemented for single-cell applications.

#### 12.2.2.2 **Recent Applications of Single-cell SIMS**

Measuring analyte distribution in cells can yield information not only on the state of the tissue but also on the role of these molecules in various processes. With the demonstration that SIMS was capable of imaging diffusible ions on a subcellular level, it was appreciated that a variety of biological processes could be monitored [61]. For example, imaging of boron has been useful in determining cellular uptake in normal and tumor tissue for applications in cancer therapy [62]. Furthermore, with the progression of SIMS instrumentation, molecular species could be detected and imaged simultaneously with atomic species on a subcellular level, which led to application in biological samples [63].

More recently, research has focused on membrane lipids and their changes during dynamic events. As one example, *Tetrahymena thermophila* is a single-cell organism that mates by adjoining two cells to allow the migration of micronuclei between them. This process requires lipid synthesis and rearrangement. Winograd and Ewing's [64] groups demonstrated that the micron-sized junction between the two cell bodies has a decrease in phosphatidylcholine relative to the rest of the cell body. Furthermore, the appearance of an unidentified peak increased in this junction, which was not present in other cell types and may be significant in forming this connection. In another application of determining analyte localization, Monroe *et al.* [65] demonstrated that vitamin E is not homogeneously distributed in an *Aplysia* neuron but is found at higher levels at the soma–neurite junction.

Similar to MALDI MS, quantitation of analytes in a tissue requires attention to detail in SIMS. The resulting signal intensity depends on the chemical composition and topography of the cell, primary ion interactions, instrumental transmission, and detector response [51]. Ostrowski and coworkers [66] developed a method to relatively quantify cholesterol between control and cholesterol-treated cells. Cholesterol has a variety of roles, including functions in metabolism, controlling phase behavior of membranes, in addition to being a precursor to hormones and vitamins [67, 68]. Also, abnormalities in cholesterol content have been observed in several diseases [69–71]. The use of an internal standard that remains constant in treated and nontreated cells is needed for normalization of the spectra; in this case, $C_5H_9^+$ was used. The control and cholesterol-treated cells were separately incubated with two different fluorophores known to adhere only to the outer membrane of the cell. To make a direct comparison, results for two cells under different treatments were analyzed in the same SIMS image to control for potential instrumental differences. Different fragments of cholesterol were produced with varying degrees of cholesterol elevations and standard deviations in signal, results which stress that fragment ions must be analyzed and chosen carefully for accurate quantitation. This methodology was able to detect a significant increase in cholesterol after incubation and could be applied for relative differences in concentration for other treatment conditions.

An enhancement to single-cell SIMS is 3D imaging. An early demonstration is the imaging of *Xenopis laevis* oocytes [58], large cells (0.8–1.3 mm in diameter) that also contain larger cellular components. These cells are also resistant to osmotic changes, which limits potential analyte diffusion [72]. Different component localizations were detected in the $x$, $y$, and $z$ planes of this model system, as shown in Figure 12.4, top panel, where some analytes have greater intensities below the cell plasma membrane. Even though these cells are resistant to osmotic changes, they did exhibit a morphological change after rinsing with water and after freeze fracture. Sample preparation will need to be optimized to maintain the original distributions of analyte and morphology in all dimensions.

In another example of 3D SIMS imaging, a new SIMS instrument design, the Ionoptika J105 3D Chemical Imager, incorporates a continuous $C_{60}^+$ ion source [54]. This results in continuous secondary ion production, which greatly enhances the duty cycle and SIMS signal. A precooled sample stage is also included for flash-frozen samples to minimize analyte redistribution. The resulting chemical image is shown in Figure 12.4, bottom panel, where adenine is localized in the center of the cell and the lipid signal is on the outer portion of the cell, as would be expected. These demonstrations of 3D-imaging yield a more complete view of cellular analyte distribution and will certainly lead to future applications in functional studies.

Further enhancements to SIMS instrumentation continue. Mass resolution improvements to the previously mentioned J105 3D chemical imager provided MS/MS capability [54], resulting in the ability to distinguish two different species with similar molecular weights. For example, analytes with $m/z$ values of 102.8 and 103.0 had different spatial distributions in cheek cells. In another advance, an Applied Biosystems QStar mass spectrometer with MS/MS capability has been modified and converted into a SIMS instrument by adding a $C_{60}$ source [55, 73]. With the development of new systems for analyte identification, SIMS implementation in bioanalytical investigations will become more frequent.

Developments in SIMS have been primarily focused on sample preparation improvements, ion source development, and demonstrations of different capabilities of instrumentation, while actual functional bioanalytical studies have been somewhat limited. Given the high spatial resolution that SIMS affords, future studies should progress the field with more functional applications. In one report, bacterial metabolism rates of carbon and ammonium were detected and quantified with SIMS on a single-cell basis via incubation of the cells with $^{13}C$ and $^{15}N$ [74]. It was demonstrated that a high variability in metabolism existed between three different strains from a common environment, each having its own purpose. Variability in metabolism was also detected in cells from the same strain. Studies such as these can further our knowledge of bacteria and how they interact with the environment.

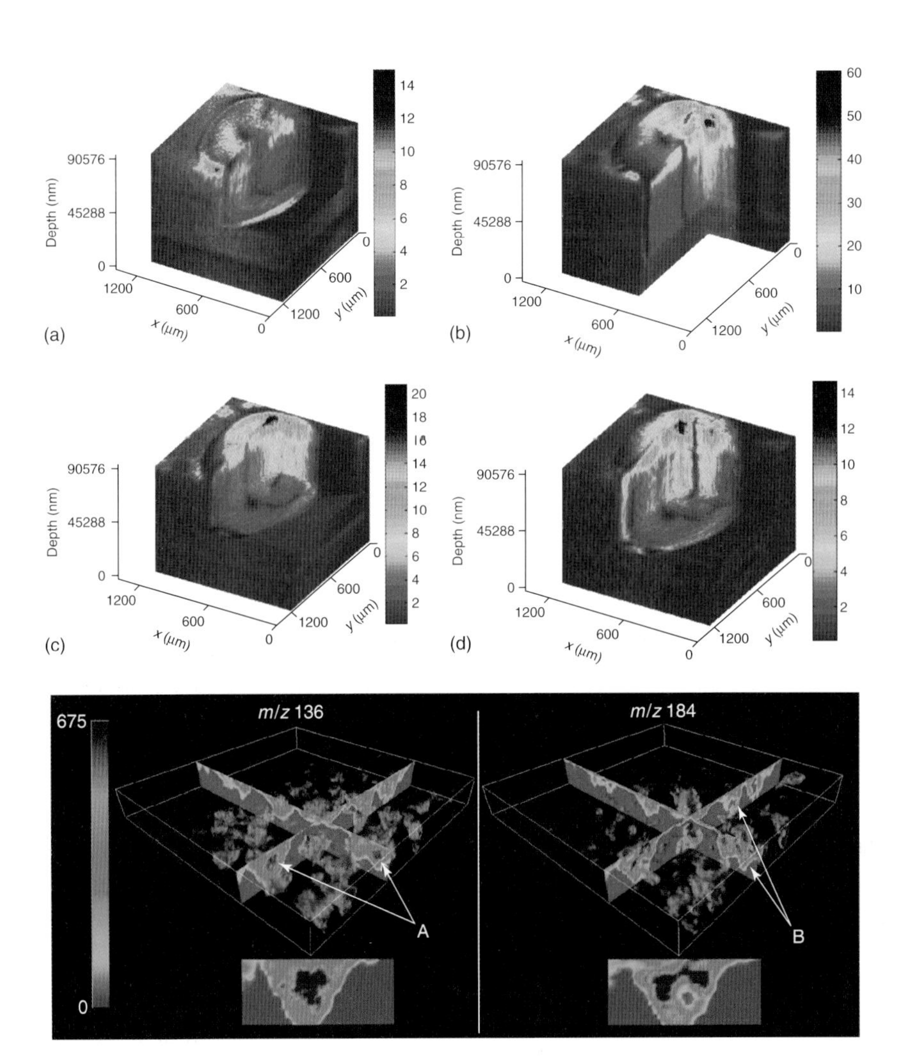


**Figure 12.4** Three-dimensional SIMS imaging. Top panel: three-dimensional biochemical images of an oocyte demonstrating different distributions of the following analytes: (a) phosphocholine peaks *m*/*z* 58, 86, 166, and 184; (b) signal summed over the *m*/*z* range 540–650; (c) signal summed over the *m*/*z* range 815–960; and (d) cholesterol peak at *m*/*z* 369. Color scale normalized for total counts per pixel for each variable (*m*/*z* range). (Reproduced with permission from the American Chemical Society [58]). Bottom panel: three-dimensional image of cells. (Left) The protonated molecular ion from adenine, which is localized to the center of the cells. (Right) Phosphocholine-containing lipids, which are observed as rings around the edge of the cell. A larger view of the orthogonal slice through a single cell is also shown for clarity. (Reproduced with permission from the American Chemical Society [54].)

Other applications, such as imaging RNA localization with respect to translation [75] or tracing copper intake of algae for environmental monitoring [76], offer data important for both fundamental and applied sciences.

### 12.2.3
### Electrospray Ionization

ESI MS has been successful in characterizing a range of samples and is the preeminent approach for proteomics. Interestingly, it is not routinely used in single-cell analysis. Unlike MALDI and SIMS, ESI involves the use of liquids and therefore requires analyte extraction from the sample prior to introduction into the mass analyzer. Therefore, sampling, extraction, and analyte dilution are important in this application. Once the sample is prepared, the analyte solution is electrosprayed (Figure 12.1c). ESI yields multiply charged species for larger molecules, with the same species being detected at different $m/z$ values, often reducing detection limits. Separation techniques, such as liquid chromatography or capillary electrophoresis (CE) are commonly hyphenated to ESI to reduce sample complexity and to concentrate analyte bands. Because ESI is not as tolerant to many inorganic salts and organic additives as are SIMS and MALDI, the separations also serve to desalt and condition the samples prior to measurement.

Nevertheless, ESI has the capability to detect a broad range of analytes from a single cell, from metabolites to large proteins, especially when optimized sample preparation approaches are implemented. There are multiple commercial nanospray ESI sources available, which are important as these minimize sample consumption and therefore fit well with single-cell ESI MS. With such approaches, the sensitivity and limits of detection for ESI MS instruments are sufficient for the detection of analytes from single cells [77].

#### 12.2.3.1 Recent Applications of Single-cell ESI

Most reports of single-cell ESI include hyphenation to CE, which separates analytes based on their size and charge with high efficiency. Using this combined approach, hemoglobin was detected in human erythrocytes, where it is found at high levels [78]. The cell was directly injected into the capillary with no prior stabilization and then lysed due to the osmotic pressure difference of the running buffer. Similar results were obtained on single erythrocytes using a different mass analyzer, thus demonstrating capability on multiple instrument platforms [79]. Other analytes, besides proteins, can be studied using single-cell ESI MS as well. Preliminary results reported by Lapainis *et al.* [80] demonstrate that characterizing a single neuron from *A. californica* is possible. To achieve these results, a home-built CE system was interfaced with ESI MS and an experimental protocol was developed. Controlled cell lysis was done within or near the separation capillary and efficient separation was achieved. Several substances were detected, including acetylcholine.

Recently, there was a report of single-cell ESI MS with mammalian cells without the need for prior separation [81]. The nanospray tip was implemented as a micropipette under a video microscope, where the cytoplasm or granules were

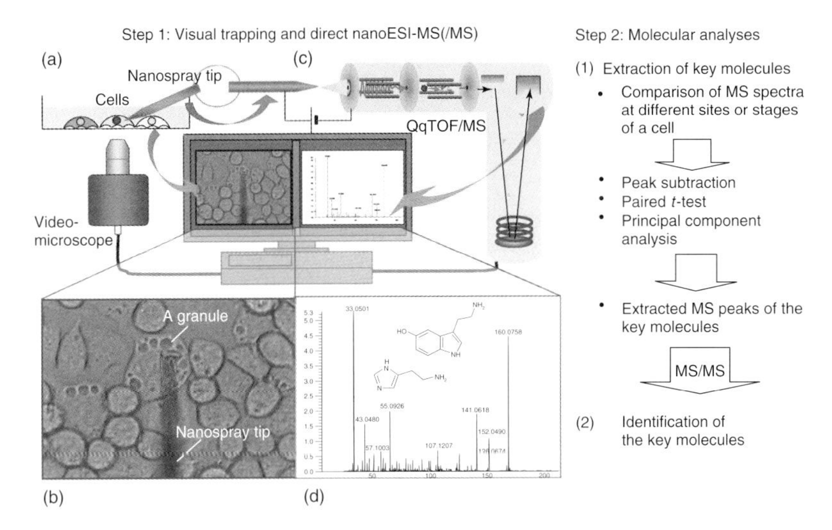


**Figure 12.5** Live single-cell video mass spectrometry. Step 1: scheme and visualization of injecting single-cell contents with resulting data. Step 2: scheme for data analysis. (Reproduced with permission from John Wiley & Sons, Ltd. [81].)

removed from the cell (Figure 12.5). Ionization solvent was added to the tip, and the sample directly introduced into the mass spectrometer. The cytoplasm, cell medium, and solvent had similar, but distinct profiles. Principal component analysis was applied to the data from each sampling, where data points from each sample set were all clearly separated (Figure 12.6). Several peaks were also identified by MS/MS, such as histamine and serotonin.

Single-cell ESI MS certainly has unrealized potential. Separations prior to ionization allow more complex samples to be measured and so protocols need to be optimized for each sample and analyte of interest. Direct infusion with ESI is advantageous because sample dilution is minimized and analysis times are much shorter. With the large range of analytes that can be detected and identified with ESI MS, efforts to improve sample introduction approaches will continue.

### 12.2.4 Other MS Approaches

MALDI, SIMS, and ESI are the most widely used ionization techniques for MS investigations of biological samples and single-cell studies. However, it is worth mentioning several of the other MS techniques that are becoming more prominent and have potential in this area. Some are related to laser desorption/ionization (LDI), which do not require a matrix for ionization. For example, with the pioneering work of Hillenkamp [82] in the development of the laser microprobe mass analysis (LAMMA), spatial resolution can be on the micron scale and a variety of small analytes can be detected [83]. Desorption/ionization on porous silicon (DIOS)

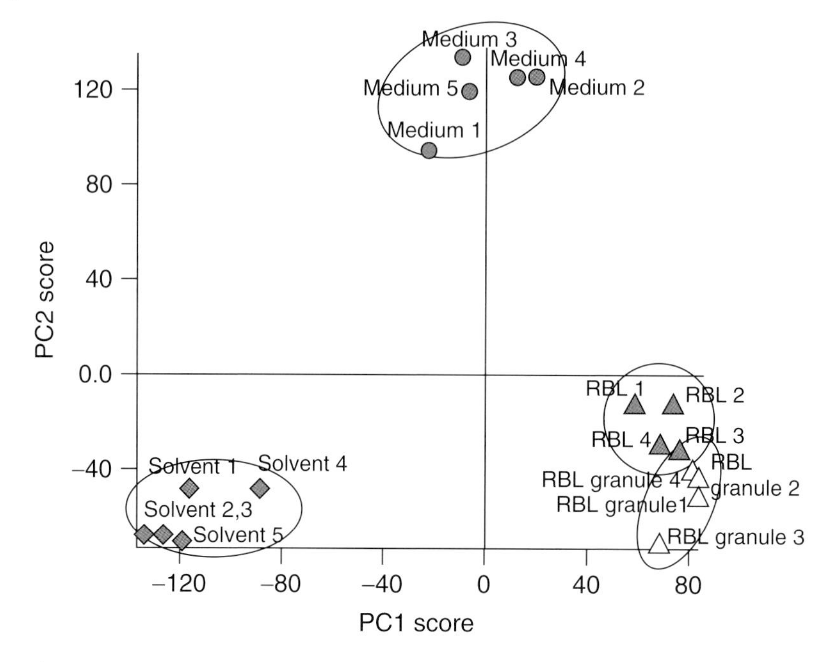


**Figure 12.6** Principal component analysis on the mass spectra of live single-cell molecular detection. Key: cytoplasm (filled triangle), granule (open triangle), cell culture medium (circle), solvent (diamond). RBL refers to RBL-2H#, a rat leukemia cell line. (Adapted with permission from John Wiley & Sons, Ltd. [81].)

can ionize peptides from single neurons and has the capability to image small molecules from single cells [59, 84]. A recent development is nanostructure-initiator mass spectrometry (NIMS), which utilizes a laser or an ion beam to ionize samples on a substrate embedded with clathrate structures to promote ionization [85]. This technique produces little fragmentation, has a lateral resolution of 150 nm, and has already achieved single-cell detection. Laser ablation electrospray ionization (LAESI) combines an ablation plume with ESI-like ionization, where plant tissue can be directly ionized and imaged [86, 87]. This technique has a 200–300 μm spatial resolution with femtomol level detection limits.

Like LAESI, there are other ambient ionization methods that have been developed as well [88]. Desorption electrospray ionization (DESI) utilizes a charged spray to extract and ionize analytes directly from the sample surface, and has already been implemented in imaging tissue sections for metabolites and lipids [89, 90]. Another example is direct analysis in real time (DART), which uses reactive ionized species to interact with the sample of interest and is beginning to be used on biological samples [91, 92]. Although this is not a comprehensive list of techniques that have the potential for single-cell MS detection, it indicates the power of these emerging methodologies for direct tissue measurements.

## 12.3
## Overall Outlook for Single-cell MS

The implementation of single-cell MS has been useful in studying neuronal systems, metabolic profiles, cell membrane components, and many other systems and analytes. MALDI, SIMS, and ESI are complementary techniques that yield a wealth of information about the presence of identified and novel analytes. With the currently available sample preparation and signal acquisition protocols, as well as instrumentation, many biological questions can be answered. Continuing to develop quantitative approaches is key to understanding the relationships between analytes. Incorporating MSI in these studies aids in determining what roles are being played by individual cells within a cellular network. Also exciting is the rapid growth of a plethora of sampling and ionization approaches that offer a new range of capabilities. The rapid evolution of technology with the increasing role of proteomic and metabolomic investigations certainly proves that single-cell MS is an area of growth. We expect that further applications and discoveries will expand during the coming decades.

### Acknowledgments

We would like to thank Zhen Li for helpful discussions during this process and Stephanie Baker for preparation of the chapter. This material is based upon work supported by the National Institute on Drug Abuse under Award No. DA 017940 and Award No. DA 018310 to the UIUC Neuroproteomics Center on Cell-Cell Signaling.

### References

1 Twerenbold, D., Gerber, D., Gritti, D., Gonin, Y., Netuschill, A., Rossel, F., Schenker, D., and Vuilleumier, J.L. (2001) Single molecule detector for mass spectrometry with mass independent detection efficiency. *Proteomics*, **1**, 66–69.

2 Garden, R.W., Shippy, S.A., Li, L., Moroz, T.P., and Sweedler, J.V. (1998) Proteolytic processing of the Aplysia egg -laying hormone prohormone. *Proc. Natl. Acad. Sci. U.S.A.*, **95**, 3972–3977.

3 Li, L., Moroz, T.P., Garden, R.W., Floyd, P.D., Weiss, K.R., and Sweedler, J.V. (1998) Mass spectrometric survey of interganglionically transported peptides in Aplysia. *Peptides*, **19**, 1425–1433.

4 Garden, R.W., Moroz, T.P., Gleeson, J.M., Floyd, P.D., Li, L., Rubakhin, S.S., and Sweedler, J.V. (1999) Formation of N-pyroglutamyl peptides from N-Glu and N-Gln precursors in Aplysia neurons. *J. Neurochem.*, **72**, 676–681.

5 Li, L., Garden, R.W., Romanova, E.V., and Sweedler, J.V. (1999) In situ sequencing of peptides from biological tissues and single cells using MALDI-PSD/CID analysis. *Anal. Chem.*, **71**, 5451–5458.

6 Jimenez, C.R., van Veelen, P.A., Li, K.W., Wildering, W.C., Geraerts, W.P., Tjaden, U.R., and van der Greef, J. (1994) Neuropeptide expression and processing as revealed by direct matrix-assisted laser desorption ionization mass spectrometry of single neurons. *J. Neurochem.*, **62**, 404–407.

7 El Filali, Z., Hornshaw, M., Smit, A.B., and Li, K.W. (2003) Retrograde labeling of single neurons in conjunction with MALDI high-energy collision-induced dissociation MS/MS analysis for peptide profiling and structural characterization. *Anal. Chem.*, **75**, 2996–3000.

8 Ma, P.W., Garden, R.W., Niermann, J.T., O'Connor, M., Sweedler, J.V., and Roelofs, W.L. (2000) Characterizing the Hez-PBAN gene products in neuronal clusters with immunocytochemistry and MALDI MS. *J. Insect Physiol.*, **46**, 221–230.

9 Neupert, S., Johard, H.A., Nassel, D.R., and Predel, R. (2007) Single-cell peptidomics of Drosophila melanogaster neurons identified by Gal4-driven fluorescence. *Anal. Chem.*, **79**, 3690–3694.

10 Neupert, S. and Predel, R. (2005) Mass spectrometric analysis of single identified neurons of an insect. *Biochem. Biophys. Res. Commun.*, **327**, 640–645.

11 Neupert, S. and Gundel, M. (2007) Mass spectrometric analysis of FMRFamide-like immunoreactive neurons in the prothoracic and subesophageal ganglion of Periplaneta americana. *Peptides*, **28**, 11–17.

12 Hager, J.W. (2004) Recent trends in mass spectrometer development. *Anal. Bioanal. Chem.*, **378**, 845–850.

13 Koppenaal, D.W., Barinaga, C.J., Denton, M.B., Sperline, R.P., Hieftje, G.M., Schilling, G.D., Andrade, F.J., and Barnes, J.H. (2005) MS detectors. *Anal. Chem.*, **77**, 418A–427A.

14 Knochenmuss, R. and Zenobi, R. (2003) MALDI ionization: the role of in-plume processes. *Chem. Rev.*, **103**, 441–452.

15 Dreisewerd, K. (2003) The desorption process in MALDI. *Chem. Rev.*, **103**, 395–426.

16 Knochenmuss, R. (2006) Ion formation mechanisms in UV-MALDI. *Analyst*, **131**, 966–986.

17 Karas, M. and Kruger, R. (2003) Ion formation in MALDI: the cluster ionization mechanism. *Chem. Rev.*, **103**, 427–440.

18 Yao, J., Scott, J.R., Young, M.K., and Wilkins, C.L. (1998) Importance of matrix:analyte ratio for buffer tolerance using 2,5-dihydroxybenzoic acid as a matrix in matrix-assisted laser desorption/ionization-Fourier transform mass spectrometry and matrix-assisted laser desorption/ionization-time of flight. *J. Am. Soc. Mass Spectrom.*, **9**, 805–813.

19 Laugesen, S. and Roepstorff, P. (2003) Combination of two matrices results in improved performance of MALDI MS for peptide mass mapping and protein analysis. *J. Am. Soc. Mass Spectrom.*, **14**, 992–1002.

20 Monroe, E.B., Koszczuk, B.A., Losh, J.L., and Sweedler, J.V. (2007) Measuring salty samples without adducts with MALDI MS. *Int. J. Mass Spectrom.*, **260**, 237–242.

21 Altelaar, A.F., Klinkert, I., Jalink, K., de Lange, R.P., Adan, R.A., Heeren, R.M., and Piersma, S.R. (2006) Gold-enhanced biomolecular surface imaging of cells and tissue by SIMS and MALDI mass spectrometry. *Anal. Chem.*, **78**, 734–742.

22 Garden, R.W., Moroz, L.L., Moroz, T.P., Shippy, S.A., and Sweedler, J.V. (1996) Excess salt removal with matrix rinsing: direct peptide profiling of neurons from marine invertebrates using matrix-assisted laser desorption/ionization time-of-flight mass spectrometry. *J. Mass Spectrom.*, **31**, 1126–1130.

23 Hatcher, N.G., Richmond, T.A., Rubakhin, S.S., and Sweedler, J.V. (2005) Monitoring activity-dependent peptide release from the CNS using single-bead solid-phase extraction and MALDI TOF MS detection. *Anal. Chem.*, **77**, 1580–1587.

24 Hatcher, N.G. and Sweedler, J.V. (2008) Aplysia bag cells function as a distributed neurosecretory network. *J. Neurophysiol.*, **99**, 333–343.

25 Li, L., Golding, R.E., and Whittal, R.M. (1996) Analysis of single mammalian cell lysates by mass spectrometry. *J. Am. Chem. Soc.*, **118**, 11662–11663.

26 Xu, B.J., Caprioli, R.M., Sanders, M.E., and Jensen, R.A. (2002) Direct analysis of laser capture microdissected cells by MALDI mass spectrometry. *J. Am. Soc. Mass Spectrom.*, **13**, 1292–1297.

27 Miao, H., Rubakhin, S.S., and Sweedler, J.V. (2005) Subcellular analysis of D-aspartate. *Anal. Chem.*, **77**, 7190–7194.

28 Rubakhin, S.S., Churchill, J.D., Greenough, W.T., and Sweedler, J.V. (2006) Profiling signaling peptides in single mammalian cells using mass spectrometry. *Anal. Chem.*, **78**, 7267–7272.

29 Rubakhin, S.S. and Sweedler, J.V. (2007) Characterizing peptides in individual mammalian cells using mass spectrometry. *Nat. Protoc.*, **2**, 1987–1997.

30 Jimenez, C.R., Li, K.W., Dreisewerd, K., Spijker, S., Kingston, R., Bateman, R.H., Burlingame, A.L., Smit, A.B. *et al.* (1998) Direct mass spectrometric peptide profiling and sequencing of single neurons reveals differential peptide patterns in a small neuronal network. *Biochemistry*, **37**, 2070–2076.

31 Neupert, S., Predel, R., Russell, W.K., Davies, R., Pietrantonio, P.V., and Nachman, R.J. (2005) Identification of tick periviscerokinin, the first neurohormone of Ixodidae: single cell analysis by means of MALDI-TOF/TOF mass spectrometry. *Biochem. Biophys. Res. Commun.*, **338**, 1860–1864.

32 Rubakhin, S.S., Garden, R.W., Fuller, R.R., and Sweedler, J.V. (2000) Measuring the peptides in individual organelles with mass spectrometry. *Nat. Biotechnol.*, **18**, 172–175.

33 Rubakhin, S.S., Greenough, W.T., and Sweedler, J.V. (2003) Spatial profiling with MALDI MS: distribution of neuropeptides within single neurons. *Anal. Chem.*, **75**, 5374–5380.

34 Rubakhin, S.S. and Sweedler, J.V. (2008) Quantitative measurements of cell-cell signaling peptides with single-cell MALDI MS. *Anal. Chem.*, **80**, 7128–7136.

35 Sweedler, J.V., Li, L., Rubakhin, S.S., Alexeeva, V., Dembrow, N.C., Dowling, O., Jing, J., Weiss, K.R. *et al.* (2002) Identification and characterization of the feeding circuit-activating peptides, a novel neuropeptide family of Aplysia. *J. Neurosci.*, **22**, 7797–7808.

36 Floyd, P.D., Li, L., Rubakhin, S.S., Sweedler, J.V., Horn, C.C., Kupfermann, I., Alexeeva, V.Y., Ellis, T.A. *et al.* (1999) Insulin prohormone processing, distribution, and relation to metabolism in Aplysia californica. *J. Neurosci.*, **19**, 7732–7741.

37 Li, L., Floyd, P.D., Rubakhin, S.S., Romanova, E.V., Jing, J., Alexeeva, V.Y., Dembrow, N.C., Weiss, K.R. *et al.* (2001) Cerebrin prohormone processing, distribution and action in Aplysia californica. *J. Neurochem.*, **77**, 1569–1580.

38 Proekt, A., Vilim, F.S., Alexeeva, V., Brezina, V., Friedman, A., Jing, J., Li, L., Zhurov, Y. *et al.* (2005) Identification of a new neuropeptide precursor reveals a novel source of extrinsic modulation in the feeding system of Aplysia. *J. Neurosci.*, **25**, 9637–9648.

39 Hummon, A.B., Kelley, W.P., and Sweedler, J.V. (2002) A novel prohormone processing site in Aplysia californica: the Leu-Leu rule. *J. Neurochem.*, **82**, 1398–1405.

40 Jakubowski, J.A., Hatcher, N.G., Xie, F., and Sweedler, J.V. (2006) The first gamma-carboxyglutamate-containing neuropeptide. *Neurochem. Int.*, **49**, 223–229.

41 Cornett, D.S., Reyzer, M.L., Chaurand, P., and Caprioli, R.M. (2007) MALDI imaging mass spectrometry: molecular snapshots of biochemical systems. *Nat. Methods*, **4**, 828–833.

42 Rubakhin, S.S., Jurchen, J.C., Monroe, E.B., and Sweedler, J.V. (2005) Imaging mass spectrometry: fundamentals and applications to drug discovery. *Drug. Discov. Today*, **10**, 823–837.

43 Seeley, E.H. and Caprioli, R.M. (2008) Molecular imaging of proteins in tissues by mass spectrometry. *Proc. Natl. Acad. Sci. U.S.A.*, **105**, 18126–18131.

44 Monroe, E.B., Jurchen, J.C., Koszczuk, B.A., Losh, J.L., Rubakhin, S.S., and Sweedler, J.V. (2006) Massively parallel sample preparation for the MALDI MS analyses of tissues. *Anal. Chem.*, **78**, 6826–6832.

45 Zimmerman, T.A., Monroe, E.B., and Sweedler, J.V. (2008) Adapting the stretched sample method from tissue profiling to imaging. *Proteomics*, **8**, 3809–3815.

46 Altelaar, A.F.M., Taban, I.M., McDonnell, L.A., Verhaert, P.D.E.M., de Lange, R.P.J., Adan, R.A.H., Mooi, W.J., Heeren, R.M.A., and Piersma, S.R. (2007) High-resolution MALDI imaging mass spectrometry allows localization of peptide distributions at cellular length

scales in pituitary tissue sections. *Int. J. Mass Spectrom.*, **260**, 203–211.

47 Altelaar, A.F., Luxembourg, S.L., McDonnell, L.A., Piersma, S.R., and Heeren, R.M. (2007) Imaging mass spectrometry at cellular length scales. *Nat. Protoc.*, **2**, 1185–1196.

48 Jo, K., Heien, M.L., Thompson, L.B., Zhong, M., Nuzzo, R.G., and Sweedler, J.V. (2007) Mass spectrometric imaging of peptide release from neuronal cells within microfluidic devices. *Lab Chip*, **7**, 1454–1460.

49 Amantonico, A., Oh, J.Y., Sobek, J., Heinemann, M., and Zenobi, R. (2008) Mass spectrometric method for analyzing metabolites in yeast with single cell sensitivity. *Angew. Chem. Int. Ed. Engl.*, **47**, 5382–5385.

50 Walker, A.V. (2008) Why is SIMS underused in chemical and biological analysis? Challenges and opportunities. *Anal. Chem.*, **80**, 8865–8870.

51 Pacholski, M.L. and Winograd, N. (1999) Imaging with mass spectrometry. *Chem. Rev.*, **99**, 2977–3006.

52 Guerquin-Kern, J.L., Wu, T.D., Quintana, C., and Croisy, A. (2005) Progress in analytical imaging of the cell by dynamic secondary ion mass spectrometry (SIMS microscopy). *Biochim. Biophys. Acta*, **1724**, 228–238.

53 Winograd, N. (2005) The magic of cluster SIMS. *Anal. Chem.*, **77**, 142A–149A.

54 Fletcher, J.S., Rabbani, S., Henderson, A., Blenkinsopp, P., Thompson, S.P., Lockyer, N.P., and Vickerman, J.C. (2008) A new dynamic in mass spectral imaging of single biological cells. *Anal. Chem.*, **80**, 9058–9064.

55 Carado, A., Passarelli, M.K., Kozole, J., Wingate, J.E., Winograd, N., and Loboda, A.V. (2008) C60 secondary ion mass spectrometry with a hybrid-quadrupole orthogonal time-of-flight mass spectrometer. *Anal. Chem.*, **80**, 7921–7929.

56 Chandra, S. and Morrison, G.H. (1992) Sample preparation of animal tissues and cell cultures for secondary ion mass spectrometry (SIMS) microscopy. *Biol. Cell*, **74**, 31–42.

57 Berman, E.S., Fortson, S.L., Checchi, K.D., Wu, L., Felton, J.S., Wu, K.J., and Kulp, K.S. (2008) Preparation of single cells for imaging/profiling mass spectrometry. *J. Am. Soc. Mass Spectrom.*, **19**, 1230–1236.

58 Fletcher, J.S., Lockyer, N.P., Vaidyanathan, S., and Vickerman, J.C. (2007) TOF-SIMS 3D biomolecular imaging of Xenopus laevis oocytes using buckminsterfullerene (C60) primary ions. *Anal. Chem.*, **79**, 2199–2206.

59 Liu, Q., Guo, Z., and He, L. (2007) Mass spectrometry imaging of small molecules using desorption/ionization on silicon. *Anal. Chem.*, **79**, 3535–3541.

60 Parry, S. and Winograd, N. (2005) High-resolution TOF-SIMS imaging of eukaryotic cells preserved in a trehalose matrix. *Anal. Chem.*, **77**, 7950–7957.

61 Ausserer, W.A., Ling, Y.C., Chandra, S., and Morrison, G.H. (1989) Quantitative imaging of boron, calcium, magnesium, potassium, and sodium distributions in cultured cells with ion microscopy. *Anal. Chem.*, **61**, 2690–2695.

62 Smith, D.R., Chandra, S., Coderre, J.A., and Morrison, G.H. (1996) Ion microscopy imaging of 10B from p-boronophenylalanine in a brain tumor model for boron neutron capture therapy. *Cancer Res.*, **56**, 4302–4306.

63 Colliver, T.L., Brummel, C.L., Pacholski, M.L., Swanek, F.D., Ewing, A.G., and Winograd, N. (1997) Atomic and molecular imaging at the single-cell level with TOF-SIMS. *Anal. Chem.*, **69**, 2225–2231.

64 Ostrowski, S.G., van Bell, C.T., Winograd, N., and Ewing, A.G. (2004) Mass spectrometric imaging of highly curved membranes during Tetrahymena mating. *Science*, **305**, 71–73.

65 Monroe, E.B., Jurchen, J.C., Lee, J., Rubakhin, S.S., and Sweedler, J.V. (2005) Vitamin E imaging and localization in the neuronal membrane. *J. Am. Chem. Soc.*, **127**, 12152–12153.

66 Ostrowski, S.G., Kurczy, M.E., Roddy, T.P., Winograd, N., and Ewing, A.G. (2007) Secondary ion MS imaging to relatively quantify cholesterol in the membranes of individual cells from differentially treated populations. *Anal. Chem.*, **79**, 3554–3560.

67 Simons, K. and Toomre, D. (2000) Lipid rafts and signal transduction. *Nat. Rev. Mol. Cell Biol.*, **1**, 31–39.

68 Rog, T., Pasenkiewicz-Gierula, M., Vattulainen, I., and Karttunen, M. (2009) Ordering effects of cholesterol and its analogues. *Biochim. Biophys. Acta*, **1788**, 97–121.

69 Wolozin, B. (2001) A fluid connection: cholesterol and Abeta. *Proc. Natl. Acad. Sci. U.S.A.*, **98**, 5371–5373.

70 Simons, K. and Ehehalt, R. (2002) Cholesterol, lipid rafts, and disease. *J. Clin. Invest.*, **110**, 597–603.

71 Lusis, A.J. (2000) Atherosclerosis. *Nature*, **407**, 233–241.

72 Kelly, S.M., Butler, J.P., and Macklem, P.T. (1995) Control of cell volume in oocytes and eggs from Xenopus laevis. *Comp. Biochem. Physiol. A Physiol.*, **111**, 681–691.

73 Piehowski, P.D., Carado, A.J., Kurczy, M.E., Ostrowski, S.G., Heien, M.L., Winograd, N., and Ewing, A.G. (2008) MS/MS methodology to improve subcellular mapping of cholesterol using TOF-SIMS. *Anal. Chem.*, **80**, 8662–8667.

74 Musat, N., Halm, H., Winterholler, B., Hoppe, P., Peduzzi, S., Hillion, F., Horreard, F., and Amann, R. *et al.* (2008) A single-cell view on the ecophysiology of anaerobic phototrophic bacteria. *Proc. Natl. Acad. Sci. U.S.A.*, **105**, 17861–17866.

75 Chandra, S. (2008) Subcellular imaging of RNA distribution and DNA replication in single mammalian cells with SIMS: the localization of heat shock induced RNA in relation to the distribution of intranuclear bound calcium. *J. Microsc.*, **232**, 27–35.

76 Slaveykova, V.I., Guignard, C., Eybe, T., Migeon, H.N., and Hoffmann, L. (2009) Dynamic NanoSIMS ion imaging of unicellular freshwater algae exposed to copper. *Anal. Bioanal. Chem.*, **393**, 583–589.

77 Valaskovic, G.A., Kelleher, N.L., and McLafferty, F.W. (1996) Attomole protein characterization by capillary electrophoresis-mass spectrometry. *Science*, **273**, 1199–1202.

78 Hofstadler, S.A., Severs, J.C., Smith, R.D., Swanek, F.D., and Ewing, A.G. (1996) Analysis of single cells with capillary electrophoresis electrospray ionization Fourier transform ion cyclotron resonance mass spectrometry. *Rapid. Commun. Mass Spectrom.*, **10**, 919–922.

79 Cao, P. and Moini, M. (1999) Separation and detection of the alpha- and beta-chains of hemoglobin of a single intact red blood cells using capillary electrophoresis/electrospray ionization time-of-flight mass spectrometry. *J. Am. Soc. Mass. Spectrom.*, **10**, 184–186.

80 Lapainis, T. and Sweedler, J.V. (2008) Contributions of capillary electrophoresis to neuroscience. *J. Chromatogr. A*, **1184**, 144–158.

81 Mizuno, H., Tsuyama, N., Harada, T., and Masujima, T. (2008) Live single-cell video-mass spectrometry for cellular and subcellular molecular detection and cell classification. *J. Mass Spectrom.*, **43**, 1692–1700.

82 Hillenkamp, F., Unsold, E., Kaufmann, R., and Nitsche, R. (1975) Laser microprobe mass analysis of organic materials. *Nature*, **256**, 119–120.

83 Verbueken, A.H., Bruynseels, F.J., and van Grieken, R.E. (1985) Laser microprobe mass analysis: a review of applications in the life sciences. *Biomed. Mass Spectrom.*, **12**, 438–463.

84 Kruse, R.A., Rubakhin, S.S., Romanova, E.V., Bohn, P.W., and Sweedler, J.V. (2001) Direct assay of Aplysia tissues and cells with laser desorption/ionization mass spectrometry on porous silicon. *J. Mass Spectrom.*, **36**, 1317–1322.

85 Northen, T.R., Yanes, O., Northen, M.T., Marrinucci, D., Uritboonthai, W., Apon, J., Golledge, S.L., Nordstrom, A. *et al.* (2007) Clathrate nanostructures for mass spectrometry. *Nature*, **449**, 1033–1036.

86 Nemes, P. and Vertes, A. (2007) Laser ablation electrospray ionization for atmospheric pressure, in vivo, and imaging mass spectrometry. *Anal. Chem.*, **79**, 8098–8106.

87 Nemes, P., Barton, A.A., Li, Y., and Vertes, A. (2008) Ambient molecular imaging and depth profiling of live tissue by infrared laser ablation electrospray ionization mass spectrometry. *Anal. Chem.*, **80**, 4575–4582.

88 Venter, A., Nefliu, M., and Cooks, R.G. (2008) Ambient desorption ionization mass spectrometry. *Trends Anal. Chem.*, **27**, 284–290.

89 Wiseman, J.M., Ifa, D.R., Song, Q., and Cooks, R.G. (2006) Tissue imaging at atmospheric pressure using desorption electrospray ionization (DESI) mass spectrometry. *Angew. Chem. Int. Ed. Engl.*, **45**, 7188–7192.

90 Wiseman, J.M., Ifa, D.R., Zhu, Y., Kissinger, C.B., Manicke, N.E., Kissinger, P.T., and Cooks, R.G. (2008) Desorption electrospray ionization mass spectrometry: imaging drugs and metabolites in tissues. *Proc. Natl. Acad. Sci. U.S.A.*, **105**, 18120–18125.

91 Cody, R.B., Laramee, J.A., and Durst, H.D. (2005) Versatile new ion source for the analysis of materials in open air under ambient conditions. *Anal. Chem.*, **77**, 2297–2302.

92 Pierce, C.Y., Barr, J.R., Cody, R.B., Massung, R.F., Woolfitt, A.R., Moura, H., Thompson, H.A., and Fernandez, F.M. (2007) Ambient generation of fatty acid methyl ester ions from bacterial whole cells by direct analysis in real time (DART) mass spectrometry. *Chem. Commun. (Camb)*, 807–809.

# 13
# Optical Sensing Arrays for Single-cell Analysis

*Ragnhild D. Whitakerand David R. Walt*

## 13.1
## Introduction to Fiber-optic Single-cell Arrays

The importance of single-cell analysis has been described in several publications [1–5] and a number of methods to analyze single cells have been published in recent years [1–13]. Single-cell methods are important because an isogenic cell population can exhibit large cell-to-cell variations with respect to cell behavior and gene expression [14–16]. Although cell-to-cell variations are masked in conventional techniques where an averaged response from many cells is measured, single-cell analyses are able to capture these differences. To obtain statistically relevant data, responses from many single cells must be simultaneously measured or within a very short time frame. In our laboratory, a method for the simultaneous analysis of thousands of single cells has been developed. The method employs a highly sensitive single-cell array platform coupled with image- and statistical analysis programs to obtain single-cell data from many cells. The single-cell platform has been employed to analyze a variety of different cell types, including bacteria, yeast, and mammalian cells. The different cell types used and the biological systems studied are chosen because the data and results obtained from the systems are masked in conventional bulk cell analysis techniques. The arrays are fabricated from commercial optical fiber bundles (Figure 13.1), in which thousands of individually clad optical fibers are fused together and then protected by an outer jacket.

The fiber bundle is first polished optically flat on a fiber polisher using diamond lapping films. The cores of the fibers are then selectively etched on one end of the fiber bundle to yield wells of uniform sizes. The glass composition enables the cores to be selectively etched while the cladding between the cores remains optically flat. The sizes of the wells can be adapted to accommodate the cell type being analyzed and to ensure that each well holds exactly one cell. After fabricating the arrays, single cells are placed in the wells and interrogated. The specific methods for array fabrication for each of the different cell types are described below. Cellular responses to a variety of stimuli are investigated and these responses are recorded using either fluorescent dyes or reporter genes that can be fluorescently interrogated. The cellular signals are detected using a

*Chemical Cytometry*. Edited by Chang Lu

ISBN: 978-3-527-32495-8

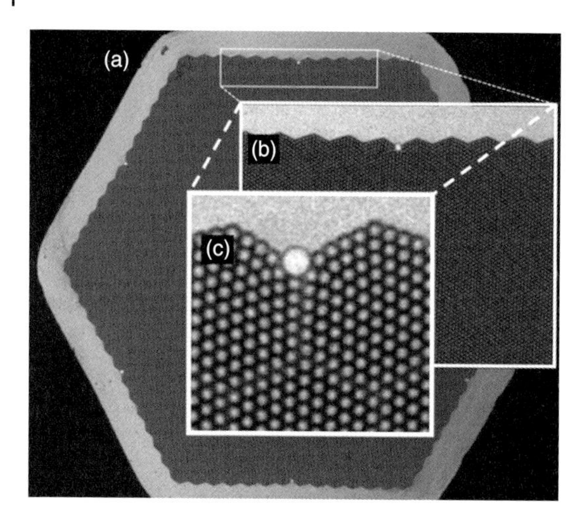


**Figure 13.1** Optical fiber bundle composed of 50 000 optical fibers, each 3.1 μm in diameter. The entire bundle is ~1.5 mm in diameter. (a–c) The fiber bundle imaged at increasing magnification (a) 5×, (b) 10×, and (c) 40× magnification. (Reproduced from [17].)

fluorescence microscope connected to a CCD camera, and the setup is coupled to an image processing program and data analysis tools to yield accurate, reproducible results (Figure 13.2).

When single cells are trapped in the wells, each individual fiber functions as a specific light guide for the single cell. By measuring fluorescence from the cells through the fibers, light from only a single cell is detected from each well by the CCD camera (Figure 13.2). The light is transferred from the cells to the detector by total internal reflection through the fiber. The loss of signal from the cell to the detector is minimal, making the detection of single-cell signals highly sensitive [17–19]. In the fiber-optic array, the cells remain in the wells throughout the experiment. By keeping the array fixed, the cells can be monitored over long time periods (up to 24 h for some cell types) and their responses to a number of perturbations or different stimuli can be recorded. In addition to recording cellular responses, the exact location of each cell in the array is determined by the use of fluorescent membrane dyes, fluorescent reporter genes, or by the use of live/dead assays.

## 13.2 Advantages of Fiber-optic Single-cell Arrays

From the chapters in this book, it is clear that many options exist when deciding on a method for single-cell experiments. The choice of a particular single-cell analysis method depends on the desired outcome of the experiments. The specific advantages of using the fiber-optic setup are

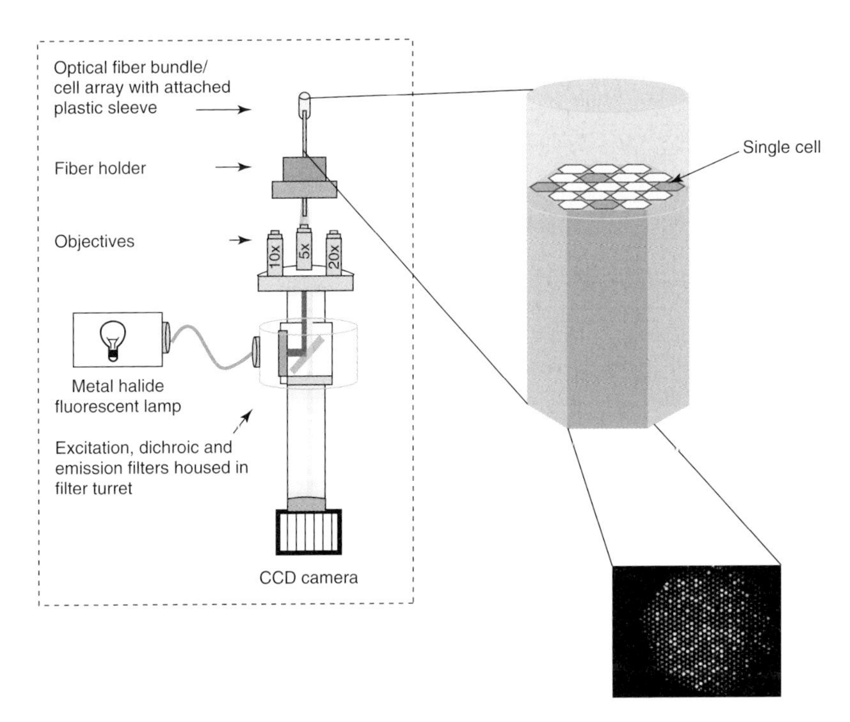


**Figure 13.2** Instrumental setup for single-cell fiber-optic array. An optical fiber array with a plastic sleeve is held in place by fiber holders. The fluorescence microscope has motor-controlled objectives and a filter turret that holds excitation, dichroic and emission filters. A metal halide fluorescent light source is used, and the fluorescence is detected using a CCD camera. Imaging is performed through the fiber.

- the ability to follow each single cell over a number of stimulations or experiments;
- the ability to determine cell viability after all stimuli and perturbations have been performed;
- the ability to determine population variance, outliers, and cellular toxicity from single-cell data;
- the ability to compare cellular responses from different strains of cell lines exposed to the exact same environment in multiplexed single-cell arrays;
- the versatility of the method allows for the analysis of a variety of biological systems in a variety of different cell types;
- the universal high density setup has minimal signal loss from the cells to the detector;
- the setup is a simple adaptation to a fluorescence microscope and should therefore be available to most research laboratories.

The fiber-optic array physically traps cells and confines them to wells throughout the experiment [11, 20, 21]. Trapping methods like the fiber-optic array are typically not ideal for selection or separation of individual cells, in which case flow-based techniques or techniques like dielectrophoresis (DEP) or optical trapping are more suitable [522–25].

## 13.3
## Fiber-optic Arrays

The optical fiber bundles used for fabricating the arrays described here are commercially available, which comprise thousands of individual fibers bundled together. In the fiber-optic bundle, each individual fiber has a light transmitting core surrounded by cladding material with a refractive index mismatch to facilitate total internal reflection at the core–clad interface. The bundle itself is protected by an outer jacket. We employ hexagonal densely packed coherent fiber bundles to fabricate our arrays (Figure 13.1) [26, 27]. Bundles with individual fiber diameters of 2.5, 3.1, 4.5, 7, 22, or 25 μm have been used. Bundles of the smaller fibers contain 24 000–50 000 individual fibers, while the bundles with larger fibers consist of ~2000 fibers. The fiber cores are made from silica doped with several elements, including barium, lanthanum, and boron. The fiber cladding consists of silica doped mainly with lead. The refractive index of the core is higher than that of the cladding (1.69 vs 1.56). When light travels from a medium with refractive index $n_1$ to another medium with lower refractive index $n_2$, the light bends away with respect to the normal of the interface surface (Figure 13.3). By increasing the incident angle with respect to the normal, the exit angle also increases until the critical angle is reached. At this point the light is reflected back to the first medium so that light travels through the fiber via total internal reflection (Figure 13.3) [19, 26, 28].

The acceptance cone of the individual fiber dictates the incident light angle that leads to total internal reflection (Figure 13.3). The collection efficiency of the fiber is defined by its numerical aperture (NA) (Equation 13.1). The higher the NA, the larger the acceptance cone, and the more light the fiber is able to collect.

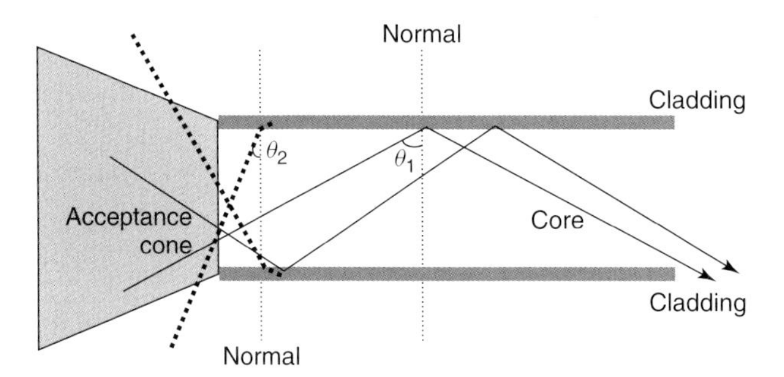


**Figure 13.3** Light transmission through a fiber caused by total internal reflection. Increasing the incident angle with respect to the normal increases the exit angle until the light is reflected. (Reproduced from [28].)

Commercial fibers have NAs ranging from 0.2 to 0.9. The NA of the fibers used in our experiments is 0.66.

$$NA = \sqrt{n_1^2 - n_2^2} \tag{13.1}$$

where $n_1$ is the refractive index of the core and $n_2$ is the refractive index of the cladding [27].

## 13.4 Single-cell Arrays for Bacteria

### 13.4.1 Array Fabrication

Bacteria are easy to cultivate and can be manipulated to detect and respond to a wide variety of external factors [29–31]. Fiber-optic arrays containing bacteria can be employed to create multiplexed bacterial biosensor arrays [32], create mercury [33] and genotoxin [34] sensors, detect promiscuous behavior in drug responses [35], and detect bacterial communication (unpublished data). When creating single-cell arrays, several criteria have to be met: the well size has to be adapted to the cell type, the cells have to be viable in the array, and any labels or expressed reporter genes must not interfere with the cellular processes being investigated. As mentioned above, single cells are isolated and confined within individual fiber-optic wells [32, 34–39]. To create single-cell arrays for bacteria, fiber bundles containing individual fibers with either 2.5 or 3.1 μm diameters are employed. The fiber bundles are first polished optically flat using a fiber polisher and diamond lapping films of decreasing coarseness (40, 30, 15, 9, 6, 3, 1, 0.5 μm). Then, 2.5-μm-deep wells are created by placing one end of the fiber in a stirred solution of dilute hydrochloric acid (HCl) for a defined time. The etching is quenched by placing the fiber end in $H_2O$, and the bundle is subsequently sonicated in order to remove debris from the surface due to the etching process. In order to place single living cells in the microarray, the distal end of the fiber is modified by gluing a 1-cm-long piece of PVC plastic tube, inner diameter 1/16 in. and outer diameter 1/8 in., to the distal end of the fiber. This plastic sleeve serves as a vessel for cell medium and allows modification of the extracellular environment as needed. In this configuration, the cells are readily accessible to stimuli injections, and the tube prevents evaporation of the medium above the cells. In order to confine the cells to the wells, a cell suspension of bacteria is placed in the vessel and the fiber is placed horizontally in a microcentrifuge and centrifuged for 1–2 min at 4000 rpm, forcing the cells into the wells. The concentration of cells in the suspension is determined by measuring the optical density (OD) of the cell suspension at 600 nm. The OD of the bacterial cell suspension is commonly between 0.1 and 0.3. The number of cells placed on the fiber must be adapted to each type of experiment. To analyze the number of wells on the fiber, which are filled with cells, fibers are dried using alcohol or critical point drying and imaged using scanning electron microscopy (SEM) [32].

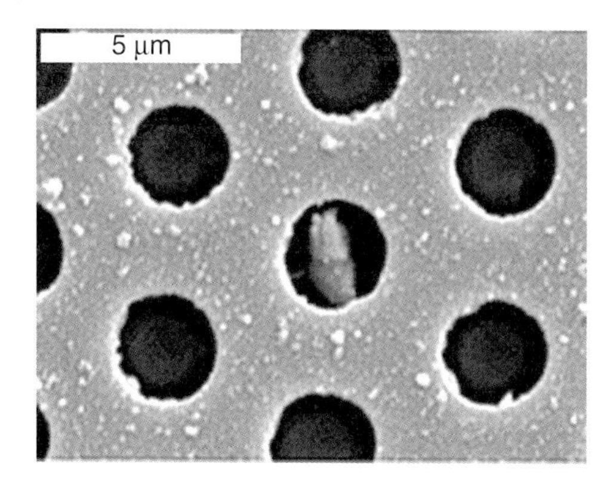


**Figure 13.4** SEM image of *E. coli* cells in 2.5-μm-diameter wells. The rod-shaped bacteria are distributed either vertically or horizontally in the array. (Reproduced from [32].)

The size of the wells is targeted to fit the bacterial cells; however, most bacteria (e.g., *Escherichia coli*) used in our experiments are rod shaped, so they do not fit perfectly into the round wells (Figure 13.4). In order to ensure that bacteria remain in the wells, the fiber surface is treated with 1% polyethyleneimine (PEI). The primary and secondary amine groups of PEI create a positive charge on the well surface, which results in the attachment of the well surface with the negatively charged bacterial cell surface through electrostatic forces. It is also possible that the sticky consistency of PEI enhances cell attachment [17, 18, 32, 39, 40]. By providing only minimal medium, confining the cells to the wells, and keeping the arrays at ambient temperature (below the cells' optimal growth temperature), cell division in the wells is avoided.

### 13.4.2 Labeling and Detection of Cellular Responses in Bacteria Arrays

Several strategies exist for labeling bacteria on the array. The bacteria are labeled to determine their exact location in the array and to detect their responses to certain stimuli. Cells in the array can be labeled using fluorescent dyes that attach to specific parts of the cell, or by employing strains that express a fluorescent reporter protein that can be detected (e.g., luminescent or intrinsically fluorescent proteins, or enzymes capable of cleaving fluorogenic substrates). In the two e xamples provided below, genetically engineered strains are employed to both optically decode cell locations and to detect cellular responses. All imaging wavelengths for fluorescent compounds described in this chapter are given in Table 13.1.

Biran *et al.* created a multiplexed live bacterial single-cell array where optical decoding was employed to differentiate between bacterial strains on the array [32]. Three *E. coli* strains were transformed using commercial plasmids to express fluorescent proteins. Each strain expressed one fluorescent protein – green fluorescent protein (GFP), red fluorescent protein (RFP), and cyan fluorescent protein (CFP).

**Table 13.1** Excitation and emission wavelengths for imaging all fluorescent compounds mentioned in this chapter.

| Fluorescent compound | Excitation wavelength (nm) | Emission wavelength (nm) | Reference |
|---|---|---|---|
| ConA–Alexa Fluor 350 | 360 | 440 | [32] |
| Cyan fluorescent protein | 440 | 490 | [32, 33, 35] |
| Ruthenium II | 455 | 613 | [38] |
| Green fluorescent protein | 480 | 530 | [32, 34, 35, 37, 41] |
| FUN 1 | 480 | 590 | [42] |
| ConA–Fluorescein | 490 | 530 | [32] |
| PKH 67 | 490 | 502 | [39] |
| DiO | 490 | 502 | [36] |
| SYTO 9 | 494 | 530 | [43] |
| $C_{12}$FDG | 494 | 530 | [32, 42] |
| Oregon green | 494 | 530 | [44] |
| Calcein AM | 495 | 515 | [39, 44] |
| EthD-1 | 495 | 630 | [39] |
| BCECF-AM | 495 | 530 | [39] |
| FITC | 495 | 530 | [39] |
| Propidium iodide | 530 | 620 | [43] |
| Red fluorescent protein | 540 | 575 | [32] |
| ConA–Tetramethylrhodamine | 540 | 580 | [32] |
| PKH 26 | 551 | 567 | [39] |
| ConA–Texas Red | 590 | 630 | [32] |
| DiD | 644 | 665 | [39] |
| ConA–Alexa Fluor 660 | 650 | 680 | [32] |

The compounds are listed in the order of excitation wavelength.

The three different strains were randomly deposited on the fiber and by imaging the array using different optical channels, the strains could be resolved and the exact location and type of each cell on the array could be determined. In this setup, Biran *et al.* created a multiplexed bacterial single-cell array. Upon decoding, this array could be further used to simultaneously detect responses from individual cells from different strains or populations. The multiplexed single-cell array provides unique insight into response differences between strains and the cell-to-cell variance within strains. The array provides information that is not available in conventional bulk assays and can significantly improve the drug discovery process [32].

Kuang *et al.* employed a multiplexed array to detect genetic noise and promiscuous drug effects in single *E. coli* cells [34, 35]. Two different *E. coli* strains were transformed with a plasmid expressing a fluorescent protein, either GFP or CFP. Upon being exposed to specific compounds, the expression of the fluorescent protein was induced. In these experiments, the cellular response to a known compound indicates the cells' exact location in the array, what type of strain it is (expressing either GFP or CFP), and the level of cellular response to stimuli [34, 35].

Kuang *et al.* monitored the protein expression in the two cell lines by imaging both GFP fluorescence and CFP fluorescence over time, obtaining temporal profiles of protein expression in response to different stimuli. Figure 13.5 illustrates the spread of responses in the two different populations when the cells were exposed to two different types of stimuli, genotoxin mitomycin C (MMC) and the synthetic allolactose analog isopropyl $\beta$-D-1-thiogalactopyranoside (IPTG).

Kuang *et al.* observed large population variance in drug responses between clonal populations. They were also able to analyze how a drug can simultaneously act on two very different cell types. The ability to simultaneously detect these promiscuous drug effects in single cells from different strains provides important information for the drug discovery process. The array provides not only data about cell-to-cell differences in each strain, but also information about how different strains react to the same exposure to a potential drug candidate [35].

In the two examples presented above, only those cells that were able to express the fluorescent reporter protein were identified and used in the analysis. Dead cells or live cells that failed to express the protein were not identified by the image processing and were therefore excluded from the analysis [32, 34, 35]. Viability assays can be employed in order to determine the viability of the cells on the array, the location of each bacterium on the fiber, and whether it is live or dead. Bacterial viability is commonly evaluated using the green variant of a SYTO nucleic acid stain together with propidium iodide [43, 45, 46]. Live cells become stained with the green SYTO dye, whereas dead cells that have compromised membrane integrity are permeable to the red fluorescent propidium iodide, which stains the nucleic acids and competitively reduces the SYTO staining in these cells [43, 45, 46]. Using these two dyes, live cells will fluoresce bright green while dead cells will fluoresce red (Table 13.1).

## 13.5 Single-cell Arrays for Yeast

### 13.5.1 Array Fabrication

Yeast are similar to bacteria in that they are easy to cultivate and easy to manipulate into detecting and responding to many different factors. Yeast are eukaryotes and their biology is therefore more relevant to more complex organisms [47]. Single-cell behavior in yeast can be studied using fiber-optic microarrays. Reporter gene expression in yeast two hybrid (Y2H) systems [32, 42], oxygen consumption in single yeast cells [38], and promoter-mediated transcriptional noise have been investigated [41]. Optical fibers (4.5 μm diameter) are used to create single-cell arrays for yeast. Array fabrication is similar to that of the bacterial arrays except that the etching time is longer, yielding wells that are ~4 μm deep in order to fit the larger yeast cells. A plastic vessel is used to hold the cells and media as described above, and yeast are placed in the wells through centrifugation at 4000 rpm. The yeast concentration is measured as OD at 600 nm and is adapted to each experiment

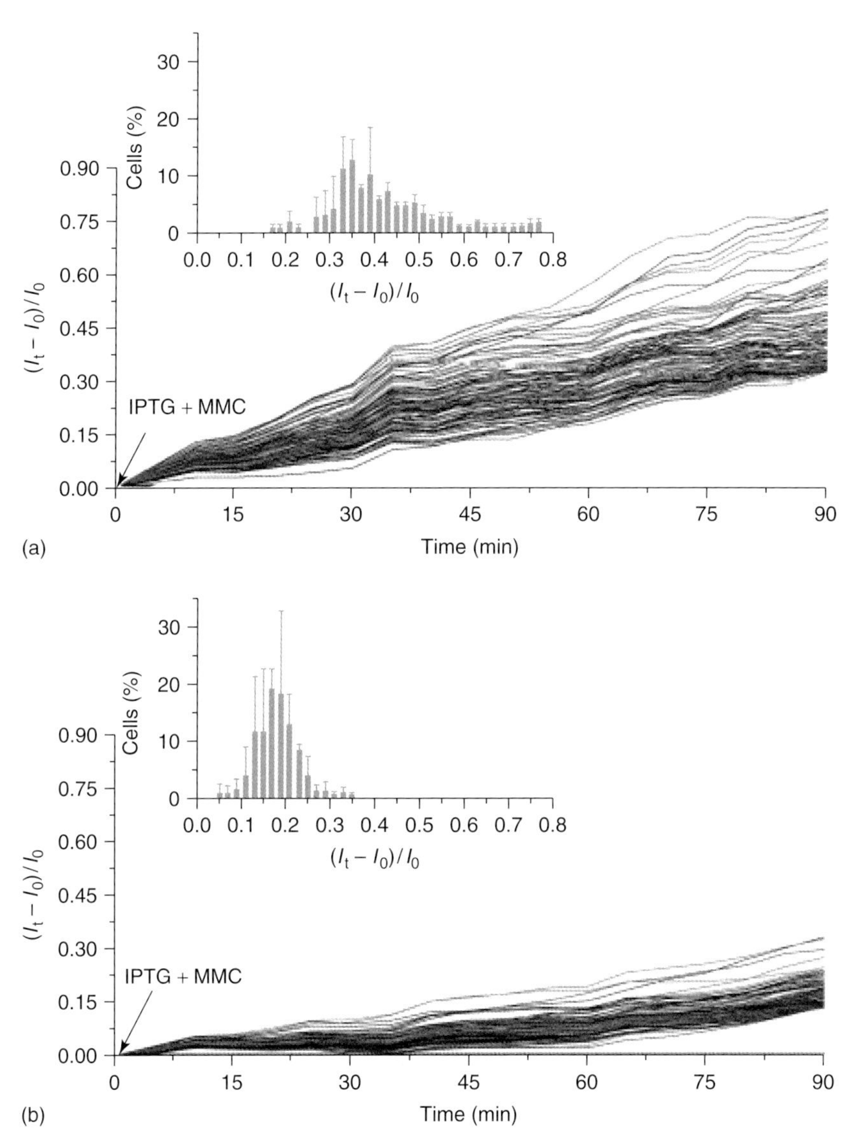


**Figure 13.5** Temporal monitoring of gene expression in different *E. coli* strains. (a) *E. coli* harboring a *lacZ::ecfp* fusion expressing CFP upon being exposed to starvation conditions and then induced with IPTG. (b) *E. coli* harboring a *recA::gfp* fusion expressing GFP upon being exposed to the genotoxin MMC. Each line in the graph represents the protein expression from a single cell, while each column in the insert illustrates the distribution of fluorescence levels in the population at one time point, 90 min after initiation. (Reproduced from [35].)

[32, 38, 41, 42]. The spherical shape of the yeast cells fits nicely into the wells, so no surface modification of the fiber is needed. The matched shapes and sizes of the cells and wells reduce the likelihood for double occupancy in the wells. Cell division is avoided due to confinement of the yeast cells and the ambient temperature during the experiment.

### 13.5.2
### Labeling and Detection of Cellular Responses in Yeast Arrays

Labeling strategies similar to those described for bacteria can be used for yeast arrays. In addition, fluorescent nanosensors have been employed to detect cellular processes in yeast. In Y2H systems, yeast strains are used to detect the interaction between two proteins that are ectopically expressed in the yeast cells. If the proteins interact, the yeast strain will express either one or more reporter genes. If the proteins do not interact, no reporter gene should be expressed. Two different arrays were employed to analyze single-cell Y2H behavior [32, 42]. Reporter gene expression in Y2H systems was interrogated through the conversion of a nonfluorescent substrate to a fluorescent product by the expressed reporter gene of $\beta$-galactosidase. The cells were exposed to cell permeable 5-dodecanoylaminofluorescein di-$\beta$-D-galactopyranoside ($C_{12}$FDG). FDG is nonfluorescent, but becomes highly fluorescent after sequential hydrolysis to 5-dodecanoylaminofluorescein (Table 13.1) [43]. In order to differentiate between different yeast strains in a multiplexed array, Biran *et al.* used the lectin Concanavalin A (ConA) bound to spectrally resolved fluorescent dyes. A different ConA–dye conjugate was used for each yeast strain. ConA binds to glycoproteins in the cell wall, providing efficient fluorescent staining of the cell wall [32, 48]. Biran *et al.* employed ConA bound to Texas Red, Tetramethylrhodamine, Alexa Fluor 660, fluorescein, and Alexa Fluor 350 (Table 13.1). Biran *et al.* were able to simultaneously detect single-cell reporter gene expression in yeast cells from different Y2H strains, while Whitaker *et al.* collected single yeast cell responses from different Y2H strains on separate fiber-optic arrays. Both experiments demonstrated that there is significant variance in responses from single yeast cells in a Y2H population. Proper characterization of protein–protein interactions in conventional bulk assays can be tedious and require several experiments. Using the fiber-optic array, it was demonstrated that by performing statistical analysis on the single-cell reporter gene expression in a population of yeast cells, the protein–protein interactions could be well characterized in a single assay, reducing the time and number of experiments needed for Y2H screens [32, 42].

Kuang *et al.* used GFP as a reporter gene to look at transcriptional noise in single yeast cells. GFP fluorescence can be monitored directly, so no addition of substrate is required. For both bacterial and yeast arrays, GFP is not a direct measure of gene expression as both the folding and half life of the protein in the cells have to be taken into consideration [49, 50]. Kuang *et al.* used the single-cell setup to analyze how promoter region differences on the yeast genome resulted in phenotypic variance between single cells in a population [41]. These analyses are important in order to fully understand the causes and effects of cell–cell variability in gene expression, and can provide very important information when evaluating whether the cell-to-cell variance provides the yeast with an adaptive advantage or if it is disadvantageous for the population [41].

Using an entirely different labeling scheme, Kuang *et al.* employed a fluorescent nanosensor strategy to measure oxygen concentration in proximity to single yeast cells [38]. The variations in oxygen concentration reflected the oxygen consumption in the yeast cells. Nanospheres that are 100 nm in diameter were first functionalized with a ruthenium(II) complex. Ruthenium fluorescence is quenched by using oxygen. So, when oxygen is consumed by the yeast cells, the concentration of oxygen in their proximity decreases and an increase in fluorescence from the ruthenium complex can be observed. The nanosensor strategy can be employed to monitor other physiologically relevant processes, such as the efflux of protons, the influx of glucose, or the concentration of carbon dioxide proximal to the yeast. In this type of array, the cells are not loaded with any dyes or transformed with plasmids, so the array provides noninvasive, reversible, and real-time measurement of cellular processes at the single-cell level. The single-cell results enable the investigation of cellular physiological responses that are unobtainable using methods that provide only an averaged cell population response and can provide unique information about how different environments affect these physiological processes in yeast [38].

Yeast viability was evaluated in some of the yeast arrays using FUN 1 (2-chloro-4-(2,3-dihydro-3-methyl-(benzo-1,3-thiazol-2-yl)-methylidene)-1-phenyl-quinolinium-iodide). This dye is cell permeable and in live cells, metabolic activity will result in intracellular compact vacuolar structures exhibiting red fluorescence. In dead cells, diffuse green fluorescence can be seen, but not red fluorescence (Table 13.1). Similar to the bacterial arrays, the viability stain enables the identification and selection of live cells in the array, allowing only live cells to be included in the image and data analysis process. Including only live cells ensures that dead cells on the array are not mistakenly identified as false negatives [42].

## 13.6 Single-cell Arrays for Mammalian Cells

Although yeast and bacterial biosensors can be models for mammalian cell responses, mammalian cells are obviously the most relevant. Fiber-optic single-cell arrays have been employed to study several processes in different mammalian cell lines. Mammalian cell lines are less straightforward to cultivate, and transfection and labeling of mammalian cells are more cumbersome than that of yeast and bacteria. Cell viability [39], cancer cell motility [36], and G-protein coupled receptor (GPCR) receptor activation [44] were all analyzed using different cell lines on fiber-optic arrays. Similar to the bacterial and yeast arrays, the results taken on a single-cell level provide information that is unobtainable using bulk cell responses. While a common scheme was used to fabricate single bacterial- and yeast arrays, each mammalian single-cell array was fabricated in a unique way.

Taylor *et al.* utilized fiber-optic bundles with 7-μm-diameter fibers to analyze NIH 3T3 mouse fibroblast cells. Wells were etched 3 μm deep at the distal end of the fiber using a solution containing hydrofluoric acid (HF) (50%) and ammonium fluoride [39]. HF is highly caustic, and extreme caution should be taken when working with HF [51]. The fiber was rinsed in deionized water and sonicated after etching to remove debris from the etching process. Prior to depositing cells on the array, the microwell array was sonicated vertically for 15 min under vacuum to remove air bubbles from the microwells. In order to place the cells in the wells, a capillary tube was filled with the cell suspension, and the etched face of the fiber was inserted into this tube. The fiber array with the tube was then incubated vertically for 1.5 h at 37 °C with 5% $CO_2$ to allow the cells to settle into the wells and adhere. No modification of the glass surface was done prior to placing the cells on the fiber, although components in the medium may have facilitated adhesion. Taylor *et al.* created a multiplexed single-cell array by labeling different cell lines with dyes that could be spectrally resolved, imaging the fluorescence signals from each dye, and overlaying the images. When the images were overlaid, a composite image of the array was created with the exact position and identity of each cell on the array. Taylor *et al.* employed two different PKH dyes – PKH 67 and PKH 26, which were incorporated into the cell membrane without affecting cell viability or proliferation [52, 53], and one 3H-Indolinium,2-[5-(1,3-dihydro-3,3-dimethyl-1-octadecyl-2H-indol-2-ylidene)-1,3-pentadienyl]-3,3-dimethyl-1-octadecyl, perchlorate (DiD) dye (a lipophilic carbocyanine), which stained the entire cell with minimal cell toxicity [43]. The dyes employed in this array were imaged using the wavelengths listed in Table 13.1. Upon verifying that the different cell lines could be identified on the fiber-optic array, Taylor *et al.* proceeded to investigate the viability of the cells on the fiber using viability stains and pH-sensitive measurements. Several different viability assays were performed (Figure 13.6). A viability/cytotoxicity kit containing Calcein acetoxymethyl (AM) ester and ethidium homodimer-1 (EthD-1) was first applied to the array. Calcein AM is cell permeable and is cleaved by esterases inside the cell, yielding a fluorescent product and causing live cells to fluoresce green. EthD-1 can only enter the cells and bind to nucleic acids if the cells have compromised membrane integrity [54, 55], causing nonviable and dead cells to fluoresce red (Table 13.1). Taylor *et al.* also evaluated cell viability on the array using 2′,7′-bis-(2-carboxyethyl)-5-(and-6)-carboxyfluorescein acetoxymethyl ester (BCECF-AM). BCECF-AM enters the cells and is cleaved via the same mechanism as Calcein AM is cleaved [43] (Figure 13.6). Finally, Taylor employed pH-sensitive nanobeads to evaluate cellular viability on the array. pH-sensitive fluorescein isothiocyanate (FITC) was trapped inside 100 nm polystyrene nanobeads and incubated with the cells (Figure 13.6). A decrease in extracellular pH accompanies cellular metabolism; consequently, the level of change in pH can be an indicator of the rate of cell metabolism and therefore cell viability. A high concentration of FITC beads in the microwells allowed for temporal monitoring of localized changes in the pH of the individual cells. The

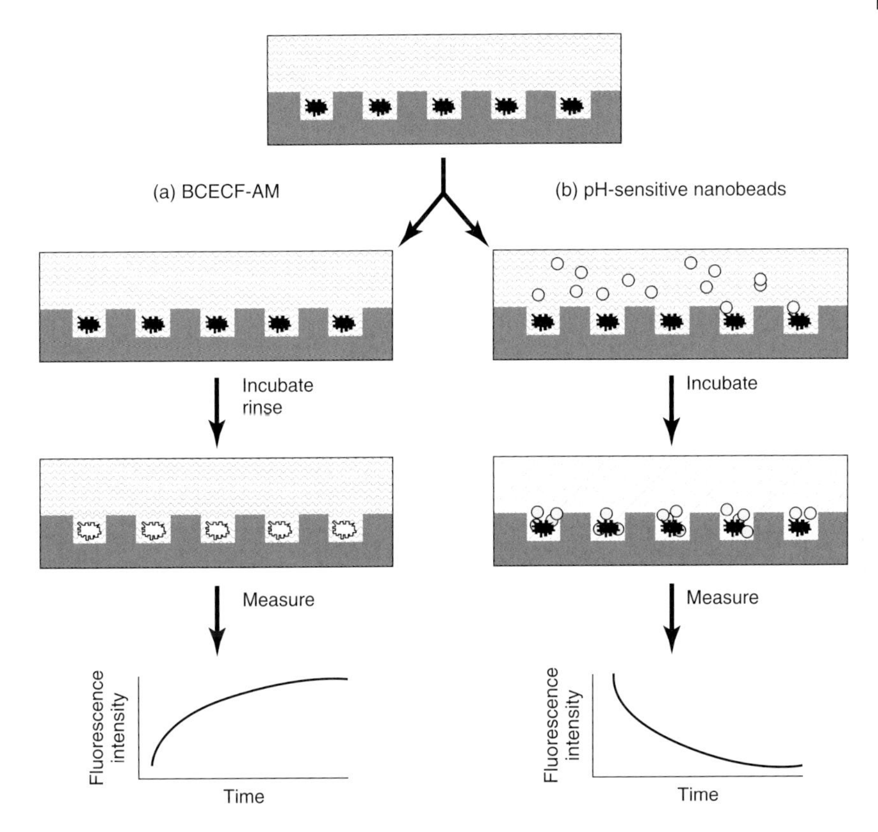


**Figure 13.6** Schematic of viability assays using BCECF-AM (a) and pH-sensitive nanobeads (b). (a) BCECF-AM is incubated with the cells, followed by rinsing. The dye is cleaved inside the cells yielding a fluorescent product; therefore, live cells exhibit increased fluorescence. (b) pH-sensitive nanobeads are immobilized in the microwell by a polymer layer. (Adapted from [35].)

beads were trapped in the wells by placing a polymer gel above the cells (Table 13.1 and Figure 13.6).

Taylor *et al.* were able to demonstrate the ability to create multiplexed mammalian cell arrays and to show that the cells remained viable for over 20 h on the array. The array was used to detect single-cell esterase activity using cytotoxicity assays and to determine the rate of cellular metabolism at the single-cell level. The results obtained with this single-cell mammalian array demonstrated how the extracellular environment affects single cells from different cell lines with respect to both their viability and rates of metabolism. These results can be used to improve many types of cellular assays where these single-cell responses are not obtainable [38].

DiCesare *et al.* used the fiber-optic array in a different fashion to investigate the migration of fibroblast cells in response to antimigratory drugs. Two different fiber-optic bundles, containing 25-μm- and 4.5-μm-diameter fibers, respectively, were polished optically flat as described above, and the flat fibers were then used in the experiments. The fibers were coated with fibronectin to promote cell adhesion

[56], and NIH/3T3 fibroblasts were labeled, placed on top of the polished flat fiber, and allowed to adhere at 37 °C in 5% $CO_2$ for 2 h. DiCesare *et al.* used a Benzoxazolium, 3-octadecyl-2-[3-(3-octadecyl-2(3H)-benzoxazolylidene)-1-propenyl]-, perchlorate (DiO) dye to label the cells (Table 13.1). DiO is very similar to the DiD dye described above and stains the entire cell with minimal leakage [36]. The dye was chosen because it did not interfere with cell migration. Other tested dyes either inhibited migration or leaked from the cells [36]. The cells were exposed to different concentrations of the antimigratory drug nocodazole [57, 58], and cell migration over the fiber was analyzed and compared with control cells that were not exposed to the drug. The experiments were performed at 37 °C. Whole cell migration was analyzed by observing the residence time of each cell on 25-μm-diameter fibers in a bundle. The fluorescence intensity measured through each fiber increased when a cell resided on it, and then decreased when the cell migrated away from it. Higher concentrations of the antimigratory drug caused the cells to exhibit increased residence time above the fibers, indicating reduced migration (Figure 13.7).

By employing smaller 4.5-μm-diameter fibers, DiCesare *et al.* also investigated the movement of subcellular structures as a result of exposure to an antimigratory drug. They found that the drug also caused the residence time of subcellular structures to increase, indicating that subcellular structures exhibited reduced motion and

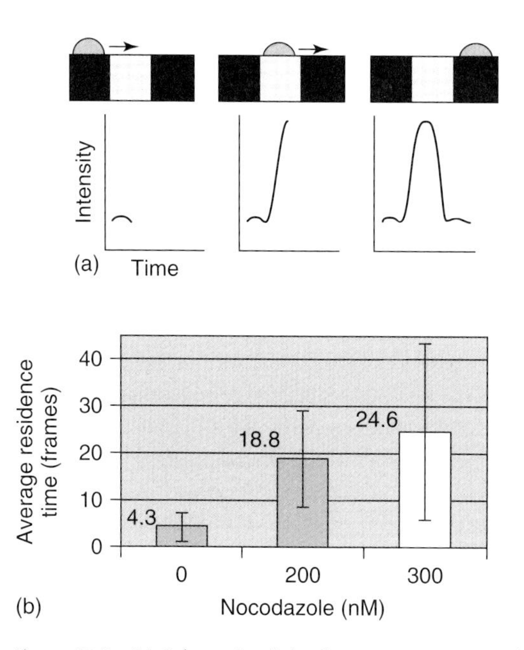


**Figure 13.7** (a) Schematic of the fluorescence measured through a fiber as a cell migrates on the fiber bundle. (b) Average residence times for the cells when exposed to different concentrations of nocodazole. (Reproduced from [36].)

perhaps were the underlying cause of the overall reduction in migratory aptitude of the cell. The fiber-optic migration arrays provide a fast and accurate way to measure the effect of antimigratory drugs. Conventional methods measuring these effects are tedious and lengthy and do not allow for single-cell analysis. Consequently, they do not allow for the determination of cell-to-cell variance to a drug exposure. The array provides valuable information that can be used for developing antimigratory drugs.

The fiber-optic microwell array was also used to investigate receptor activation in mammalian cells. Fibers that are 22 μm in diameter were employed to analyze the activation of GPCRs ectopically expressed in Chinese hamster ovarian (CHO) cells [44]. The mammalian single-cell array setup in these experiments was very similar to the bacteria and yeast cell arrays described above. Deep wells of 15 μm in size were etched in the distal end of the fiber, and the glass surface of the fiber was modified with fibronectin to promote cell adhesion [56]. A plastic sleeve was attached to the well end of the fiber bundle and a cell suspension was pipetted into this sleeve and incubated overnight to allow the cells to settle in the wells and adhere. A calcium-sensitive dye was employed to analyze GPCR activation in single CHO cells [44]. Three different cell lines were investigated separately using the fiber-optic array, and their responses to different agonists were recorded. When GPCR receptors are activated, the concentration of intracellular calcium is modulated and intracellular calcium can therefore be used as an indicator of receptor activation. Calcium-sensitive dextran-bound Oregon green BAPTA-1 was used to measure cytosolic calcium modulations. The dye was chosen because the large bound dextran molecule ensures that the dye stays in the cytosol. The dye does not leak out of the cell or sequester into intracellular calcium storages. Leakage and sequestration of the dye are undesirable; leakage causes decrease in the fluorescence over time while sequestration causes the dye to reflect the calcium levels in the cell storages, which can mask the calcium modulations caused by receptor activation. The large dye molecule is cell impermeable and was loaded into the cell using a pinocytic loading protocol [59]. The receptors were stimulated with several different agonists and nonstimulating compounds, and the intracellular calcium concentration was recorded by imaging the dextran-bound dye every 3 s for 4 min for each agonist (Table 13.1). A Calcein AM viability assay was performed at the end of the experiment and only live cells were included in the subsequent data analysis. The single-cell GPCR receptor activation assay allowed for determination of response distribution in cellular populations, analysis of drug responses, and investigation of outliers in clonal populations. In addition to these results, the temporal patterns of intracellular calcium modulation were analyzed using machine learning techniques including principal component analysis (PCA) and k-nearest neighbor (k-NN). These techniques allow all the temporal response patterns to be compared. The data analysis demonstrated that when a promiscuous receptor is stimulated with several different agonists, the patterns of cytosolic calcium oscillations in the population are specific for each agonist. Although calcium response specificity had been suggested previously in the literature [60–62], by simultaneously investigating hundreds of single cells, we were able to confirm this idea [44]. For the mammalian cell arrays, only normal growth medium was used when culturing the cells on the array, but due to the confinement of the cells in the wells, cell division was not observed.

## 13.7 Image and Data Analysis for Single-cell Arrays

All single-cell arrays described in this chapter were imaged using a fluorescence microscope coupled to a CCD camera. Image acquisition and image processing is performed using IPLab software. Different optical channels are selected by employing excitation and emission filters. Grayscale images and the pixel intensities represent a particular fluorophore. The wells on the array are uniform in shape and they all project onto the same number of pixels in the image. When creating a decoding image for cells in a multiplexed array, wells are selected by segmenting using the imaging program. Images taken at different optical channels (representing different strains/cell lines) are overlaid so that a decoding image is created showing all the different cells and their positions on the array (Figure 13.8). When temporal data from cells are obtained, all the images are stacked, creating a series of images with each image representing a time point in the experiment. Desired wells are selected either from a decoding image or from a live/dead assay and overlaid onto the time stack of images. The fluorescence intensities from desired wells are extracted and a time trace for each cell is obtained (Figure 13.5).

Upon extracting fluorescence data from each cell at different time points, subsequent data analysis techniques are performed. Statistical analyses including *t*-test, ANOVA, and normal fitting of reporter gene expression in a population are performed using Matlab computing software. PCA and k-NN analyses are performed in the machine learning program Weka. The post image analyses enable determination of population variances and mathematical comparisons between all the single-cell responses in the same population or comparisons of population single-cell responses between experiments [63, 64]. In addition to the commercial image analysis and machine learning programs used in most analyses, custom image and statistical analysis programs have been developed for specific experiments. The custom written programs enable automated segmentation and image overlay, extraction of cellular data, and post image statistical analysis, thereby decreasing the time requirements for image and data analysis. Advantages of custom image analysis programs are the speed and increased data processing ability; drawbacks are the time requirements for program development and the suitability of the program to a limited set of experiments.

## 13.8 Summary

Fiber-optic single-cell arrays and their application to a variety of biological systems have been discussed in this chapter. The arrays are fabricated from commercial

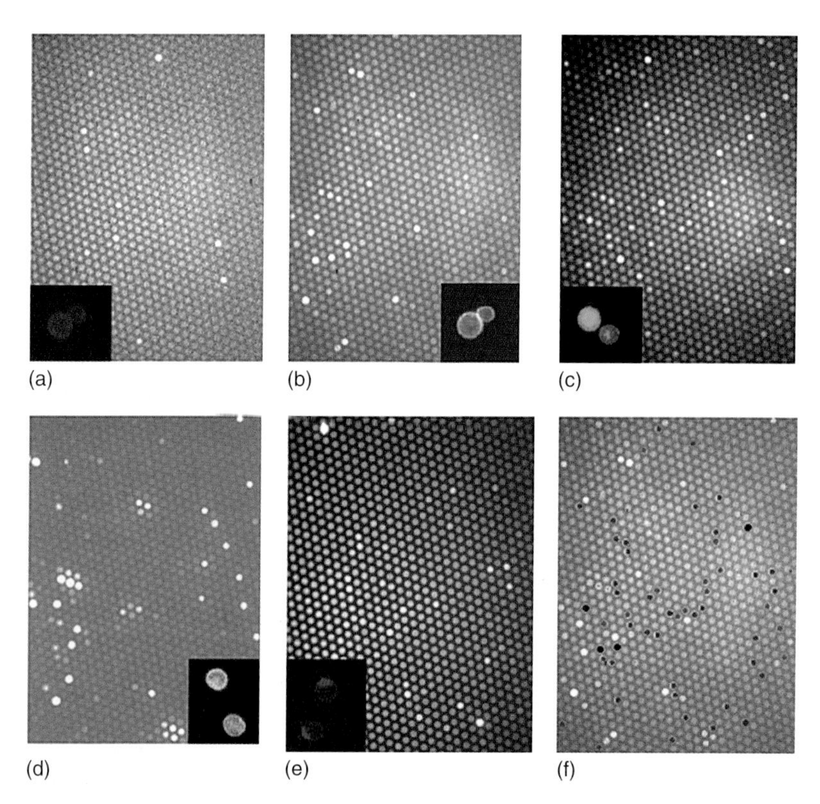


**Figure 13.8** Decoding images of five different yeast strains on the fiber-optic array. (a) Yeast cells labeled with ConA–Texas Red. (b) Yeast cells labeled with ConA–Tetramethylrhodamine. (c) Yeast cells labeled with ConA–Alexa fluor. (d) Yeast cell labeled with ConA–Fluorescein. (e) Yeast cells labeled with ConA–Alexa fluor 350. The inserts shows fluorescence images of the encoded yeast that are pseudocolored with the corresponding color label. (f) All the segments from a to e are overlaid and a decoding image for all the yeast strains on the array is created. (Reproduced from [32].)

fiber-optic bundles and are adaptable to a wide variety of cell types. Cellular responses are detected using a standard fluorescence microscope connected to a CCD camera, employing instruments that are common equipment in many laboratories. Fiber-optic cell arrays were made with bacteria, yeast, and mammalian cell lines, and were used to screen for toxic compounds, monitor environmental changes, analyze drug responses, investigate physiological processes, and study cell-to-cell variability. The high density of the array enables monitoring of hundreds to thousands of cells simultaneously. This ability provides information that is not obtainable using conventional cellular analysis techniques where an averaged response from a cell population is obtained. The flexibility of the single-cell array makes it amenable to study a wide variety of cellular processes and the simple instrumental setup makes the array accessible to many laboratories.

## References


1 Buse, P., Tran, S.H., Luther, E., Phu, P.T., Aponte, G.W., and Firestone, G.L. (1999) *J. Biol. Chem.*, **274**, 7253–7263.

2 Curtis, J.E. and Grier, D.G. (2003) *Phys. Rev. Lett.*, **90**, 133901/1–133901/4.

3 Edwards, B.S., Oprea, T., Prossnitz, E.R., and Sklar, L.A. (2004) *Curr. Opin. Chem. Biol.*, **8**, 392–398.

4 Feldhaus, M. and Siegel, R. (2004) *Methods Mol. Biol. (Totowa)*, **263**, 311–332.

5 Fuchs, A.B., Romani, A., Freida, D., Medoro, G., Abonnenc, M., Altomare, L., Chartier, I., Guergour, D., Villiers, C., Marche, P.N., Tartagni, M., Guerrieri, R., Chatelain, F., and Manaresi, N. (2006) *Lab Chip*, **6**, 121–126.

6 Gerdts, G. and Luedke, G. (2006) *J. Microbiol. Methods*, **64**, 232–240.

7 Grier, D.G. (2003) *Nature (London)*, **424**, 810–816.

8 Grier, D.G. (2003) *Nature*, **424**, 810–816.

9 Kacmar, J., Zamamiri, A., Carlson, R., Abu-Absi, N.R., and Srienc, F. (2004) *J. Biotechnol.*, **109**, 239–254.

10 Koo, M.K., Oh, C.H., Holme, A.L., and Pervaiz, S. (2007) *Cytometry A*, **71A**, 87–93.

11 Kovac, J.R. and Voldman, J. (2007) *Anal. Chem. (Washington)*, **79**, 9321–9330.

12 Zhang, T. and Fang, H.H.P. (2004) *Biotechnol. Lett.*, **26**, 989–992.

13 Zheng, H.-Z., Liu, H.-H., Chen, S.-X., Lu, Z.-X., Zhang, Z.-L., Pang, D.-W., Xie, Z.-X., and Shen, P. (2005) *Bioconj. Chem.*, **16**, 250–254.

14 Di Carlo, D., Aghdam, N., and Lee, L.P. (2006) *Anal. Chem.*, **78**, 4925–4930.

15 Di Carlo, D. and Lee, L.P. (2006) *Anal. Chem.*, **78**, 7918–7925.

16 Lidstrom, M.E. and Meldrum, D.R. (2003) *Nat. Rev. Microbiol.*, **1**, 158–164.

17 Epstein, J.R. and Walt, D.R. (2003) *Chem. Soc. Rev.*, **32**, 203–214.

18 Epstein, J.R. and Walt, D.R. (2002) *Proc. SPIE Int. Soc. Opt. Eng.*, **4578**, 89–95.

19 Walt, D.R. (2000) *Science (Washington)*, **287**, 451–452.

20 Rettig, J.R. and Folch, A. (2005) *Anal. Chem.*, **77**, 5628–5634.

21 Xia, Y. and Whitesides, G.M. (1998) *Annu. Rev. Mater. Sci.*, **28**, 153–184.

22 Albrecht, D.R., Underhill, G.H., Wassermann, T.B., Sah, R.L., and Bhatia, S.N. (2006) *Nat. Methods*, **3**, 369–375.

23 Chiou, P.Y., Ohta, A.T., and Wu, M.C. (2005) *Nature*, **436**, 370–372.

24 Eriksson, E., Enger, J., Nordlander, B., Erjavec, N., Ramser, K., Goksoer, M., Hohmann, S., Nystroem, T., and Hanstorp, D. (2007) *Lab Chip*, **7**, 71–76.

25 Jeffries, G.D.M., Edgar, J.S., Zhao, Y., Shelby, J.P., Fong, C., and Chiu, D.T. (2007) *Nano Lett.*, **7**, 415–420.

26 Hecht, E. (1998) *Optics*, 3rd edn, Addison Wesley Longman, Inc.

27 Ming-Kang Liu, M. (1996) *Principles and Applications of Optical Communications*, Irwin Professional Publishing.

28 Ball, D. (2006) *Spectroscopy*, **21**, 19–22.

29 Bjerketorp, J., Hakansson, S., Belkin, S., and Jansson, J.K. (2006) *Curr. Opin. Biotechnol.*, **17**, 43–49.

30 Gu, M.B., Mitchell, R.J., and Kim, B.C. (2004) *Adv. Biochem. Eng. Biotechnol.*, **87**, 269–305.

31 Wolf, B., Brischwein, M., Grothe, H., Stepper, C., Ressler, J., and Weyh, T. (2007) *Microsystems*, **16**, 269–307.

32 Biran, I. and Walt, D.R. (2002) *Anal. Chem.*, **74**, 3046–3054.

33 Biran, I., Rissin, D.M., Ron, E.Z., and Walt, D.R. (2003) *Anal. Biochem.*, **315**, 106–113.

34 Kuang, Y., Biran, I., and Walt, D.R. (2004) *Anal. Chem.*, **76**, 2902–2909.

35 Kuang, Y. and Walt, D.R. (2005) *Anal. Biochem.*, **345**, 320–325.

36 DiCesare, C., Biran, I., and Walt, D.R. (2005) *Anal. Bioanal. Chem.*, **382**, 37–43.

37 Kuang, Y., Biran, I., and Walt, D.R. (2004) *Anal. Chem.*, **76**, 6282–6286.

38 Kuang, Y. and Walt, D.R. (2007) *Biotechnol. Bioeng.*, **96**, 318–325.

39 Taylor, L.C. and Walt, D.R. (2000) *Anal. Biochem.*, **278**, 132–142.

40 Bencic-Nagale, S. and Walt, D.R. (2005) *Anal. Chem.*, **77**, 6155–6162.

41 Blake, W.J., Balazsi, G., Kohanski, M.A., Isaacs, F.J., Murphy, K.F., Kuang, Y.,

Cantor, C.R., Walt, D.R., and Collins, J.J. (2006) *Mol. Cell*, **24**, 853–865.

42 Whitaker Ragnhild, D. and Walt David, R. (2007) *Anal. Biochem.*, **360**, 63–74.

43 Invitrogen *www.invitrogen.com*

44 Whitaker, R.D. and Walt, D.R. (2007) *Anal. Chem.*, **79**, 9045–9053.

45 Miyanaga, K., Takano, S., Morono, Y., Hori, K., Unno, H., and Tanji, Y. (2007) *Biochem. Eng. J.*, **37**, 56–61.

46 Selvaraju, S.B., Khant, I.U.H., and Yadav, J.S. (2008) *Lett. Appl. Microbiol.*, **47**, 451–456.

47 Baronian, K.H.R. (2004) *Biosens. Bioelectron.*, **19**, 953–962.

48 Tkacz, J.S., Cybulska, E.B., and Lampen, J.O. (1971) *J. Bacteriol.*, **105**, 1–5.

49 Lippincott-Schwartz, J. and Patterson, G.H. (2003) *Science (Washington)*, **300**, 87–91.

50 Wielgus-Kutrowska, B., Narczyk, M., Buszko, A., Bzowska, A., and Clark, P.L. (2007) *J. Phys. Condens. Matter*, **19**, 285223/1–285223/8.

51 Material_Safety_Data_Sheet. (MSDS) *http://fscimage.fishersci.com/msds/11171.htm*

52 Rousselle, C., Barbier, M., Comte, V.V., Alcouffe, C., Clement-Lacroix, J., Chancel, G., and Ronot, X. (2001) *In Vitro Cell. Dev. Biol. Anim.*, **37**, 646–655.

53 Barbier, M., Laurier, J.-F., Seigneurin, D., Ronot, X., and Boutonnat, J. (2002) *C. R. Biol.*, **325**, 393–400.

54 Wang, X.M., Terasaki, P.I., Rankin, G.W. Jr., Chia, D., Zhong, H.P., and Hardy, S. (1993) *Hum. Immunol.*, **37**, 264–270.

55 Kaneshiro, E.S., Wyder, M.A., Wu, Y.P., and Cushion, M.T. (1993) *J. Microbiol. Methods*, **17**, 1–16.

56 Raghavan, S. and Chen, C.S. (2004) *Adv. Mater. (Weinheim)*, **16**, 1303–1313.

57 De Brabander, M.J., Van de Veire, R.M.L., Aerts, F.E.M., Borgers, M., and Janssen, P.A.J. (1976) *Cancer Res.*, **36**, 905–916.

58 Hoebeke, J., Van Nijen, G., and De Brabander, M. (1976) *Biochem. Biophys. Res. Commun.*, **69**, 319–324.

59 Okada, C.Y. and Rechsteiner, M. (1982) *Cell*, **29**, 33–41.

60 Berridge, M.J. (1992) *Adv. Second Messenger Phosphoprotein Res.*, **26**, 211–223.

61 Sanchez-Bueno, A. and Cobbold, P.H. (1993) *Biochem. J.*, **291**, 169–172.

62 Savineau, J.-P., Guibert, C., and Marthan, R. (2005) *Lung Biol. Health Dis.*, **197**, 167–183.

63 Bakken, G.A. and Jurs, P.C. (2000) *J. Med. Chem.*, **43**, 4534–4541.

64 Witten, I.H. and Frank, E. (2005) *Data Mining*, 2nd edn, Morgan Kaufmann, San Francisco.

# Index



*Chemical Cytometry*. Edited by Chang Lu

ISBN: 978-3-527-32495-8

## *f*

***n***